Edible Oleogels

Structure and Health Implications

Editors
Alejandro G. Marangoni and Nissim Garti

AOCS PRESS
Urbana, Illinois

ISBN 978-0-12-810222-0

Library of Congress Cataloging-in-Publication Data
Edible oleogels : structure and health implications / editors, Alejandro
Marangoni, Nissim Garti.
p. cm.
Includes bibliographical references and index.
1. Oils and fats, Edible--Health aspects. 2. Essential fatty acids in
human nutrition. 3. Lipids in human nutrition. I. Marangoni, Alejandro G.,
1965- II. Garti, Nissim, 1945-
QP752.E84E35 2011
612.3'97--dc22

2010052396

15 14 13 2 3 4 5

Contents

Preface

The dangers of consuming high amounts of saturated and *trans* fats have been well documented in the past 50 years. The replacement of these fats in manufactured food products has been very slow. It was not until mandatory labeling of "*trans* fat" content became a reality, that serious efforts took place to remove dangerous *trans* fats from our diets. Surprisingly, even after much negative publicity and consumer rejection of such food ingredients, we can still find food products in the market containing *trans* fats. Any amount of *trans* fat used in food products is bad—there is no lower safe limit. Five years ago, replacement of *trans* fats meant addition of saturated fat stocks including palm fat and fully hydrogenated stocks. Even though the issue of which saturated fatty acids are worse for you than others is an active one, as a general public health policy, this is a mute point—the consumer will never differentiate between types of saturated fatty acids. Public health policy does not work in this way.

An exciting and novel strategy to eliminate *trans* fats from our diets and avoid excessive amounts of saturated fat lies in the ability to structure oil and make it behave like a solid crystalline fat, without being one. This has been in an area of very active research in the last five years since the first publications on organogels appeared in the Journal of the American Oil Chemists' Society. This area of research endeavor has steadily grown from obscurity to the point where the concept of an organogel or oleogel has entered the consciousness of the lipid scientist and technologist. It is thus our view, that it is time to highlight research in this area that has taken place in the past five years.

This book will review recent advances in the structuring of edible oils to form new mesoscale and nanoscale structures, including nanofibers, mesophases, and functionalized crystals and crystalline particles. Many of the approaches proposed rely heavily on our understanding of self-assembly in non-aqueous systems and a healthy dosing of luck. All of these systems share the commonality of being dimensionally constrained in their growth mode—from nanofibers and mesofibers to nanoplatelets and mesoplatelets, the special structuring ability of the molecules discussed is a function of the unique structures they can form, which arises from unique growth patterns.

The materials studied constitute a unique class of soft materials and their study is worthwhile on this fact alone, but the possible development of *trans* fat free, low saturate functional shortenings for the food industry will make a major impact on the health characteristics of the foods we consume. The explosion in obesity and Type-II diabetes we are currently experiencing provides a further impetus to accelerate and foster this research. As scientists we have a moral responsibility to try and help mitigate this serious problem in our society.

This book will be of interest to lipid scientists and technologists who work on crystallization, fats and oils, food science, interfacial processes, dispersion technology, rheology, food polymers, as well as medical doctors and health practitioners.

Alejandro G. Marangoni
Nissim Garti

1

An Overview of the Past, Present, and Future of Organogels

Alejandro G. Marangoni[1] and Nissim Garti[2]

[1]*Canada Research Chair in Food and Soft Materials Science, Guelph-Waterloo Physics Institute, Department of Food Science, University of Guelph, Guelph, Ontario, Canada and* [2]*Ratner Chair of Chemistry, Casali Institute of Applied Chemistry, Institute of Chemistry, Hebrew University of Jerusalem, Edmond J. Safra Campus, Givat Ram, Jerusalem, Israel*

Introduction

For over twenty years, clinical and epidemiological research has suggested that there is a positive relationship between *trans* fatty acid intake and a *decrease* in serum HDL ("good" cholesterol) in combination with an *increase* in serum LDL ("bad" cholesterol) (Mensink & Katan, 1990; Zock & Katan, 1992; Judd et al., 1994; Ascherio et al., 1999; Institute of Medicine, 2002; Mensink et al., 2003). Together these effects increase the risk of coronary heart disease. Both the Institute of Medicine and the American Heart Association recommend a reduction in the intake of *trans* fatty acids, preferably eliminating them from the diet (Institute of Medicine, 2002; American Heart Association, 2004). However, most consumers would find this difficult to achieve without appropriate labeling of staple foods. Prompted originally by the Center for Science in the Public Interest in 1994, the U.S. Food and Drug Administration (FDA) decreed that as of January 2006, food manufacturers must include the *trans* fatty acid content in product labels (US FDA 2003a; US FDA, 2003b).

This FDA ruling is an important milestone for public health, but it creates serious technological hurdles for the food manufacturing industry—it is difficult to eliminate *trans* fats from a food formulation. At the core of the problem is the ability to transform an oil (liquid at room temperature) to a fat ("solid" at room temperature). The difference between an oil and a fat is subtle. Oils and fats are mostly composed of molecules called triaclyglycerols: three fatty acids esterified onto a glycerol backbone. Whether such material is solid or liquid at a particular temperature depends on the chemical nature and physical properties of the constituent fatty acids.

Hydrogenation has been successfully commercially used for almost 100 years to transform oils into fats. From the time the British patent on liquid phase hydrogenation was issued to Norman in 1903 and its introduction in the U.S. in 1911, few chemical processes have made as great an economic impact on any industry.

Hydrogenation opened new markets for vegetable-oil-based specialty products (O'Brien, 2003). Three reactions take place during hydrogenation: (1) the saturation of carbon-carbon double bonds, (2) the conversion of *cis* geometric isomers into more stable *trans* isomers, and (3) the creation of new positional isomers, where double bonds are shifted to new positions along the fatty acid chain. Both the saturation of double bonds as well as the *cis* to *trans* isomerization of double bonds will increase the melting point of a fat. Thus, cooling of this hydrogenated fat below the melting point of the newly created triacylglycerol species containing saturated and *trans* fatty acids will lead to the partial crystallization of the material. The resulting semisolid fat matrix will be a network of fat crystal aggregates with liquid oil trapped inside (Narine & Marangoni, 1990a). The solid-like characteristics of this material are due to this underlying fat crystal network (Narine & Marangoni, 1990b; Marangoni, 2000; Marangoni & Rogers, 2003). The presence or absence of this network of crystallized fat determines whether the material is a fat or an oil, respectively. Thus, the only previously known way to provide structure an oil, and thus convert it into a plastic fat, is by addition of high-melting temperature saturated or *trans* fats. This represents a major problem since fats containing high amounts of *trans* and saturated fatty acids are known to be atherogenic—they contribute to the build-up of cholesterol and other substances in artery walls. The American Heart Association discourages the consumption of excessive amounts of saturated animal and vegetable fats such as milk fat, palm oil, palm kernel oil, and coconut oil, as well as *trans* fats (American Heart Association, 2004). A new strategy for structuring edible oils is thus required.

In recent years, scientists have modified the physical properties of oils, which have a low viscosity and no elasticity, to resemble those of fats, which have a solid-like character and are elasto-plastic. Many food products that require a specific texture and rheology can be made with these novel oil-based materials without causing significant changes to final product quality. The major approach to forming these novel oil-based materials is to incorporate oil components that by various molecular interactions will alter the physical properties of the oil so that its fluidity will decrease and the rheological properties will be similar to those of fats. The continuous phase of these oil gels is lipidic, and they exhibit the characteristic physical properties of hydrogels. To distinguish these food oil gels from traditional "organogels", which are usually gels of organic solvents used in various industrial applications in the chemical industry, we call these edible lipid oil gels, oleogels.

Several approaches have been taken to construct oleogels from vegetable oils and the research experience has brought us to a new area in which our food products will be healthier with no *trans* and minimal saturated fats. Before introducing the contributors to this book, we will briefly review the definition of a "fat" and some of the strategies that can be used to gel oils in the context of specific examples. Then, some of the applications of organogels will be summarized. Finally the chapters presented in this book will be introduced.

Fat Crystal Networks

For semisolid fat products (i.e., ice cream, margarine, and chocolate), the solid lipids, which normally exist as a three-dimensional (3D) colloidal fat crystal network, determine the physical properties of the product. Upon crystallization, hardstock triacylglycerols aggregate to form fat crystals similar to colloidal gels. The resulting clusters aggregate into flocs. The weak links between the flocs in the final macroscopic network are portrayed in Fig. 1.1.

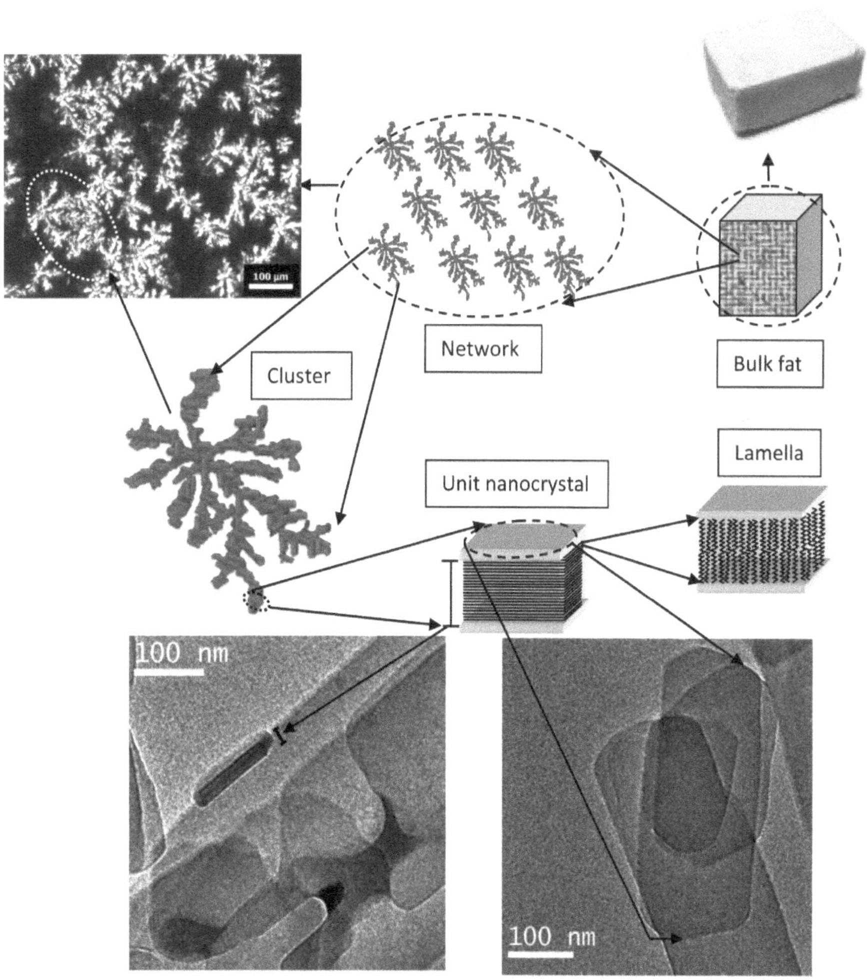

Fig. 1.1. Structural hierarchy in colloidal fat crystal networks. Source: Acevedo & Marangoni, 2010.

Heat, mass, and momentum transfer conditions have significant effects on the final microstructure and resultant macroscopic physical properties (i.e., hardness, yield stress, and compressibility) of fat products. For example, many studies have shown that the hardness of a fat crystal network is directly correlated to hardness determined by sensory analysis, and thus the sensory perception of the food product (Rousseau & Marangoni, 1998).

Soft, plastic materials, including fats, have different levels of structure present and each level influences the macroscopic properties of the material (Tang & Marangoni, 2006). The rheological properties of fats are the result of the combined effects of solid fat content (SFC), polymorphism, and fat crystal network microstructure, including the shape, size, area fraction, and the distribution pattern of the fat crystals (Tang & Marangoni, 2006). Since the hardstock triacylglycerols (TAGs) are responsible for the network structure, it is often difficult or impossible to eliminate these ingredients to improve the health aspects of the product without sacrificing some of the characteristic properties of the product. For example, the TAG profile is responsible for the narrow melting range of chocolate and the broad melting profile of margarine. An alternative to structuring lipid-based materials with or without reduced levels of the traditional saturated- and trans-containing hardstocks is to replace the colloidal fat network with organogels or oleogels.

Gels, Hydrogels, and Oleogels

An organogel or oleogel is a gel where the liquid phase is oil, as opposed to a hydrogel, which is a gel that has a continuous liquid water phase. Before we proceed to discuss different strategies for gelling oils, we first need to define what a "gel" is, which is more difficult than it might initially seem. The definition of a "gel" has been evolving since 1861 when Thomas Graham attempted a loose definition. Sixty-five years later, Dr. Dorothy Jordan Lloyd stated "the colloid condition, the 'gel', is one which is easier to recognize than to define" (Jordan Lloyd, 1926). Her major contribution was in recognizing that all gels must be comprised of at least two components, a liquid and a gelling substance (i.e., a solid), and that the entire system must have the mechanical properties of a solid (Jordan Lloyd, 1926). The major downfall to this definition is that not all colloids are gels and not all gels are colloids. Over the next twenty years the definition progressed to the point where Hermans proposed that gels are coherent colloid dispersed systems of at least two components. They exhibit mechanical properties that are consistent with the solid state, and both the dispersed (gelator) phase and dispersion medium must extend continuously throughout the whole system (Hermans, 1949). Because of the exclusivity of this definition, Ferry offered a more descriptive definition of a gel: "a gel is a substantially diluted system which exhibits no steady state flow" (Ferry, 1961). Another common definition of a gel is "if it looks like Jello, it's a gel" (Jordan Lloyd, 1926). From this point, a gel must contain two features: (1) it has a continuous microscopic structure with macroscopic dimensions that is permanent on the time scale of an analytical experiment, and (2) it is solid-like in its rheological behavior, despite being comprised mostly of liquid (Weiss & Terech, 2006).

Organogelling Strategies

Organogels or oleogels offer an alternative to the use of "bad" fats such as saturated and *trans* fats. The building blocks of these organogels can vary widely, mainly falling under the following categories:

(1) Crystalline Particles:

A network formation of colloidal crystalline triacylglycerol (TAG) particles traps the liquid TAG phase inside, thus causing gelling. The size and shape of the crystals and the interactions between them determine the mechanical properties of the network (Bot et al., 2007; Marangoni, 2004). It is also possible to structure similar edible oils with diacylglycerols (DAG), monoacylglycerols (MAG), and fatty acids (Pernetti et al., 2007a; Ojijo et al., 2004; Wright & Marangoni, 2006). Structuring can also be achieved with wax esters/sorbitan monostearate, ceramides (Daniel & Rajasekharan, 2003; Rogers, 2009), and lecithin/sorbitan tristearate (Pernietti, 2007b).

(2) Crystalline Fibers:

Self-assembled fibrillar networks (SAFIN) of low molecular weight gelators (LMOG) of phytosterols with oryzanol (Bot & Agterof, 2006; Bot et al., 2008), 12-hydroxystearic acid, and ricinoleic acid have been reported (Rogers et al., 2008; Rogers et al., 2009), where the structuring agent forms helical and twisted crystalline ribbons hundreds of micrometers long.

(3) Polymeric Strands:

Polymers promote gelation of a medium. These gels can be of two types: gels formed by covalent bonding (chemical gel) or gels formed by self-assembly (e.g., Van der Waal's force or H-bonding) (physical gel). Formulation of fatty acid based co-monomers and vegetable oil based polymers (Lu & Larock, 2009; Yilmaz & Kusefoglu, 2005; Sharma & Kundu, 2006) has been reported in the context of sustainable resources, but these systems lack the potential for food applications. An abundance of literature can be found on gelation of biopolymers such as cellulose, starch, gelatin, etc. in aqueous solutions as well as in dispersions (Ruan et al., 2008; Ostrovskii et al., 1999; Frohoff-Hülsmann et al., 1999, Brouillet-Foumann et al., 2003; Lee et al., 2008; Gilsenan & Ross Murphy, 2000; Borchard & Burg, 1990), but similar studies in edible oil medium are scarce (Aiache et al., 1992; Ruiz Martinez et al., 2003).

(4) Particle-filled Networks:

A lipid-based product is structured by dispersing a large amount of inert filler particles in the continuous lipid phase. The inert particles may be a solid (suspension) or a liquid (emulsion), and must be at sufficiently high concentrations to exceed the close packing fraction. This allows mechanical contact between particles forming the network. One such example is peanut butter where peanut oil (which consists of 1–2% hardstock) is structured using 50% solid non-fat particles (Pernetti et al., 2007a). The major disadvantage to using a dispersed system is that it requires large amounts of structurant and is not suitable for most food products.

(5) Liquid Crystalline Mesophases:

This idea is based on forming semi crystalline scaffolds in the oil with gel-like properties. The scaffolds are composed of liquid crystalline mesophases using the oil as the continuous phase spacing the meso structures. Liquid crystals (LCs) are a state of matter with intermediate properties between a liquid and a solid crystal. In this mesophase, the molecules arrange into a short and/or long order structured lattice whereas the lattices are partially ordered in space. The lattices may display 1D, 2D, or 3D order. LCs exhibit different physical properties (optical, electrical, and visco-elastic) that distinguish them from one another, depending on their composition and structure. Liquid crystalline mesophases are categorized into three main classes: thermotropic, metallotropic, and lyotropic. The lyotropic liquid crystals commonly used as fat replacers consist of cylinders or channels filled with hydrophilic liquids such as water or polyols. (Amar-Yuli et al., 2009; Ben-Ishai et al., 2009; Efrat et al., 2010). See Fig. 1.2.

The field of organogelation in food is in its infancy and requires further research by the fats and oils community. Presented below is a short introduction to some of the most promising food-grade oleogel systems currently known. The purpose of this monograph is to expand this list significantly and provide further insight into the structure and functionality of this novel class of soft materials.

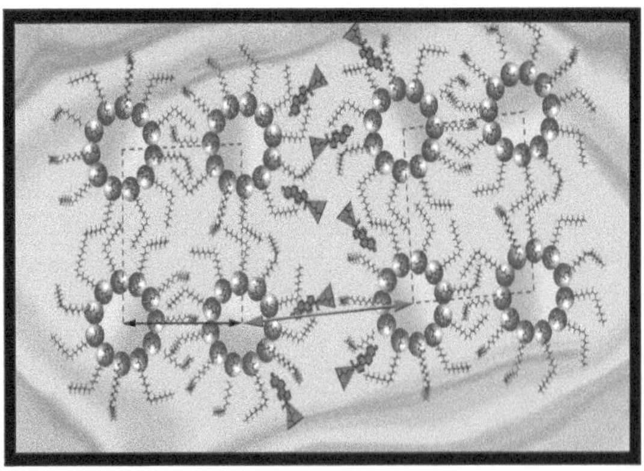

Fig. 1.2. Lyotropic liquid crystals forming a channel of hydrophillic liquid in an organic medium.

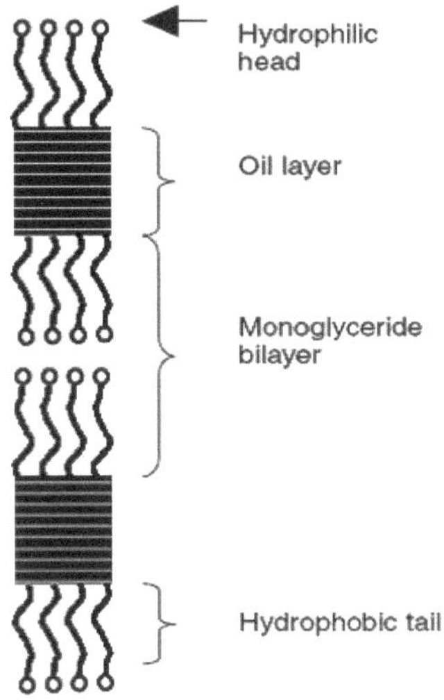

Fig. 1.3. Schematic representation of monoglyceride/vegetable oil lamellar mesophase. Source: Ojijo et al., 2004.

Monoacylglycerides (MAGs)

MAGs have many different self-assembly structures due to their amphiphilic character. These structures include lamellar, micellar, cubic, and hexagonal mesophases (Michel & Sagalowicz, 2006). MAGs can structure both lipid (Ojijo et al., 2004) and aqueous phases (Sein et al., 2002). In lipids, gel formation by MAGs is believed to occur via the formation of reverse bilayers (Fig. 1.3) (Ojijo et al., 2004). MAGs (i.e., monopalmitate and monostearate) showed gel-like behavior above 1.3%; below this concentration an entangled network resulted (Ojijo et al., 2004). At low cooling rates the MAGs align with a lamellar structure acting as a "backbone" for crystallization. Conversely, at high cooling rates, numerous fine crystals are formed in a single crystallization step. A major advantage to these molecules is that they are able to incorporate a large amount of water while maintaining a fat-like consistency (Heertje et al., 1998; Marangoni et al., 2007).

Wax Esters

Food grade waxes, in low concentrations, may also be used to structure edible oils (Daniel & Rajasekharan, 2003). For example, Candelilla wax has been demonstrated to structure safflower oil at 1% (Toro-Vazquez et al., 2007). Candelilla wax is comprised predominately of n-alkanes (~50%), which range from 29 to 33 carbons (Toro-Vazquez et al., 2007). It also contains 20–29% high molecular weight esters, 12–14%

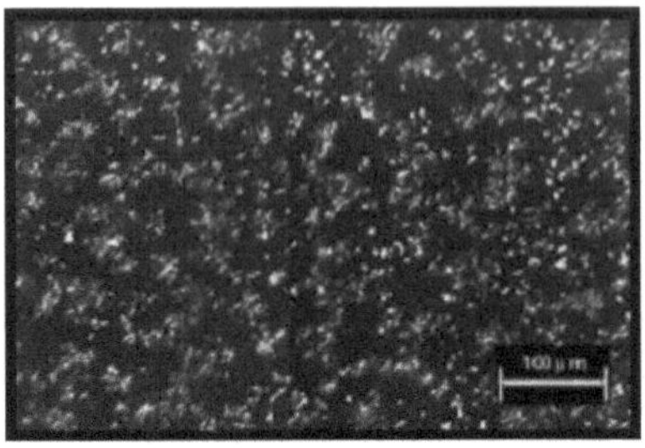

Fig. 1.4. Polarized light micrograph of 3% Candelilla wax organogel after 14 days of storage. Source: Toro-Vazquez et al., 2007.

alcohols and sterols, and 7–9% free acids (Toro-Vazquez et al., 2007). Organogelation of the vegetable oil is achieved by the n-alkanes forming small microcrystalline particles of colloidal dimensions that aggregate in time to form a 3D network entrapping the liquid oil (Fig. 1.4) (Toro-Vazquez et al., 2007).

Lecithin and Sorbitan Tristearate

Lecithin forms an unstable organogel with small amounts of a polar solvent (Kumar & Katare, 2005). As the amount of polar additive is increased, the reverse micelles begin to form flexible, long tubular micelles of 2.0 to 2.5 nm in radius and hundreds to thousands of nanometers in length (Kumar & Katare, 2005). After reaching a critical length, these extended micelles overlap, forming a transient, entangled 3D network with corresponding increases in the viscosity and viscoelastic properties. The hydrogen-bonded network is built up by the polar head groups of the lecithin molecules, and the glycerol residue stabilizes the micellar aggregates in the organic solvent. As the concentration of polar solvent increases, more solvent is incorporated into the spherical lecithin micelles, increasing the cross-sectional area of the lecithin polar region in which the solvent is arranged (Kumar & Katare, 2005). Further addition of water leads to packing constraints in the spherical micelles, which are diminished during the transition from spherical micelles to cylindrical tubules with a smaller surface curvature (Fig. 1.5) (Kumar & Katare, 2005).

Lecithin organogels have limited potential for food use due to their extreme sensitivity to water; however, sorbitan tristearate (STS) and lecithin combined in certain ratios may be a suitable replacer for hardstock fats (Pernetti et al., 2007b). Individually, neither lecithin nor sorbitan tristearate (STS) can gel oil at concentrations between 6 and 20% in the absence of small amounts of polar solvents (Pernetti et al., 2007b). Individually they form viscous pourable solutions. However, if specific ratios of lecithin:STS are added between 40:60 and 60:40, firm gels are obtained at concentrations as low as 4 wt% (Pernetti et al., 2007b). Lecithin:STS has limited use as

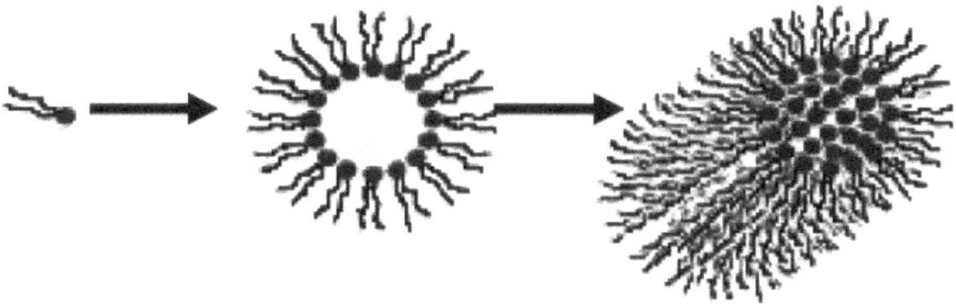

Fig. 1.5. Formation of a 3D network of reverse cylindrical micelles in lecithin organogels. Source: Kuman & Katare, 2005.

a hardstock replacer because the gel collapses at approximately 30°C and significant softening begins at 15°C (Pernetti et al., 2007b).

Phytosterols and Oryzanol

Some of the most exciting organogelation strategies involve the use of the previously mentioned low molecular weight organogelators (LMOGs), which self-assemble from the melt and grow specifically to form a 3D network that bears no resemblance to a colloidal fat crystal network, but has the ability to bind up to 99% oil in its structure.

LMOGs form gels, which are small molecules that assemble into fibers, rods, liquid crystals, micelles, or ribbons to immobilize oil. A self-assembled fibrillar network (SAFiN) is a specific type of LMOG that forms crystalline fibers to entrap liquid oil. Molecular gels occur when the LMOG undergoes supramolecular aggregation and the corresponding SAFiN forms in a solution (sol) at low concentrations of gelator molecules (~2%) (Weiss & Terech, 2006). In brief, as the sol is cooled, the solution becomes super-saturated, causing a chemical potential that is the driving force for phase separation and nucleation. Gelator molecules begin to self-assemble in stochastic nucleation events that have highly specific interactions, promoting one-dimensional growth. Crystallographic mismatches vary the degree of branching in the network structure. Fibers grow along one dimension and interact with each other to form a three-dimensional network. Fig. 1.6 shows the resulting micrographs from a system that has a high degree of crystallographic mismatches (left) and a low degree of mismatches (right).

Some materials that form these types of structures include phytosterols and oryzanol mixtures, 12-hydroxystearic acid, and 12-hydroxy ricinoelaidic acid. Only a limited number of molecules can self-assemble into these structures.

Organogels of vegetable oil have been investigated using γ-oryzanol and phytosterols at concentrations of 2–4% structurant (Bot & Agterof, 2006). Bot and

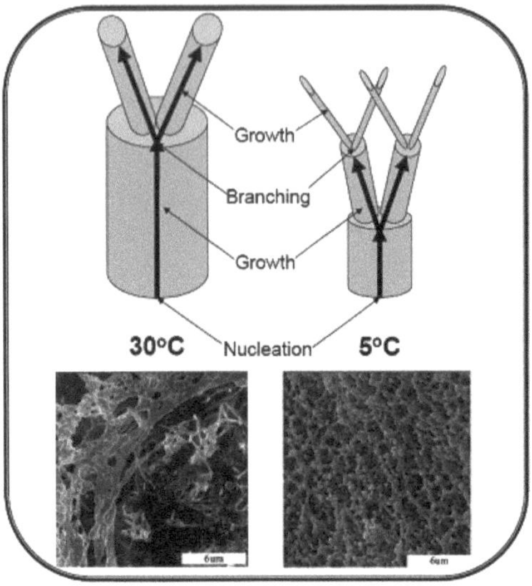

Fig. 1.6. Crystallographic mismatching causes SAFiN fibers to branch.

Agterof examined a series of phytosterols and found that dihydrocholesterol, cholesterol, β-sitosterol, and stigmasterol produced firm gels (2006). They reported that stigmasterol was the only compound that did not form a translucent gel (Bot & Agterof, 2006). Upon examination of the phytosterol, the researchers concluded that the presence of the hydroxyl group was critical for the formation of a gel. Gel formation accelerated when the ring structure did not have double bonds present, and if more than one double bond was present, a gel did not form. The composition of the alkyl group did not appear to have a major impact on the gelling ability of the phytosterol (Bot & Agterof, 2006). This group proposed a mechanism for the formation of the gel matrix (Fig. 1.7) (Bot & Agterof, 2006). The position of the hydroxyl group is critical because it limits the solubility, hence, upon cooling, the gelator molecules undergo phase separation. The phase separation minimizes the surface free energy, causing the hydroxyl groups to align, which reduces contact with the apolar solvent.

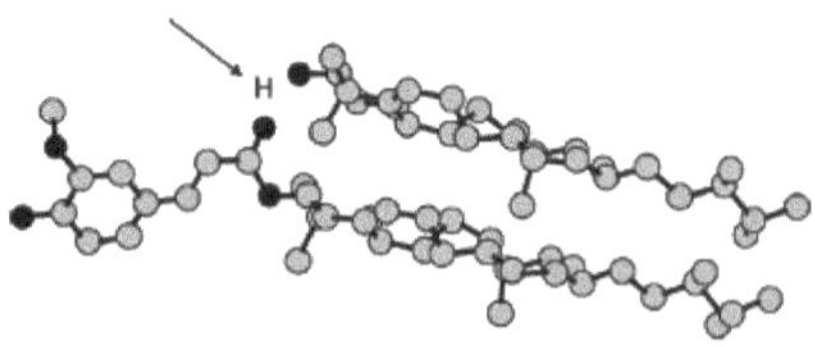

Fig. 1.7. Proposed stacking of sitosterol and oryzanol molecules, based on a preliminary energy minimization calculation (RM Hermant, Unilever R&D, unpublished data). An H-bridge can be formed near the arrow, resulting in a slightly wedge-shaped stack. Source: Bot & Agterof, 2006.

Long Chain Fatty Acids, Fatty Alcohols, and Their Mixtures

12-hydroxystearic acid (12HSA) was the first long chain substituted fatty acid reported to thicken oil-based food products (i.e., peanut butter) (Elliger et al., 1972). Since then saturated fatty alcohols have been shown to be less efficient at gelling oil than saturated fatty acids (Daniel & Rajasekharan, 2003). In addition, dicarboxylic acids (adipic, suberic, and sebacic acids) are reportedly more efficient than monocarboxylic acids of the same chain length (Daniel & Rajasekharan, 2003). A possible gelation mechanism for fatty acids is that they align themselves head-to-tail in a linear fashion. The acyl chains may form an irregular network, stabilized by intermolecular hydrogen bonding, which could entrap organic liquids. Alternatively, they could form a large spherical monolayer that entraps liquid oil in the interior hydrophobic environment, resulting in gelation (Daniel & Rajasekharan, 2003).

In a more systematic study, a series of fatty alcohols and fatty acids (FA) with chain lengths ranging from 16 to 22 carbons were evaluated for their oil-structuring potential (Gandolfo et al., 2004). The individual compounds and their mixtures structured several vegetable oils at levels as low as 2% (Gandolfo et al., 2004). Mixtures of the same chain length fatty alcohols and fatty acids resulted in a synergistic effect with gelation observed below 20°C at 7:3 and 3:7(w/w) fatty alcohol/fatty acid ratios. The synergistic effect in the 7:3 stearyl alcohol/stearic acid mixture was due to the microstructure of the composite material. The larger number of small crystals observed in the mixture was attributed to effects on the crystallization kinetics resulting from a minimal interfacial tension at the specific 1:3 and 3:1 molecular ratios (Gandolfo et al., 2004). Monolayer studies of stearic acid/stearyl alcohol showed a minimum area per molecule at the 1:3 and 3:1 molecular ratios (Patist et al., 1999), leading to a decrease in interfacial tension at these particular ratios and therefore a smaller critical radius and hence a greater rate of nucleation. Thus the crystalline network was comprised of numerous small crystals, allowing for more crystal-crystal contacts in the 3D network and enhanced oil entrainment (Schaink et al., 2007).

Applications of Organogels

The potential varied applications for organogels in the food, pharmaceutical, cosmetics, and petrochemical industries have led to an increased interest in these soft materials. The excitement is partly due to the large diversity of microscopic and mesoscopic structures possible (Terech & Weiss, 1997). In the food industry, there is potential to use organogels to minimize oil migration in multi-component foods, such as cream-filled chocolates, and to provide structure to edible oils, thereby reducing the need for saturated and *trans* fatty acids. Other applications include the stabilization of water-in-oil emulsions and controlled release media for nutraceuticals and pharmaceuticals (Hughes et al., 2009). Organogels are also being investigated for topical applications in drug delivery (Giorgano et al., 1998). Investigations are underway in the petrochemical industry to study the prospect of containing oil spills using phase selective gelation with organogelators (Bhattacharya & Krishnan-Ghosh, 2001), in addition to gelling flammable solvents for storage and transport (Abdallah & Weiss, 2000b). Some organogels have the ability to form aerogels that occur when the solvent is removed and the network structure is unaltered. Interest in SAFiNs, which include organogelator molecules, exists in a wide range of research fields and industries. Furthermore, there are many SAFiNs present in nature, including fibrous actin (Greer, 2002), clathrin (Kirchhausen, 2000), tubulin (Oakley & Akkari, 1999), keratin (Fuchs, 1995), insulin (Waugh, 1946), collagen (Caria et al., 2004), silk (Jin & Kaplan, 2003), and amyloid fibrils that are found in Alzheimer's and other neurodegenerative diseases (Lui et al., 2004). Hence, with such diverse applications, an intimate understanding of the physical chemistry of these systems is of paramount importance. The potential of organogels as novel soft materials has been underutilized thus far, on account of the inability to modify or predict the microstructure of the fibular network (Wang et al., 2006a; 2006b).

This Book

The book describes the current state of the field through the variety of possible oleogels that have been studied recently for food applications. We have brought together pioneers from different areas of the oleogel endeavor, spanning most of the different types of structuring known to date. An introduction to the whole area of oleogels is given in Chapter 2 by Zetzl and Marangoni from the University of Guelph. Here the authors provide a general overview of edible oil gels. Some of the earliest work on oleogels dates back to the use of monoglycerides to structure oils. It is quite remarkable MAGs can structure oils at very low volume fractions of solids. While they are prone to polymorphic, or more appropriately mesomorphic, transformations in time, which causes the release of oil, many valuable lessons can be learned to help guide future endeavors. Dr. Terentjev from Cambridge University reviews past work and his current work on MAG oleogels in Chapter 8. Dr. Arjen Bot (Unilever Research Laboratory), Dr. Amanda Wright (University of Guelph), and Dr. Michael Rogers (Rutgers University) present their research on low-molecular weight organogelators, namely the famous phytosterol-oryzanol system of Bot (Chapter 3), the curious unsaturated hydroxylated fatty acid organogelator system of Wright (Chapter 4), and the in-depth analysis of one of the earliest organogelators known, 12-hydroxystearic acid by Rogers (Chapter 5). These LMOGs are particularly interesting since 1% structurant is usually enough to gel 99% oil phase, which is quite remarkable. The LMOG organogelators are the ultimate example of powerful self-assembly in an organic medium, with stunning mesoscale structures being formed. Interestingly, edible waxes have been found to gel oil as well. Dr. Toro-Vazquez from the Universidad Autonoma de San Luis Potosi and Dr. Sato from Hiroshima University discuss the structure and properties of Candelilla wax organogels

(Toro-Vazquez, Chapter 6) and Rice-Bran wax (Sato, Chapter 7). The concentration of waxes required for proper structuring is higher than that for LMOGs, (5–10%), but similar to that of MAGs. Dima Libster and Nissim Garti from the Hebrew University of Jerusalem present their work on lyotropic liquid crystals in Chapter 11. This unique approach to gelling oil relies on the self-assembly of amphiphillic molecules in the presence of small amounts of polar additives. This is a new area of research, which promises some interesting insight into the structuring of oils. Tania Dey, Dennis Kim (Kraft R&D), and Alejandro Marangoni from the University of Guelph present the use of polymer-surfactant combinations to structure oils into strong gels in Chapter 13. Raffaelle Mezzenga from the ETH presents a unique method for structuring oils using proteins in Chapter 12. He managed to induce the self-assembly of protein into a cellular solid structure, with oil trapped inside. This unique structure is considerably solid-like and behaves similarly to a fat. Dr. Kiyotaka Sato contributes a second chapter, this time on 'beta gels' (Chapter 9). This is interesting work—Sato's group managed to induce massive nucleation of a pure triglyceride into an alpha form, followed by a controlled transformation into a beta form, without any re-crystallization taking place. The result was the creation of a large number of small platelets, which could gel oils below 5% concentration. This work is exciting since Sato is using only a natural triglyceride. Perhaps this is the way of the future—we need to learn how to crystallize fats such that we will need 75% less saturated fat to do the structuring job. Naomi Hughes, James Rush, and Alejandro Marangoni (H.C. Brill, University of Waterloo, and University of Guelph) summarize in Chapter 14 the first and only clinical study on edible oil oleogels. Here they report the effects of ingesting oleogels vs. free oil or margarine or butter.

This collection of papers represents the latest cutting-edge research in the area. It is the sincere hope of the editors of this book that our knowledge in this area grows exponentially, with the potential of being able to make more healthful manufactured food products, decreasing saturated fats, and eliminating *trans* fats.

References

Abdallah, A.J.; Weiss, R.G. n-Alkanes gel n-Alkanes (and many other organic liquids. Langmuir 2000(b), 16, 352–355.

Acevedo, N.; Marangoni, A.G. Characterization of the nanoscale in triacylglycerol crystal networks. Crystal Growth and Design 2010, 10, 3327–3333.

Aiache, J.M.; Gauthier, P.; Aiache, S. Int. J. Cosmetic Sci., 1992, 14, 228–234.

Amar-Yuli, J.; Libster, D.; Aserin, A.; Garti, N. Solubilization of food bioactives within lyotropic liquid crystalline mesophases, Curr. Opin. Colloid Interface Sci. 2009, 14, 21–32.

American Heart Association. Fat. <http://www.americanheart.org/presenter.jhtml?identifier=4582> (accessed March 26, 2004).

Ascherio, A.; Stampfer, M.J.; Willett, W.C. Trans fatty acids and coronary heart disease. <http://www.hsph.harvard.edu/reviews/transfats.html> (1999).

Ben-Ishai, P.; Libster, D.; Aserin, A.; Garti, N.; Feldman, Y. Molecular interactions in lyotropic reverse hexagonal liquid crystals: A dielectric spectroscopy study. J. Phys. Chem. B 2009, 113, 38, 12639–12647.

Bhattacharya, S.; Krishnan-Ghosh, Y. First report of phase selective gelation of oil from oil/water mixtures. Possible implication toward containing oil spills. ChemComm 2001, 185–186.

Borchard, W.; Burg, B. Molecular mechanisms during the thermoreversible gelation of gelatin-water-systems. Progr. Colloid Polym. Sci. 1990, 83, 200–210.

Bot, A.; Agterof, W.G.M. Structuring of edible oils by mixtures of γ-oryzanol with β-sitosterol or related phytosterols. J. Amer. Oil Chem. Soc. 2006, 83, 513–521.

Bot, A.; Flöter, E.; Lammers, J.G.; Pelan, E.G. Understanding and controlling the microstructure of complex foods, McClements, D.J., Ed. Woodhead Publishing: Cambridge, 2007; 575–599.

Bot, A.; den Adel, R.; Roijers, E.C. Fibrils of γ-oryzanol + β-sitosterol in edible oil organogels. J. Am. Oil. Chem. Soc. 2008, 85, 1127–1134.

Brouillet-Fourmann, S.; Carrot, C.; Mignard, N. Gelatinization and gelation of corn starch followed by dynamic mechanical spectroscopy analysis. Rheol Acta, 2003, 42, 110–117.

Caria, A.; Bixio, L.; Kostyuk, O.; Ruggiero, C. Elastic Scattering and Light Transport in Three-Dimensional Collagen Gel Constructs: A Mathematical Model and Computer Simulation Approach. IEEE Trans. Nanobiosci. 2004, 3, 85–89.

Daniel, J.; Rajasekharan, R. Organogelation of plant oils and hydrocarbons by long-chain saturated FA, fatty alcohols, wax esters, and dicarboxylic acids. J. Amer. Oil Chem. Soc. 2003, 80, 417–421.

Elliger, C.A.; Guadagni, D.G.; Dunlap, C.E. Thickening action of hydroxytearates in peanut butter. J. Amer. Oil Chem. Soc. 1972, 49, 536–537.

Efrat, R.; Abramov, Z.; Aserin, A.; Garti, N. Nonionic-anionic mixed surfactants cubic phases. Part I: phase Structural Chaotropic and Kosmotropic Effect. J. Phys. Chem. B 114(33), 2010, 10709–10716.

Ferry, J.D. Viscoselastic Properties of Polymers. Wiley: New York, 1961; 391.

Frohoff-Hülsmann, M.A.; Lippold, B.C.; McGinity, J.W. Aquaeous ethyl cellulose dispersion containing plasticizers of different water solubility and hydroxypropyl methyl-cellulose as coating material for diffusion pellets II: properties of sprayed films. Eur. J. Pharm. Biopharm. 1999, 48, 67–75.

Fuchs, E. Keratins and the Skin. Ann. Rev. Cell Dev. Biol. 1995, 11, 123–153.

Gandolfo, F.G.; Bot, A.; Flöter, E. Structuring of edible oils by long-chain FA, fatty alcohols and their mixtures. J. Amer. Oil Chem. Soc. 2004, 81, 1–5.

Gilsenan, P.M.; Ross-Murphy, S.B. Viscoelasticity of thermoreversible gelatin gels from mammalian and piscine collagens. J. Rheol. 2000, 44, 871–883.

Giorgano, J.; Daleo, C.; Stacks, S.M. Topical ondansetron attenuates nociceptive and inflammatory effects of intradermal capsaicin in humans. Eur. J. Pharm. 1998, 354, 1, R13.

Greer, S.C. Reversible polymerizations and aggeragations. Ann. Rev. Phys. Chem. 2002, 53, 173–200.

Heertje, I.; Roijers, E.C.; Hendrickx, H.A.C.M. Liquid crystalline phases in the structuring of food products. Lebensmittel Wissenschaft und Technologie 1998, 31, 387–396.

Hermans, P.H. Colloid Science II. Kruyt, H.R., Ed.; Elsevier: Amsterdam, 1949; 484.

Hughes, N.E.; Marangoni, A.G.; Rogers, M.A; Rush, J.W.E. Potential food applications of edible oil organogels. Trends in Food Sci. Tech. 2009, 8, 470–480.

Institute of Medicine. Letter report on dietary reference intakes for trans fatty acids. <http://www.iom.edu/report.asp?id=5410> (July 10, 2002).

Jin, H.-J.; Kaplan, D.L. Mechanism of silk processing in insects and spiders. Nature 2003, 424, 1057–1061.

Jordon Lloyd, D. Colloid Chemistry. Alexander, J. Ed.; The Chemical Catalog Co.: New York, 1926; 1, 767–782.

Judd, J.T.; Clevidence, B.A.; Muesing, R.A.; Wittes, J.; Sunkin M.E.; Pordczasy, J.J. Dietary trans fatty acids: effects on plasma lipids and lipoproteins of healthy men and women. Am. J. Clin. Nutr. 1994, 59, 861–868.

Kirchhausen, T. Clathrin. Ann. Rev. Biochem. 2000, 69, 699–727.

Kumar, R.; Katare, O.P. Lecithin organogels as a potential phospholipid-structured system for tropical drug delivery: A review. Pharm. Sci. Tech. Today 2005, 6, E298–E310.

Lee, K.Y.; Kim, Y-R.; Park, K.H.; Lee, H.G. Rheological and gelation properties of rice starch modified with 4-alpha-glucanotransferase. Int. J. Biol. Macro. 2008, 42, 298–304.

Lu, Y.; Larock, R.C. Novel Polymeric Materials from Vegetable oils and Vinyl Monomers: Preparation, Properties, and Applications. ChemSusChem 2009, 2, 136–147.

Lui, W.; Prausnitz, J.M.; Blanch, H.W. Amyloid Fibril Formation by Peptide LYS (11–36) in Aqueous Trifluoroethanol. Biomacromolecules 2004, 5, 1818–1823.

Marangoni, A.G. Elasticity of high-volume-fraction fractal aggregate networks: a thermodynamic approach. Phys. Rev. B 2000, 62, 13951–13955.

Marangoni, A.G.; Rogers, M.A. Structural basis for the yield stress in plastic disperse systems. Appl. Phys. Lett. 2003, 82, 3239–3241.

Marangoni, A.G. Fat Crystal Networks. Marcel Dekker: New York, 2004.

Marangoni, A.G.; Idziak, S.H.J.; Vega, C.; Batte, H.; Ollivon, M.; Jantzi, P.S.; Rush, J.W.E. Encapsulation-structuring of edible oil attenuates acute elevation of blood lipids in insulin in humans. Soft Matter 2007, 3, 183–187.

Mensink, R.P.; Katan, M.B. Effect of dietary trans fatty acids on high-density and low-density lipoprotein cholesterol levels in healthy subjects. N. Eng. J. Med. 1990, 323, 439–445.

Mensink, R.P.; Zock, P.L.; Kester, A.D.M.; Katan, M.B. Effects of dietary fatty acids and carbohydrates on the ratio of serum total to HDL cholesterol and on serum lipids and apolipoproteins: a meta-analysis of 60 controlled trials. Am. J. Clin. Nutr. 2003, 77, 1146–1155.

Michel, M.; Sagalowicz, L. Probing Food Structure. Food Material Science Principles and Practices. Aguilera, J.M., Lillford, P.J., Eds.; Springer: New York, 2008; 203–226.

Narine, S.S.; Marangoni, A.G. Fractal nature of fat crystal networks. Phys. Rev. E. 1999a, 59, 1908–1920.

Narine, S.S.; Marangoni, A.G. Mechanical and structural model of fractal networks of fat crystals at low deformations. Phys. Rev. E. 1999b, 60, 6991–7000.

Nyberg, L.; Nilsson, A.; Lundgren, P.; Duan, R-D. Localization and capacity of sphingomyelin digestion in the rat intestinal tract. J. Nutr. Biochem. 1997, 8, 112–118.

Oakley, B.R.; Akkari, Y.N. γ-Tubulin at ten: progress and prospects. Cell Struc. and Funct. 1999, 24, 365–372.

O'Brien, R.D. Fats and Oils, 2nd edition. CRC Press: Boca Raton, FL, 2003; 92–104.

Ojijo N.K.O.; Neeman I.; Eger, S.; Shimoni, E. Effects of monoglyceride content, cooling rate and shear on the rheological properties of olive oil/monoglyceride gel networks. J. Food Sci. Agr. 2004, 84, 1585–1593.

Ostrovskii, D.; KjØniksen, A.-L.; Nyström, B.; Torell, L.M. Macromolecules 1990, 32, 1534–1540.

Patist, A.; Devi, S.; Shah, D.O. Importance of 1:3 Molecular Ratio on the Interfacial Properties of Mixes Surfactant Systems. Langmuir 1999, 15, 7403–7405.

Pernetti, M.; van Malssen, K.F.; Floter, E.; Bot, A. Structuring edible oils by alternatives to crystalline fat. Curr. Opin. Colloid Interface Sci. 2007a, 84, 989–1000.

Pernetti, M.; van Malssen, K.; Kalnin, D.; Flöter, E. Food Hydrocolloids 2007b, 21, 855–861.

Rogers, M.A.; Wright, A.J.; Marangoni, A.G. Engineering the oil binding capacity and crystallinity of self-assembled fibrillar networks of 12-hydroxystearic acid in edible oils. Soft Matter 2008, 4, 1483–1490.

Rogers, M.A. Novel structuring strategies for unsaturated fats—Meeting the zero-trans, zero-saturated fat challenge: A review. Food Res. Int. 2009, 42, 747–753.

Rogers, M.A.; Wright, A.J.; Marangoni, A.G. Nanostructuring fiber morphology and solvent inclusions in 12-hydroxystearic acid/canola oil organogels. Curr. Opin. Colloid Interface Sci. 2009, 14, 33–42.

Rousseau, D. Marangoni, A.G. The effects of interesterification on physical and sensory attributes of butterfat and butterfat-canola oil spreads. Food Res. Int. 1998, 31, 381–388.

Ruan, D.; Lue, A.; Zhang, L. Gelatin behavrios of cellulose solution dissolved in aqueous NaOH/thiourea at low temperature. Polymer 2008, 49, 1027–1036.

Ruíz Martínez, M.A.; Muñoz de Benavides, M.; Morales Hernández, M.E.; Gallardo Lara, V. Influence of the concentration of a gelling agent and the type of surfactant on the rheological characteristics of oleogels. Il Farmaco, 2003, 58, 1289–1294.

Schaink, M.H.; van Malssen, K.F.; Morgado-Alves, S.; Kalnin, D.; van der Linden, E. Crystal network for edible oil organogels: Possibilities and limitations of the fatty acid and fatty alcohol systems. Food Res. Int. 2007, 40, 1185–1193.

Sein, A.; Verheij, J.A.; Agterof, W.G.M. Rheological characterization, crystallization, and gelation behavior of monoglyceride gels. J. Colloid Interface Sci. 2002, 249, 412–422.

Sharma, V.; Kundu, P.P. Condensation polymers from natural oils. Prog. Polym. Sci. 2006, 31, 983–1008.

Tang, D.; Marangoni, A.G. Quantitative study on the microstructure of colloidal fat crystal networks and fractal dimensions. Adv. Colloid Interface Sci. 2006, 128–130, 257–265.

Terech, P.; Weiss, R.G. Low molecular mass gelators of organic liquids and the properties of their gels. Chem. Rev. 1997, 97, 3133–3159.

Toro-Vazquez, J.F.; Morales-Rueda, J.A.; Dibildox-Alvarado, E.; Charo-Alonso, M.; González-Chávez, M.; Alonzo-Macias, M.M. Thermal and textural properties of organogels developed by candelilla wax in safflower oil. J. Amer. Oil Chem. Soc. 2007, 84, 989–1000.

U.S. Food and Drug Administration. Federal Register—68 FR 41433 July 11, 2003: Food Labeling: Trans Fatty Acids in Nutrition Labeling, Nutrient Content Claims, and Health Claims <http://www.cfsan.fda.gov/~lrd/fr03711a.html> (July 11, 2003a).

U.S. Food and Drug Administration. Food labeling: trans fatty acids in nutrition labeling, nutrient content claims, and health claims. Small entity compliance guide <http://www.cfsan.fda.gov/~dms/transgui.html> (August 20, 2003b).

Wang, R.; Lui, X-Y.; Xiong, J.; Li, J. Real-time observation of fiber network formation in molecular organogel: supersaturation-dependent microstructure and its related rheological property. J. Phys. Chem. B 2006a, 110, 7275–7280.

Wang, R.; Liu, X.Y.; Narayanan, J.; Xiong, J.Y.; Li, J.L. Architecture of fiber network: from understanding to engineering of molecular gels. J. Phys. Chem. B 2006b, 110, 25797–25802.

Waugh, D.F. A fibrous modification of insulin. I. The heat precipitates of Insulin. J. Amer. Chem. Soc. 1946, 68, 247–250.

Weiss, R.G.; Terech, P. Introduction. Molecular Gels Materials with Self-Assembled Fibrillar Networks. Weiss, R.G. Terech, P., Eds.; Springer: Dordrecht, Netherlands, 2006; 1–13.

Wright, A.J.; Marangoni, A.G. Formation, structure and rheological properties of ricinelaidic acid—vegetable oil organogels. J. Amer. Oil Chem. Soc., 2006, 83, 497–503.

Yilmaz, M.; Kusefoglu, S.H. Gelation of soybean oil with polybutadiene. J. Appl. Polym. Sci. 2005, 96, 2240–2246.

Zock, P.L.; Katan, M.B. Hydrogenation alternatives: effects of trans fatty acids and stearic acid versus linoleic acid on serum lipids and lipoproteins in humans. J. Lipid Res. 1992, 33, 399–410.

2

Novel Strategies for Nanostructuring Liquid Oils into Functional Fats

Alexander K. Zetzl and Alejandro G. Marangoni
Department of Food Science, University of Guelph, Guelph, Ontario, Canada

Introduction

The study of fats and oils and their effect on human health has been one of the most studied topics in human nutrition to date. Over decades of study, several indicators of human cardiovascular health have been identified, including blood serum cholesterol, triacylglycerols, and free fatty acid levels. Blood serum cholesterol levels have been extensively studied and are believed to be one of the most effective of these indicators. One of the first reviews on diet and its effect on serum cholesterol was conducted by Keys et al. in 1965. This group showed that serum cholesterol is greatly affected by the amount and composition of the fat that is consumed in one's diet. The group also believed that an individual's serum cholesterol could be predicted fairly accurately based on their diet (Keys et al., 1965). In the 40 years after the publication of this groundbreaking study, a much more detailed understanding of fat and its affect on serum cholesterol has been attained. Total fat can be broken down into saturated, *cis* monounsaturated, *trans* monounsaturated, and *cis* polyunsaturated fatty acids (Mensink et al., 2003). Serum cholesterol also consists of separate fractions, low density lipoprotein (LDL) and high density lipoprotein (HDL). Low density lipoprotein is commonly referred to as the "bad cholesterol" and has been used for many decades as the specific indicator for coronary artery disease (CAD) (Mensink et al., 2003). Saturated fatty acids were shown to greatly increase our levels of LDL and because of this were believed to have a deleterious effect on our cardiovascular health (Mensink et al., 2003). A temporary solution for the partial replacement of saturated fats appeared to be *trans* fatty acids. They provided a very similar functionality to saturated fats and were derived from highly unsaturated vegetable oils, which gave us the belief that they would be much healthier. Ironically, however, *trans* fatty acids have since been shown to have an even greater negative effect on our cardiovascular health than saturated fats (Mensink et al., 2003). This can be partially explained by the effects that *trans* fatty acids have on our serum cholesterol levels.

It has been recently shown that the ratio of total cholesterol to HDL may be a more effective and specific indicator for CAD than simply looking at the total change in cholesterol, or the LDL levels alone (Mensink et al., 2003). Mensink et al. compiled the results from selected studies between the years 1970 and 1997 to create one of the most extensive looks at serum cholesterol to date. From their results it was clear that the ΔTotal:HDL cholesterol ratio for saturated fatty acids increased slightly, while *trans* fatty acids increased greatly. Saturates do induce an increase in LDL levels, however, they also induce an increase in the amount of HDL. Contrary to LDL, increased concentrations of HDL in the blood through one's diet has actually been shown to decrease the risk of CAD. As the large increase in LDL may be offset by the corresponding increase in HDL, the effects of saturated fatty acids on CAD are not as pronounced when considering these two specific indicators simultaneously. *Trans* fatty acids on the other hand show an even greater increase in LDL levels than the saturated fats, yet they cause a decrease in HDL levels instead of an increase. This leads to a significantly greater increase in the ΔTotal:HDL cholesterol ratio and provides insight as to why *trans* fats have a larger deleterious effect on cardiovascular health than the saturated fats (Mensink et al., 2003). *Cis* mono- and polyunsaturated fatty acids caused a decrease in the amount of LDL and an increase in the amount of HDL. As mono- and polyunsaturated fatty acids cause increases in the levels of HDL and lead to decreases in the ΔTotal:HDL cholesterol ratio, they are beneficial to cardiovascular health and may prevent CAD (Mensink et al., 2003). Polyunsaturated fatty acids induced a slightly greater decrease in the amount of LDL compared to monounsaturated fatty acids and were shown to be the most beneficial in reducing the risk of CAD. This study showed that LDL alone should not be used as a marker of CAD but rather the corresponding change in both types of cholesterol should be noted. An increase in LDL may be offset by a corresponding increase in HDL levels, showing that the total serum cholesterol to HDL ratio might be the most effective method for determining the true risk that each fat, oil, or fatty acid imposes on cardiovascular health (Mensink et al., 2003).

Based on the total evidence from decades of research it can be concluded that diets high in *trans* and saturated fats have numerous deleterious health effects. These range from adverse effects on lipoprotein (cholesterol) profiles, increased incidence of heart disease and metabolic syndrome (Aro et al., 1997). By 2010, it is projected that in excess of 230 million people will be affected by metabolic syndrome, a diet-related disease, associated with an elevated risk of developing type II diabetes mellitus, cardiovascular disease (CVD) and premature death (Isomaa et al., 1993). Negative health effects associated with consuming *trans* and saturated fats may be reversed by altering the intake of these heart unhealthy fats and replacing them with polyunsaturated fats. It is estimated that replacing 5% of our daily energy intake from saturated fats with either equivalent energy from carbohydrates, mono-, or polyunsaturated fats would be associated with a decreased risk of CVD in the range of 22 to 37% (Roche, 2005).

In an attempt to curb these diet-related epidemics, governments across the globe are passing aggressive legislation to limit and in some cases ban the use of *trans* fats. Denmark was the first country which restricted the amount of *trans* fatty acids in industrially produced fats and oils to 2% (w/w) or less (Danish Order No. 160 of March 2003). In January 2006 the United States Food and Drug Administration and the Canadian Food Inspection Agency enacted mandatory labeling of *trans* fatty acid content in food products containing more than 0.5 g of *trans* fatty acids. Also, California and Connecticut recently passed state legislation banning all artificial *trans* fats from food products (CA AB 97, 2008, enacted, Chapter No. 207). Florida has proposed restrictions on *trans* fats in schools via the "Florida Healthier Child Care and School Nutrition Act," and one state after another is considering enacting "hard-line" legislation restricting or banning *trans* fat (National Conference of State legislatures, 2008). The American Heart Institute believes even greater restrictions are necessary, advocating that no more than 10% of daily energy should be consumed in the form of *trans* AND saturated fats combined (American Heart Association 2009). In 2007, Health Canada stated that *trans* fat should consist of no more than five percent of a food product's total fat content, and when dealing with cooking oils and margarines, it should be no more than two percent. The levels of *trans* fats in specific goods was monitored by Health Canada every six months for two years. The recently released results reported that the levels of *trans* fat in our diet is still dangerously high. Health Canada reported that 26% of chicken products, 21% of French fries, and 50% of baked good still contain higher levels of *trans* fat compared to their recommendations in 2007 (Health Canada, 2009).

With this in mind, it becomes more evident than ever that novel technologies need to be implemented when attempting to curb the associated epidemics and to address the resultant gigantic food manufacturing problem.

To fulfill these new legislative requirements, the food industry must vigorously investigate alternatives to traditional triacylglyceride (TAG) structuring. TAGs provide structure in numerous food products including ice cream, cheese, butter, lard, etc. Unfortunately, it is the *trans* and saturated lipids that provide the structure and solid-like properties of these foods. Although the structure they confer on products is desirable, and indeed required in many products, both types of fatty acids have been shown to negatively influence human health. Since it is the hardstock TAGs that are responsible for network structure, it is often difficult or impossible to eliminate these ingredients to improve the health characteristics of a food product without sacrificing some of the quality associated with that particular food product. Food manufacturers are very reluctant to change the characteristics of successful industrially produced food products to any extent! Thus, the task of replacing a major ingredient responsible for many of the quality attributes of a food product is an extremely difficult task, and an even harder sell. Pernetti eloquently states that finding alternatives to TAGs with healthy properties, versatility, and performance is a tremendous challenge (Pernetti et al., 2007a).

Without being overly critical of the food manufacturing industry, the "solution" to replacing *trans* fats adopted by industry has been the replacement of partially hydrogenated fats with saturated fats from palm "oil". Palm oil is a misnomer since it should, by definition, be referred to as palm "fat", since it is solid at room temperature (20–25°C). This is obviously true for palm kernel "fat" and coconut "fat" as well. Calling these materials "oils" is incorrect and should be phased out of our language. For the record, the fatty acid composition of palm fat is included in Table 2.A below.

Lauric acid (12:0) has been shown to cause a health-beneficial decrease in the ΔTotal:HDL cholesterol ratio when consumed, while stearic acid (18:0) shows a neutral effect on the ΔTotal:HDL cholesterol ratio. Palmitic acid (16:0) on the other hand has been shown to be the most atherogenic saturated fatty acid (Mensink et al., 2003; Mensink, 2005). Palmitic acid shows an opposite effect upon our ΔTotal:HDL cholesterol ratio compared to lauric and stearic acid and increases the ratio substantially. What is concerning is that palm fat contains no lauric acid, minimal amounts of stearic acid, and high levels of the 16:0 palmitic acid.

A recent study by the University of Texas Southwestern Medical Center reported that palmitic acid may also block signals by the hormone leptin (Hannam, 2010). Leptin is responsible for telling the brain that the stomach is full. If this signal is partially blocked or impaired, individuals will tend to overeat. Snacks consumed over the weekend as a reward for 5 days of healthy eating have been shown to keep extra weight off over the long term (Hannam, 2010). The problem is that these snacks have now been shown to affect the amount we eat after the weekend is over. Many snacks that are commonly consumed as "treats", are high in saturated fat, which in turn is high in palmitic acid. The palmitic acid blocks the signaling of the leptin hormone, and causes us to feel less full on the Monday after the weekend. This phenomenon has created the term "Monday Diet Hangover", and can fully be attributed to the effect of palmitic acid (Hannam, 2010). Other fats however, do not have any affect upon leptin signals. Monounsaturated fats, specifically oleic acid (omega-9s),

Table 2.A. Fatty Acid Composition (% w/w) of Palm Fat.

Fatty Acid	Concentration
14:0	1.8
16:0	55
18:0	2.5
18:1	30
18:2	10
18:3	0.40
20:0	0.10
Total Saturates	60

Source: Firestone, Physical and Chemical Characteristics of Oils, Fats, and Waxes, AOCS Press, 1999 (Firestone, 2006).

Table 2.B. Novel Strategies For Structuring Liquid Oils.

Oil Structuring Mechanism	Required Structuring Agent %
Structured Emulsions	3–6 %
Fat Crystal Networks of Stearic Acid Rich Fats in High Oleic Oils	> 40 %
Wax Ester Organogels	0.5–10 %
Polycrystalline Fibrillar and Particle Organogels	1–8 %
Polymer Organogels	3–10 %

do not impair leptin signaling and help us feel full. This is another advantage of the monounsaturated fats; not only are they more heart healthy, but we tend to eat less of them due to the neutral effect on leptin's signals to the brain (Hannam, 2010).

This means that palm fat is clearly not a healthy alternative to *trans* fat, which was in turn, shown to be a very unhealthy alternative to saturated fat. If anything, palm fat is a temporary fix to the *trans* fat dilemma. Even the "fix" is only in consumer consent, as the general public is not yet fully aware of these newer health studies, and is for the most part still open to alternatives like palm fat, as long as the saturates and *trans* fats are reduced or eliminated from the food product. In an ever growing health conscious society however, this may rapidly change, indicating the urgent need for the development of a functional *and* healthy substitute to traditionally used fats in food products. To create a "healthy", functional fat, the health advantageous characteristics of monounsatured and polyunsaturated fats needs to be utilized. The aim of this review is to examine strategies for replacing colloidal fat crystal networks comprised of *trans* and saturated fats. These strategies include small-molecule organogels of unsaturated oils, structured oil-in-water emulsions, and polymer gels. This chapter is not meant to be an exhaustive review, but an overview of the work in this area. Table 2.B lists the novel structuring mechanisms that will be described in this chapter.

Structured Emulsions

At the core of this technology is the creation of an oil-in-water emulsion stabilized by hydrated saturated monoglyceride multilayers. These multilayers are composed of monoglyceride bilayers with large amounts of water "sandwiched" in between.

The monoglyceride bilayers are formed at temperatures above their Krafft temperatures and are in the L_α state. Upon cooling, the fatty acid chains of the hydrated monoglycerides crystallize in a hexagonal conformation (α subcell polymorph) to an orthorhombic or triclinic conformation (β' or β polymorph). The material composed of hydrated alpha polymorphic monglyceride is referred to as an alpha gel while the material composed of hydrated beta prime or beta polymorphic monoglyceride is referred to as the coagel or beta gel. There is no separate nomenclature for the beta prime structured material; it is grouped with the beta gel.

A co-surfactant is added along with the saturated monoglyceride to enhance the formation and stabilization of this, so called, α-gel state. While these hydrated monoglyceride multilayers are formed, oil is added to the system, which is then exposed to an external shear field. A monoglyceride monolayer then surrounds the oil droplets upon which the hydrated saturated monoglyceride multilayers can deposit. Thus, in the end, the oil becomes encapsulated within hydrated monoglyceride multilayers, forming an oil-in-water emulsion. Upon crystallization of these multilayers, the oil becomes effectively microencapsulated within hydrated crystalline spheres.

By adjusting the surface charge of the vesicles by judicious addition of a charged co-surfactant and/or medium pH adjustments, it is possible to fine-tune vesicle-vesicle interactions to create a range of rheological behaviors, from liquid to solid, and also prevent the alpha gel to coagel transformation (Marangoni et al., 2007; Batte et al., 2007a; Batte et al., 2007b; Marangoni et al., 2008). The surface charge can be adjusted so as to create a semisolid material of high viscosity that structurally resembles a cellular solid, as shown in Figure 2.1. As well, refer to Figs. 2.2 and 2.3.

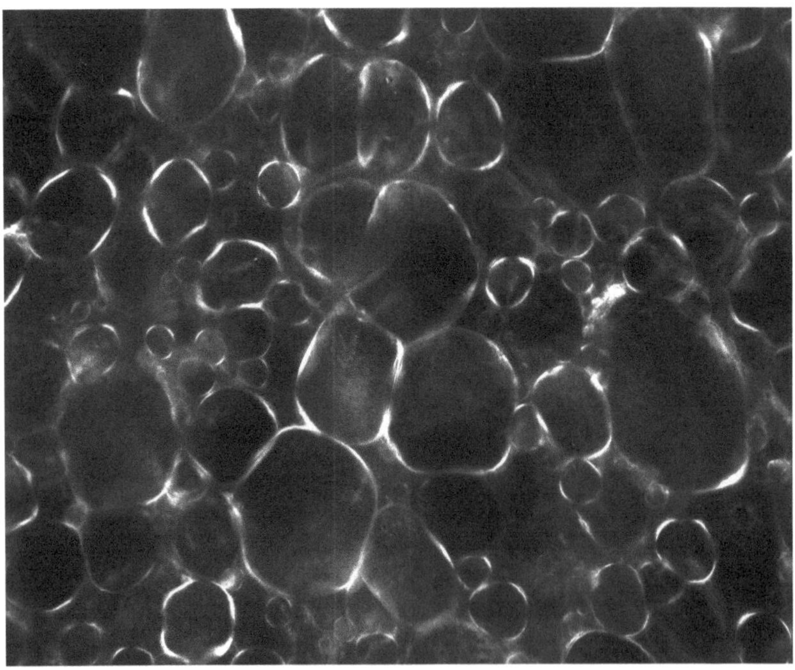

Fig. 2.1. Polarized light micrograph of a cellular solid composed of hydrated crystalline monoglyceride walls where liquid oil is encapsulated within the cellular lumen.

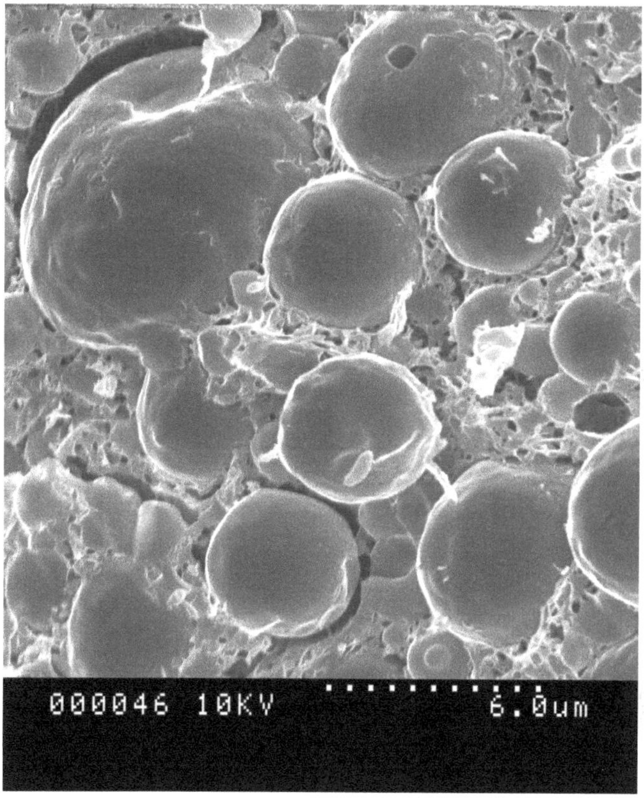

Fig. 2.2. Scanning Electron Micrograph of a structured oil-in-water emulsion similar to CoaSun™.

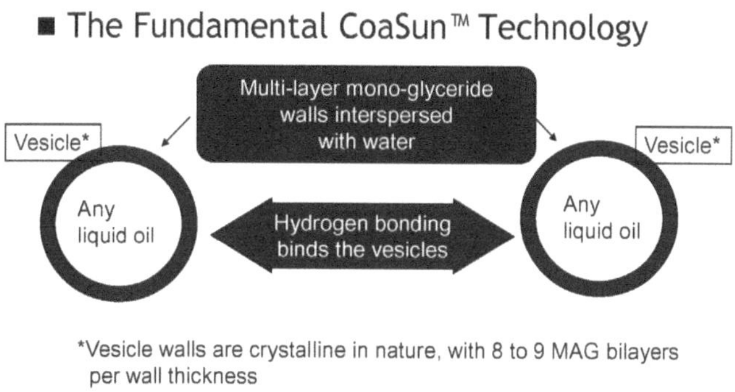

Fig. 2.3. Principle behind the technology involved in the creation of a structured oil-in-water emulsion that can be effectively used as an all-purpose shortening.

This structured emulsion, now a commercial product marketed under the brand CoaSun™, can be effectively used to manufacture a wide variety of bakery products, particularly cookies, cakes, scones, muffins, and waffles. The great advantage of these materials is their low saturated fatty acid content, which makes it possible to both manufacture healthier food products and also make claims on food labels. If prepared with canola oil, CoaSun™ will contain between 7 and 8% total saturates and can usually replace traditional shortenings on a 1:1 basis (Huschka et al., 2010a; Huschka et al., 2010b, Marangoni, 2007).

Another nutritional advantage of structured emulsions such as the shortening-like CoaSun™ product is that they contain 30–40% water. As stated previously this water lies between monoglyceride bilayers that surround the oil droplets. Fig. 2.4 shows how the monoglyceride bilayers would form into droplets; the outer walls of the droplet consist of multiple bilayers, each holding a layer of water between the hydrophilic head groups.

CoaSun™ therefore contains less fat due to the addition of water between the bilayers and a reduction in the total amount of saturated fatty acids, as only unsaturated vegetable oils are used to make the shortening. The addition of a large percentage of water to make CoaSun™ can significantly cut production costs, however, baking parameters must be adjusted accordingly.

In a controlled dietary study performed by Rush et al. in 2009, meals of butter, margarine, canola oil, and CoaSun™ (referred to as MAG GEL in the study) were consumed to determine their effects on postprandial triglyceride levels. Three to six hours after consumption of the CoaSun™ meal, individuals' postprandial triglyceride levels were significantly lower compared to when they consumed the meals of butter, margarine, or canola oil. The surprising result to this study was the significant

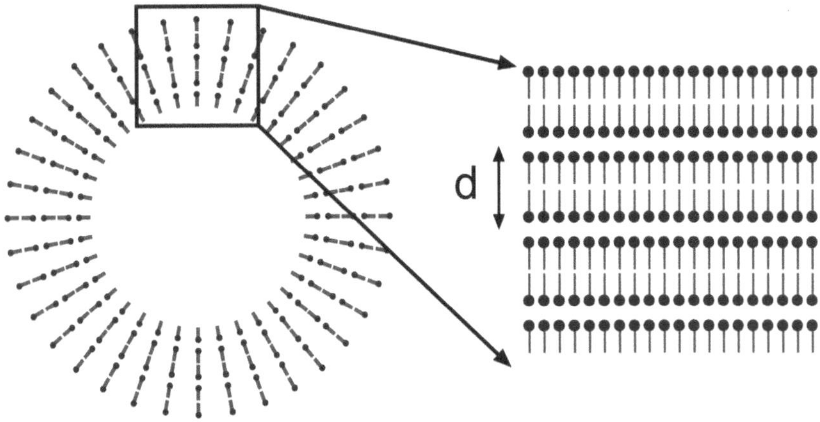

Fig. 2.4. Schematic diagram depicting a structured emulsion droplet consisting of hydrated crystalline monoglyceride bilayers.

rise in triglyceride levels caused by the consumed oil, compared to the meal of CoaSun™, which had the same fatty acid composition. There was also no significant difference in the triglyceride responses among the butter, margarine, and oil trials, even though they contained a wide range of saturated fat. This clearly indicates that it is the unique structure of CoaSun™ that is responsible for the beneficial metabolic effects of the treatment (Rush et al., 2009). This is confirmed by a previous study by Rush et al., which showed that the Coasun™ and oil trial differences in metabolic response were eliminated when CoaSun's unique structure was destroyed. This study also showed that ingestion of CoaSun in the form of a margarine-like spread resulted in beneficial suppression of blood lipid and insulin responses in humans compared to compositionally equivalent controls lacking the monoacylglyceride structure (Rush et al., 2008).

Another benefit of this structuring mechanism is the ability to encapsulate and partially protect omega-3 fatty acids, phytosterols, phytstannols, fat-soluble antioxidants, and many other bioactive compounds. Omega-3 fatty acids in particular have become a highly sought addition to a variety of food products, including yogurts and juices, due to their requirement for normal functional development of the retina and brain, particularly in premature infants. Omega-3 fatty acids also increase bleeding time; decrease platelet aggregation, blood viscosity, and fibrinogen; and they increase erythrocyte deformability, thus decreasing the tendency of thrombus formation. In patients with hyperlipidemia, omega-3 fatty acids decrease low-density-lipoprotein cholesterol if the saturated fatty acid content is decreased. They also consistently lower serum triglycerides in normal subjects and in patients with hypertriglyceridemia, whereas the effect on high-density lipoprotein (HDL) varies from no effect to a slight increase. However these fatty acids are highly labile to oxidation, which is why encapsulation is not only advantageous but in many cases necessary to attain any of their benefits (Simopoulos, 1991).

Phytosterols have recently become of great interest due to their cholesterol fighting effects, which will be further described later in this chapter. To take advantage of these effects, CoaSun™ can be made with phytosterol/stanol enriched oil. Cookies made using this enriched oil are shown in Fig. 2.5. As can be seen by their nutritional facts label, the cookies have 0 g of *trans* fat in addition to 0 g of saturated fat, which is in itself novel. Add to this the fact that the cookies have the additional health benefits of phytosterols and it is a truly groundbreaking product!

Overall, there are many benefits to a structured emulsion product like CoaSun™. These benefits include a reduced total fat content, a reduction in the amount of saturated fat, regulation of blood lipid and insulin responses in humans, and lower postprandial triglyceride levels; however, the higher water content in this product means that modifications to current production lines may be required, potentially adding significant implementation costs. Also, while the shortening-like product is suitable for many baking applications, it does not provide the proper textural parameters required for the production of laminates or fillings.

Nutrition Facts

Serving Size (35g)
Servings Per Container

Amount Per Serving

Calories 120	Calories from Fat 25
	% Daily Value*
Total Fat 3g	5%
Saturated Fat 0g	0%
Trans Fat 0g	
Cholesterol 5mg	2%
Sodium 75mg	3%
Total Carbohydrate 23g	8%
Dietary Fiber 1g	4%
Sugars 10g	
Protein 2g	

Vitamin A 0% • Vitamin C 0%
Calcium 0% • Iron 4%

*Percent Daily Values are based on a 2,000 calorie diet. Your daily values may be higher or lower depending on your calorie needs:

	Calories:	2,000	2,500
Total Fat	Less than	65g	80g
Saturated Fat	Less than	20g	25g
Cholesterol	Less than	300mg	300 mg
Sodium	Less than	2,400mg	2,400mg
Total Carbohydrate		300g	375g
Dietary Fiber		25g	30g

Calories per gram:
Fat 9 • Carbohydrate 4 • Protein 4

Fig. 2.5. Phytosterol enriched oatmeal-raisin cookie using the structured emulsion product CoaSun™. These cookies contain 0 g of saturated fat and 0 g of *trans* fat.

Interesterified High-stearic Acid Fats

Interesterification, either chemical or enzymatic, can improve the functionality and physical properties of fats and oils (List et al., 1995). This method has been a primary tool in the creation of plastic fats for baked goods, and the reduction of *trans* fatty acids in these products (Dijkstra et al., 2008). Reducing the dietary intake of *trans* fatty acid has been recommended by various scientific studies, as well as public and regulatory policy (Berger et al., 2005; Eckel et al., 2007). For example, the American Heart Association recommends that the intake of *trans* fatty acids be less than 1% of total daily energy (Lichtenstein et al., 2006). A recent study showed that a 2% increase in energy intake from *trans* fatty acids forms was associated with a 23% increase in the risk of incidence of coronary heart disease (Eckel et al., 2007).

To achieve similar functionality as partially hydrogenated fat in a fat product, a saturated hard stock such as fully hydrogenated rapeseed oil or fully hydrogenated soybean oil rich in stearic acid can be used. Early studies noted that stearic acid (18:0) has a neutral effect on serum cholesterol levels (Hegsted et al., 1965; Keys et al., 1965). Another study examined the effect of cocoa butter and structured fat with high stearic acid content and showed that stearic acid elicits a hypocholesterolemic effect compared with other saturated fatty acids (Bonanom & Grundy, 1988). More recent work strongly suggests that stearic acid does not affect lipoprotein metabolism to a significant extent (Mensink et al., 2003). Furthermore, consumption of oleic acid has been shown to have beneficial effects on human health, such as decreased vulnerability of plasma low density lipoprotein (LDL) to oxidation (Wahrburg, 2004). Moreover, oleic acid is more stable toward oxidation relative to linoleic and linolenic acids (Kris-Etherton et al., 1999; Prudy, 1985).

Thus, a possible strategy for structuring food products with less harmful fats is to use mixtures of fully hydrogenated vegetable stocks with high-oleic acid oils. However, these fats tend to have poor functionality (high melting point, waxy mouthfeel) and thus interesterification using sodium methoxide or lipases is necessary. Fig. 2.6 shows the solid fat content vs. temperature profiles for non-interesterified (NI), chemically interesterified (CI) and enzymatically interesterified (EI) mixtures of 30% fully hydrogenated canola oil in high oleic sunflower oil. The enzymatic interesterification was carried out using a random lipase from *Candida antartica.*

The SFC-temperature profile of the interesterified samples suggests that the interesterfied are a much more functional fat. Moreover, the NI crystals were in the β polymorphic form, while the EI and CI samples were in the β′ form, which results in an overall reduction in crystal cluster size (Ahmadi et al., 2008a; 2008b; 2009; Ahmadi & Marangoni, 2009). This can easily be seen in Fig. 2.7, where the NI fat groups in large crystal aggregates at 30°C, while the CI and EI samples do not.

It is evident that these mixtures require a chemical randomization of the fatty acids on the TAG backbone to be functional in food applications. This randomization results in a wider melting range of the fat, which is highly desirable. For years the β′ has been the polymorphic form that is most sought for use in margarines and shortenings. The β′ polymorph is believed to be the most functional polymorphic form for these products because it usually contains crystals of small size and needle-like shape. Compared to large crystal aggregates, these small needle-like crystals provide a smoother texture and better mouth-feel (Ghotra et al., 2002). The larger melting range caused by a mixed triglyceride composition also becomes obvious when thinking

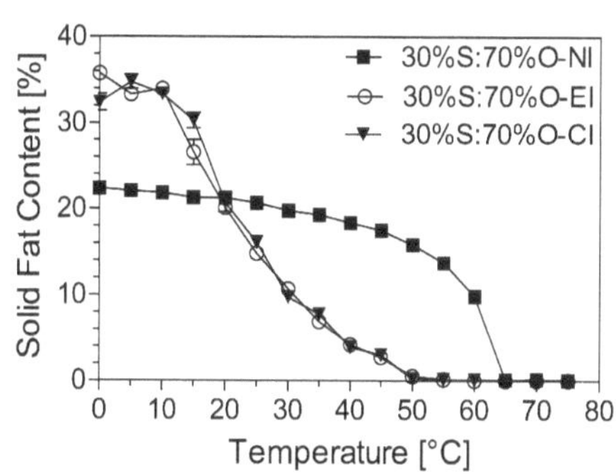

Fig. 2.6. Solid fat content vs. temperature profiles for 30% mixtures of fully hydrogenated canola oil in high oleic sunflower oil (CI: chemically interesterified, EI: enzymatically interesterified; NI: non-interesterified).

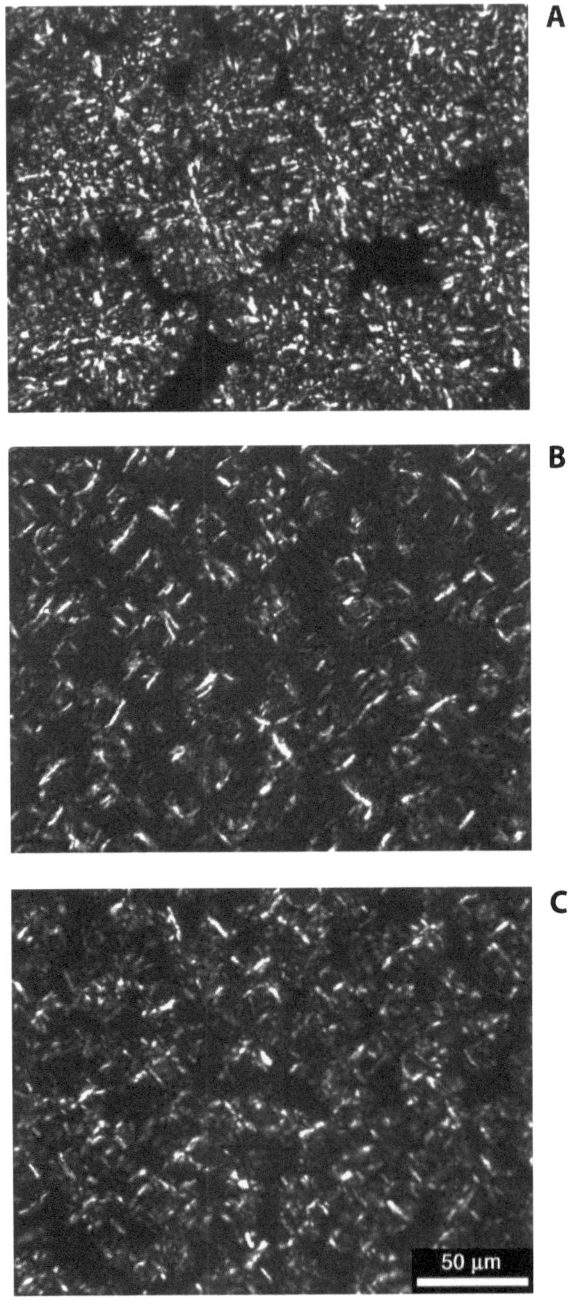

Fig. 2.7. Polarized light micrograph of 40% hydrogenated canola oil in high oleic sunflower oil crystallized at 30°C. A) Non Interesterified, B) Chemically Interesterified, C) Enzymatically Interesterified.

of the functional properties of a product like margarine. Margarine has a small melting range, quickly changing from hard to melted, which is clearly undesirable. By combining the small crystal size of the β′ form with a mixed triglyceride profile, a superior margarine-like fat can be produced.

Fig. 2.8 shows the changes in the storage modulus G′ (A) and Yield Force (B) upon chemical and enzymatic randomization of some of the mixtures. The increase in G′ and yield force of the chemically interesterified samples was due to a decrease in crystal size (Ahmadi & Marangoni, 2009). This increase is beneficial since similar structuring can thus be achieved with less structuring material (i.e., less fat).

This family of fats is highly functional and can be used to manufacture shortenings used in baking applications, such as soft cookies, without any changes in formulations or process (Ahmadi & Marangoni, 2009). Thus, if the motivation of food

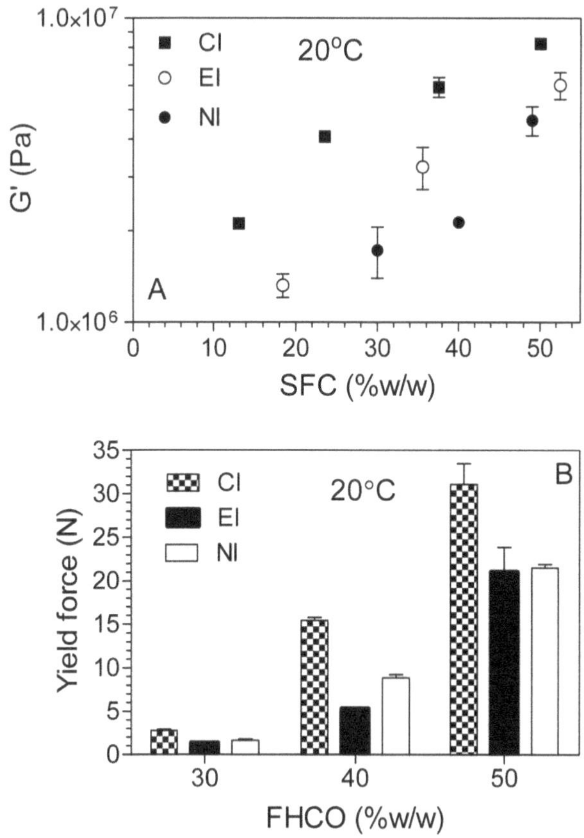

Fig. 2.8. A) Changes in the storage modulus (G′) as a function of solid fat content (SFC), and (B) yield force as a function of the relative proportion of fully hydrogenated canola oil (FHCO) in high-oleic sunflower oil upon chemical (CI) and enzymatic (EI) interesterification of non-interesterified (NI) mixtures.

manufacturers is merely the removal of *trans* fatty acids, without any need to reduce the saturated fatty acid content at the same time, these high-stearic acid fats represent an interesting option.

Through the reduction of *trans* fatty acids, the health characteristics of these mixtures may be greatly improved, however, the word "hydrogenated" would still be included on ingredients labels. This word is avoided by many companies wherever possible because there is a widely held public belief that hydrogenated fats are detrimental to cardiovascular health. Though this is only the case in specific situations, it will be difficult to re-educate the general public as to when exactly hydrogenation can cause heart detrimental fats and oils. This is the main reason why other novel structuring methods should be pursued, as many of these do not have the same public stigma.

Organogels

A substance is a gel if it has a continuous structure with macroscopic dimensions that is permanent on the time scale of an analytical experiment and is solid-like in its rheological properties (Flory, 1953). The formation of a gel is influenced by several factors, any of which could change the mechanical properties of the gel, or completely prevent its formation. These factors include the following: temperature, molecular weight of the polymer, structure of the solvent, presence of surfactant, concentration, incorporation method, and whether there are any interactions with other components. It is difficult to optimize these conditions to create a product that matches a set of requirements. Tiny alterations in one factor can radically change the resulting gel.

Some patterns have been identified when trying to understand which conditions can be used to obtain a gel with specific characteristics, but often an initial trial and error approach must be taken to try and produce the gel.

An organogel can be defined as an organic liquid entrapped within a thermoreversible, three-dimensional gel network. This gel network is formed by the self-assembly of a relatively low concentration of organogelator molecules into long crystalline fibers (Fig. 2.9), and are thus called Self-assembled fibrillar networks or SAFINs (Hughes et al., 2009; Rogers et al., 2007).

SAFINS are found naturally and include the following materials: fibrous actin, clathrin, tubulin, keratin, insulin, collagen, silk, and amyloid fibrils that are found in Alzheimer's and other neurodegenerative diseases (Rogers et al., 2007). The difference between SAFINS and platelet crystals is that SAFINS show only one-directional growth whereas platelet crystals show two-dimensional growth. This is clearly illustrated in Fig. 2.9 where many of the SAFIN fibers posses a similar orientation because most of them would have originated at the same nucleation point (Weiss & Terech, 2006).

Organic solvents can actually be gelled at organogelator concentrations as low as 0.5%. Depending upon the chemical properties of the organogelator, gels can be formed from organic solvents (benzene or hexane) or liquid oils.

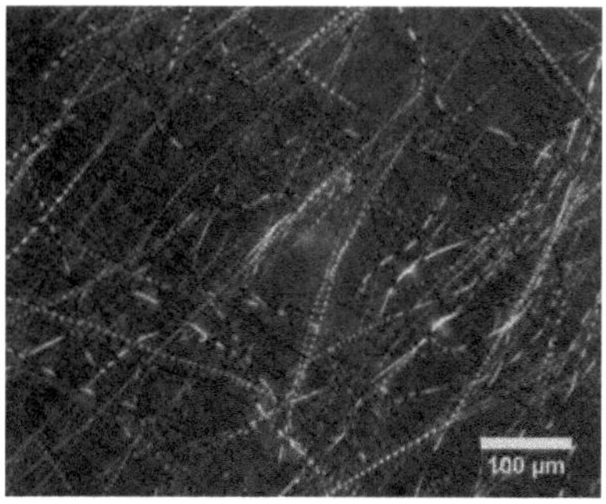

Fig. 2.9. Polarized light micrograph of a 12-hydroxystearic organogel at a 2.5% (w/w) concentration in rapeseed oil. Notice the alternating dark and bright areas indicative of a helical twist in the fibers.

The wide variety of both current and future applications of organogels has been outlined in two comprehensive reviews (Terech & Weiss, 1997; Abdallah & Weiss, 2000). While some industries are already making use of organogel technology, many of its potential applications are still in the research and development phase.

The potential applications of organogels in food systems are numerous and offer a revolutionary option for nano and microstructuring edible oils into functional fats. A number of organogelator systems have the ability to gel edible oils at low concentrations (0.5–2.0% wt). General categories of network-forming edible oil structurants include Triacylglycerols (TAGs), Diacylglycerols (DAGs), Monoacylglycerols (MAGs), fatty acids, fatty alcohols, waxes, wax esters, sorbitan mono-stearate, as well as the following mixtures: fatty acids and fatty alcohols, lecithin and sorbitan tri-stearate, and phytosterols and oryzanol (Pernetti et al., 2007a). More specifically, the following organogelators or mixtures thereof are well-known for their ability to structure edible oils: 12-hydroxystearic acid (Rogers et al., 2007; Terech & Weiss, 1997; Rogers et al., 2008a; Rogers & Maragoni, 2008a; Rogers et al., 2008b, Rogers & Marangoni, 2008b), ricinelaidic acid (Wright & Marangoni, 2006; Wright & Marangoni, 2007), candelilla wax (Toro-Vazquez et al., 2007a), mixtures of β-sitosterol and γ-oryzanol (Bot & Agterof, 2006), mixtures of stearic acid and stearyl alcohol (Gandolfo et al., 2004; Schaink et al., 2007), and mixtures of lecithin and sorbitan tri-stearate (Pernetti et al., 2007b), and more recently mixed ceramides (Rogers et al., 2009). Regardless of the network composition, organogels have numerous potential functionalities in food products including (but not limited to) the restriction of oil mobility and migration, the replacement of saturated and *trans* fats, the stabilization of emulsions, and the ability to control the rate of nutraceutical release. Fig. 2.10 shows one of the most promising applications of these organogel systems, the purely physical stabilization of water-in-oil emulsions. This emulsion gel was stable for over a month at room temperature conditions regardless of the large size of the water droplets. The birefringent network of 12-HSA fibers can be appreciated throughout the continuous oil gel.

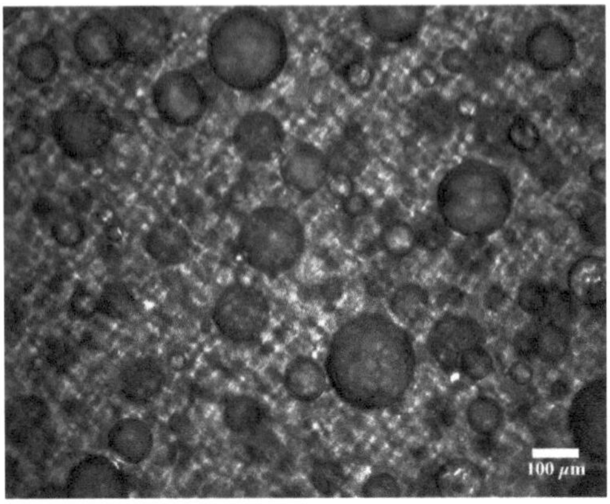

Fig. 2.10. Polarized light micrograph of a 12-hydroxystearic organogel at 2% (w/w) concentrations stabilizing large water droplets. Notice the birefringent fibers in the continuous oil gel phase.

When creating an organogel using 12-HSA and canola oil, as in Fig. 2.11, a change in storage temperature from 5 to 30°C results in drastically different gel structures even though all other variables are kept constant. The gel stored at 5°C is fairly opaque while the gel stored at 30°C is quite transparent.

Although 12-HSA has not yet been approved as a food-grade additive, it is derived from a natural product, castor seed oil. Due to 12-HSA's ability to produce strong, thermoreversible gel networks at low concentrations (0.5–2.0% w/w), it was chosen by Hughes et al. in 2009 to test the potential food application of edible oil organogels as a replacement for conventional fat-based spreads (Hughes et al.,

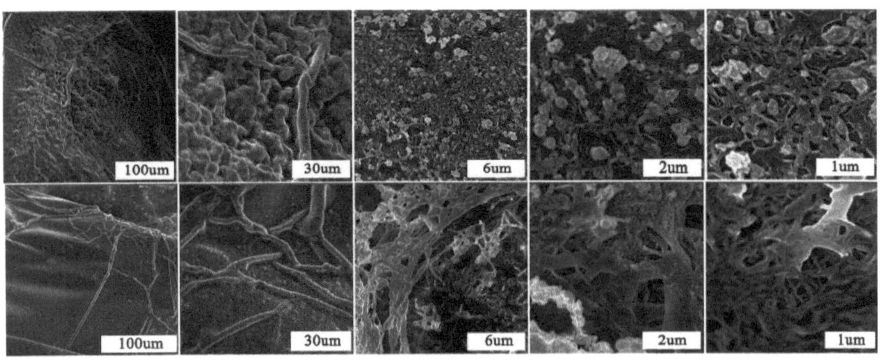

Fig. 2.11. Cryo-SEM of 12-HSA organogels imaged at 5°C (top panels) and 30°C (bottom panels). Images at 100 μm and 30 μm used osmium tetroxide fixation and isobutanol washing, while the images with scale bars of 6, 2, and 1 μm were produced by washing with a hexane/acetone solution. Figure adapted from Rogers et al., 2007.

2009). This study was conducted in a very similar manner to that of Rush et al. in 2009, however, the MAG Gel Coasun™ contained a large proportion of water while the 12-HSA gel in Hughes' study contained no water. In Hughes' study, the nutritional response to the acute ingestion of the 12-HSA organogel, butter, margarine, and canola oil was measured. The results showed that the 12-HSA organogel, which consisted primarily of canola oil (98% w/w), caused mean postprandial serum triacylglycerol levels that were significantly lower than those corresponding to butter and margarine, yet were not significantly different from the results corresponding to canola oil. The mean postprandial serum free fatty acid levels corresponding to the organogel were also significantly lower than those of butter and margarine but were not significantly different from canola oil. Though the 12-HSA organogel provides the same response benefits as canola oil, the structured gel network prevents the oil from leaking into the food product on which it is applied, and also provides a more spread-like appearance when compared to an unstructured oil (Hughes et al., 2009). If a certified food grade organogelator could be used in place of 12-HSA, it is probable that it would provide the same spread-like structure while maintaining the reduced postprandial serum free fatty acid and triglyceride levels.

The health attributes of SAFINS will be further discussed in the chapter entitled "Clinical Study on 12-hydroxystearic Acid Organogel Ingestion."

The greatest challenge in this field is to find an organogelator that is not only food grade, but offers a functional advantage as well. The phytosterol-oryzanol system developed in Bot's group at Unilever R&D is the most promising system in this respect. On one hand, the material forms optically transparent gels structured by crystalline nanotubes of 7 nm diameter (Bot et al., 2008), while also having the added value of the proven cholesterol-reducing properties of phytosterols.

As can be seen in Fig. 2.12, the chemical structure of β-sitosterol and 5α-cholestan-3β-ol are very similar to that of cholesterol. β-sitosterol differs only based on its alkyl side chain while the absence of a double bond in 5α-cholestan-3β-ol's ring structure is the only difference between it and cholesterol (Bot et al., 2008). Phytosterols such as these are able to compete with cholesterol for absorption sites in the small intestine, reducing the total cholesterol uptake by the body. As an added bonus, the phytosterols themselves are not absorbed by these sites and do not contribute to the calorie content of a food (Bot et al., 2008).

Unfortunately, none of structures from Fig 2.12 have the ability to structure oil individually. However, as shown by Bot's group, phytosterol-oryzanol mixtures can have this ability. Dr. Bot's latest work illustrates the need for a greater understanding of the ratios required to form an effective phytosterol-oryzanol gel. Several mixtures were prepared in an attempt to characterize phytosterol:γ-oryzanol systems and to try and better understand the self-assembly process. Prepared mixtures used heavy mineral oil and various combinations of the structures from Fig. 2.12. These included 8:8 wt%:wt% cholesterol:γ-oryzanol, 8:8 wt%:wt% 5α-cholestan-3β-ol:γ-oryzanol, and 8:8 wt%:wt% β-sitosterol:γ-oryzanol. Several other ratios of cholesterol:γ-oryzanol

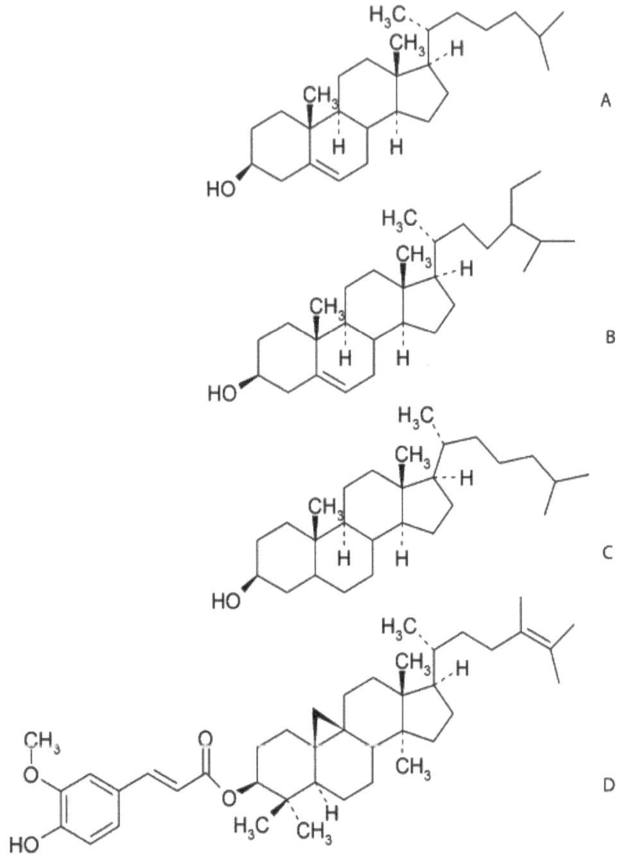

Fig. 2.12. Chemical structures for cholesterol (A), β-sitosterol (B), 5α-cholestan-3β-ol (C), and γ-oryzanol (D). Source: Rogers et al., 2010.

were tested, however, optically transparent gels are only formed when the amount of phytosterol is equal to the amount of γ-oryzanol. A minimum activation energy of nucleation is observed when there is a 8:8 wt%:wt% ratio of cholesterol and γ-oryzanol, which results in the formation of fine tubules that refract minimal light. Ratios other than 8:8 wt%:wt% cause an increase in the activation energy of the system, which subsequently increases the average crystal size and allows for the formation of more junction zones, which are capable of diffracting light. If optically transparent gels are desired, this greatly reduces the potential combinations in which phytosterol:γ-oryzanol gels can be created. If more opaque gels are useable or required, there appears to be a fairly large combination of mixture concentrations that can be used to create physically similar gels (Rogers et al., 2010).

This work also included several other key discoveries. A hydroxyl group on the phytosterols is essential for assembly into tubules; an absence of double bonds in the phytosterol rings increases the rate of assembly; and thirdly the aliphatic side chain has a moderate effect on the occurrence of gelation (Rogers et al., 2010). Even though the phytosterol-oryzanol system is only at the preliminary stages of being understood, our group has developed a similar system using ceramides that follows in the footsteps of Bot's phytosterol-oryzanol work—self-assembled fibrillar network structuring and a powerful pharmacological effect.

Ceramides are a polar class of lipids and are capable of gelling edible oil at concentrations as low as 2% structuring agent. Ceramides possess a great variability in regard to chain length, saturation, and chemical substitution. Pure ceramides with a two carbon side-chain form long, thin fibers several hundred microns in length. Mixed ceramide systems, however, which can be derived from milk or eggs, have long carbon chains (i.e., C16–C24) and produce small maltese-cross crystals, which are able to effectively immobilize oil (Rogers, 2009).

Several studies have shown that feeding ceramides and sphingolipids (which contain a ceramide group) to rats reduces the total serum cholesterol by 30% and improves the chemical composition of serum lipoproteins (Gellen et al., 1995; Kobayashi et al., 1997).

This means that ceramides are the first food grade organogelator to provide a powerful pharmacological effect, immobilize oil at low concentrations, and possess the ability to be chemically modified to alter their physical properties.

Though ceramides are food grade, the use of pure ceramides in industry would not be feasible due to their extreme cost. Rather than using the pure ceramides, mixed system ceramides can be created using milk and eggs through the use of the enzyme phospholipase C. This enzyme is used to hydrolyze sphingomyelin between the primary hydroxyl group of the ceramide and the choline phosphate ester. Though the conversion from sphingomyelin to ceramide is performed highly efficiently by this enzyme, it is also quite expensive, though insignificantly so in comparison to pure ceramide. Unfortunately, there is also a solubility issue that must be overcome. Sphingomyelin is insoluble in water while phospholipase C is insoluble in organic solvents. This means that complex media must be used to ensure complete hydrolysis of the sphingomyelin (Zhang et al., 2006). This means that, while ceramide systems appear to be model organogelators, their overall cost renders them impractical for widespread use.

Though scanning electron microscopy (SEM) has been used to image ceramide organogels, it is difficult to prepare the sample and not change the gel structure in some way. This means that imaging the "native state" gel is extremely challenging and makes understanding the interactions of this native gel equally difficult. The forces that govern these gels have been proposed, but until we determine more effective ways of imaging the unaltered gel, we will be unsure if these proposed forces are completely correct.

Wax Ester Organogels

A wax is a fatty material that consists of a long hydrocarbon chain, with or without a functional group. Common functional groups found in waxes include alcohol, ketone, aldehyde, and ester. One of the groups of highest interest for industrial and potential food applications is the ester. Esters provide the characteristic smell of most fruits, and therefore are commonly used in perfumes. Currently wax esters are used in cosmetics, as lubricants and polishes, surface coatings, inks, and also as an ingredient in select food products (Akoh & Min, 2002; Ghosh & Banyopadhyay, 2005). Wax esters are derived predominantly from plants; however, they can also be derived from insects and marine animals. Plants, as the most widely available and easily attained source, would be ideally suited to provide wax esters in high quantities.

Candelilla wax contains 20–29% high molecular weight esters, and has demonstrated the ability to structure safflower oil at concentrations as low as 1% (Toro-Vazquez, 2007b). Carnauba wax, which consists of fatty esters (80–85%), free alcohols (10–15%), acids (3–6%), and hydrocarbons (1–3%), has also demonstrated the ability to structure oil at concentrations similar to that shown by candelilla wax. Candelilla wax appears to be slightly more effective at structuring oil compared to carnauba wax, as candelilla is able to induce gelation at concentrations as low as 2% (w/w), while carnauba is not able to form a gel until a concentration of 4% (w/w) has been reached (Dassanayake et al., 2009).

Recently, a wax derived from rice (*Oryza sativa*) bran has shown gelling potential that far exceeds even candelilla wax (Dassanayake et al., 2009).

Rice bran wax is another natural wax and is derived from the bran of milled rice kernels. The high production of rice, particularly in eastern Asia, allows for high availability and low cost of rice bran to be used for rice bran wax production. The difference between rice bran wax and many other waxes, such as candelilla and carnauba, is that it forms long (20–50 μm), needle-like crystals, which are easily able to form organogels. In comparison, candelilla and carnauba form spherulitic crystals, which usually have a diameter less than 10 μm. Crystals of this size and type are not conducive to the formation of organogels. While neither candelilla or carnauba wax are able to form a gel at concentrations lower than 2%, rice bran wax was shown to gel oil at concentrations as low as 0.5% (w/w). The long needle-like crystals of the rice bran wax, as shown in Fig. 2.13, form strong crystal matrices, which enables large volumes of liquid oil to become entrapped between the crystalline strands.

When Fig 2.13 is compared to the SAFIN organogel in Fig. 2.9, it can be seen that the rice bran wax does not possess the helical twist that is characteristic of the fibers in SAFIN organogels. Fig. 2.13 also shows a fairly uniformly distributed mesh of crystals compared to the fibers in Fig. 2.9 that all originate from key nucleation points. Rice bran wax was reported to form an organogel at concentrations as low as 0.5% (w/w), however, the gel still remains fairly oily in appearance at concentrations

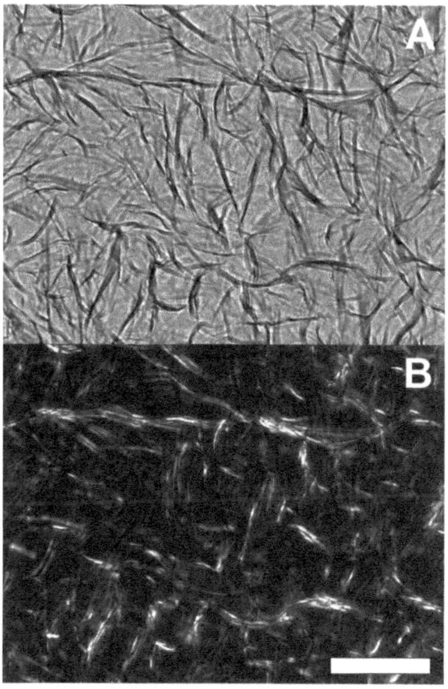

Fig. 2.13. Non-polarized (A) and polarized light (B) micrographs of rice bran wax in canola oil at a concentration of 5% (w/w). Magnification bar = 50 μm.

as high as 5%. The gel is also considerably more opaque and is not as jello-like in appearance compared to the SAFIN organogels.

Though rice bran wax could be produced in high quantities with low production costs, it may be difficult to perfect this organogel for use in foods. The oily appearance of the wax at concentrations as high as 5% and the typical wax mouthfeel and taste would make this product a difficult ingredient to use in most foods. Also, rice bran wax organogels do not provide the same added health benefits that some other structured oil systems provide.

Polymer Organogels

In our opinion, this is the next frontier of liquid oil structuring. There is a wide availability of food-grade polymers (polysaccharides and proteins) that could fulfill the promise of gelling oil. Being a frugal industry by nature, this "cheap" structuring of oils should help the implementation of these new technologies in the marketplace.

Currently, many biopolymers are being assessed in our laboratory for their potential as organogelators at concentrations below 10%. Recently, our group managed to gel vegetable oil using ethylcellulose and other components in the absence

Fig. 2.14. Polymer organogel of ethylcellulose in vegetable oil.

of water. This clear, elastic gel, and others like this, may represent the future of oil (nano)structuring into functional fats. See Fig. 2.14.

Little published work is available on this topic. We could only identify one paper on the gelation of ethylcellulose-oil mixtures into a plastic solid by addition of olive oil-derived surfactants (Martinez et al., 2003).

The solubility of the ethylcellulose in oil is highly dependent on the number of hydroxyl to ethoxy substitutions per glucose molecule in the ethylcellulose chain. There are only three hydroxyl groups total per glucose molecule. Only when there are 2.4–2.5 hydroxyl-ethoxy substitutions will the ethylcellulose be soluble in polar and non-polar organic solvents such as a vegetable oil. When the degree of substitution is 1.5 or lower the ethylcellulose is water soluble, as shown in Fig. 2.15 (Koch, 1937).

The viscosity of the resulting gel or ethylcellulose-oil solution is greatly affected by the length of the ethylcellulose chains. When molecular weight is high, this allows the formation of much firmer gels or much more viscous ethylcellulose-oil solutions. Ethylcellulose is commercially available with widely varying molecular weights, or viscosities. This means that the ethylcellulose (molecular weight) can be customized depending on the desired application.

To create the polymer organogel, the crystalline ethylcellulose is added to the liquid oil that is to be gelled. This mixture must then be heated to above the glass transition temperature of the ethylcellulose, which occurs at approximately 130°C. At this point the ethylcellulose unfolds and, upon cooling, forms a backbone for the cross-linked gel network. Fig. 2.16 clearly shows a network of unfolded ethylcellulose polymer strands that are responsible for the oil entrapment.

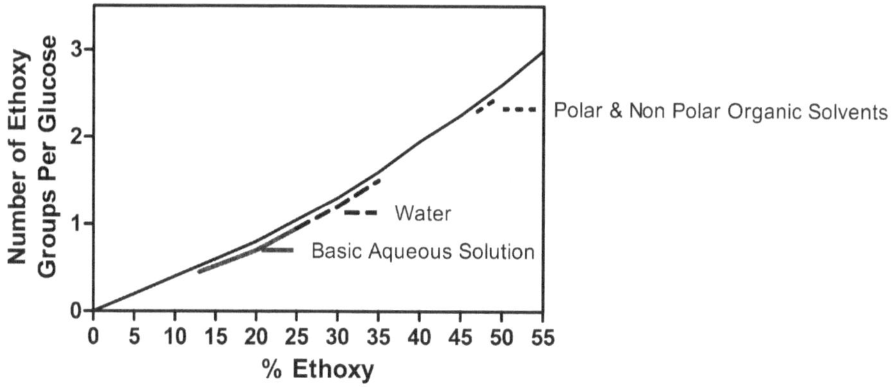

Fig. 2.15. Differences in solubility of ethylcellulose in various liquids based on the hydroxyl-ethoxy substitutions. Specific solubility areas are highlighted. Source: adapted from Koch, 1937.

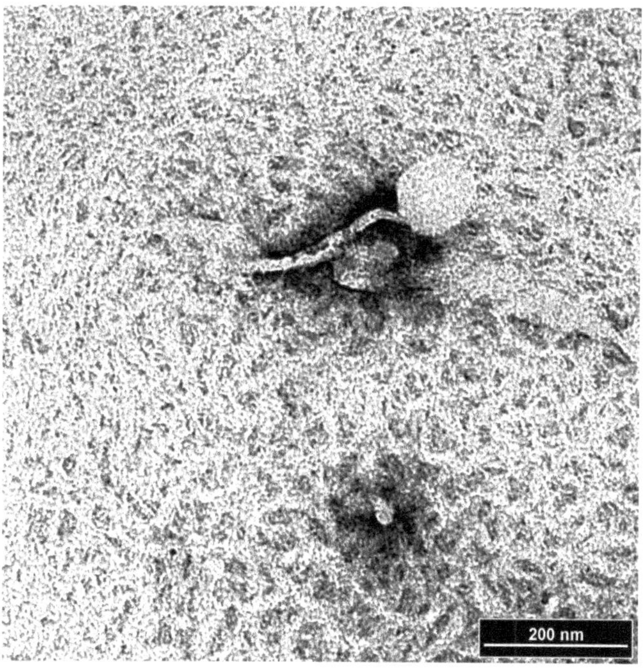

Fig. 2.16. Transmission electron micrograph showing the unfolded polymer strands that comprise the cross-linked network of an ethylcellulose gel.

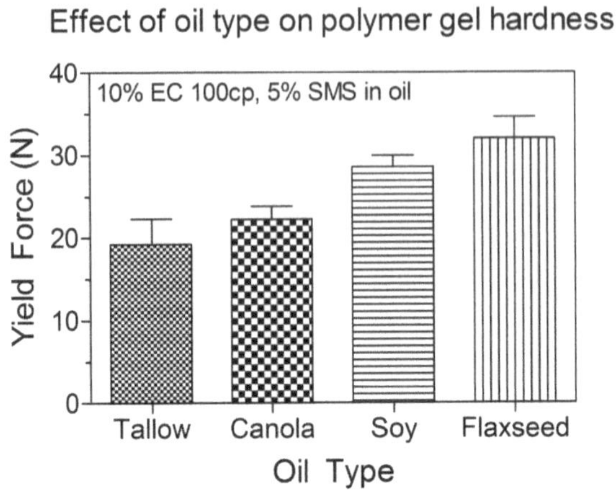

Fig. 2.17. Relative hardness of ethylcellulose gels made with beef fat or various oils. A texture analyzer was used to compress samples of size 2 cm × 2 cm × 1 cm at a temperature of 25°C. Source: Laredo et al., 2011.

The forces that govern the stability of these types of gel are currently being studied. We believe that hydrogen bonding among ethylcellulose stands has a major effect upon gel stability. It is also possible that hydrophobic interactions may occur between the acyl chains of the oil and the methyl groups of the ethylcellulose side chains. There may also be interactions among the methyl groups on the ether side-chains of the ethylcellulose polymer. Furthermore, there is evidence that the unsaturations in the triacylglycerol molecules influence the strength of the gel. This is most likely because of a hydrophobic interaction between the double bonds in the triacylglycerols of the oil and the ethylcellulose side-chains. By using different oils with varying saturation levels, it can clearly be seen in Fig. 2.17 that the yield force (hardness) varies substantially among the different oils.

The unsaturation levels of the different oils also causes large differences in the amount of oil that will migrate from a gel sample to a piece of filter paper. As shown in Fig. 2.18, the gel made with the more highly unsaturated flaxseed oil demonstrates a reduced migration level of approximately 35%, while the gel consisting of canola oil exhibited a migration level of 45%.

The great advantage of these newly discovered polymer gels, and the polymer organogelator ethylcellulose, is the significant cost reduction compared to other organogelators. Ethylcellulose is readily available commercially in a variety of forms and molecular weights, is fairly inexpensive (especially when compared to traditionally used organogelators), and, best of all, is also certified food grade and is essentially fiber. This allows for the replacement of traditional, highly saturated fats, with polymer organogels in a variety of potential food products. Our lab is currently investigating the replacement of animal fat with polymer organogels in emulsion-type meat products. Products containing vegetable oil organogels display textural attributes

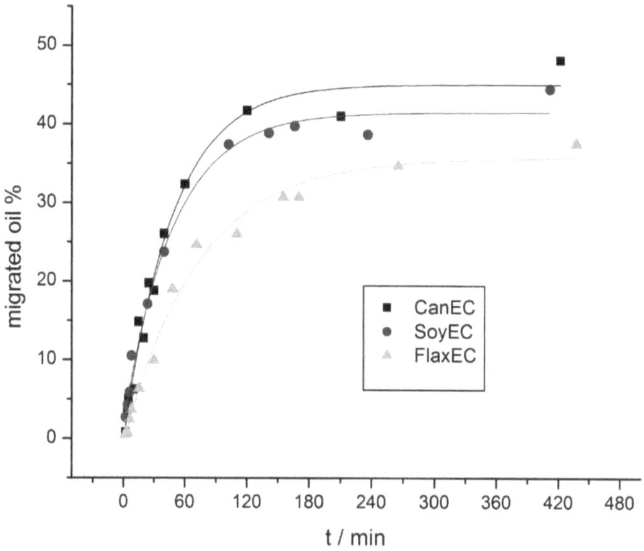

Fig. 2.18. Percent migrated oil onto Whatman #2 filter paper from organogel samples using various types of oil. Source: Laredo et al., 2011.

matching those of traditional products made with animal fat. When an un-gelled vegetable oil is used to replace the animal fat, the resulting products tend to be much firmer and do not receive the same level of consumer appeal. If the organogels can be used in these products, there can be significant reduction in the amount of saturated fat and cholesterol while retaining the same textural attributes and, hopefully, consumer appeal. Similar development is taking place for a variety of other products and will hopefully lead to a much healthier future!

Ethylcellulose is currently approved for indirect food uses in North America, yet it has been approved for food applications in Europe since the end of 2006. Some of the most common current uses of the polymer include the following: a component of paper and paperboard that is in contact with fatty or water-based foods, use in inks that are used to mark fruits and vegetables, and use to help control the release of nutrients in animal feed pellets. In North America, it is likely that ethylcellulose will receive full approval for use in food products within the next few years.

Conclusions

In this work we have presented some of the most promising strategies for structuring liquid oils. The motivation for this work is to create a semisolid structure with the functionality of fats but with the nutritional profile of liquid oil—low in saturates, devoid of *trans* fatty acids, and possibly containing some nutritionally beneficial fatty acids. Structured emulsions, SAFINs, high-stearic acid fats, and polymer organogels will all be part of the solution to the problem of improving on the nutritional quality of our manufactured food products in consideration of sensory quality and price.

Acknowledgments

The authors acknowledge the financial assistance of the Natural Sciences and Engineering Research Council of Canada, the Ontario Ministry of Agriculture and Food, the Advanced Food Materials Network, the Canadian Foundation for Innovation and the Canada Research Chairs Program.

References

Abdallah, D.J.; Weiss, R.G. Organogels and Low Molecular Mass Organic Gelators. Adv. Mater. 2000, 12, 1237–1247.

Ahmadi, L.; Wright, A.; Marangoni, A.G. Chemical and enzymatic interesterification of tristearin-triolein-rich blends I: chemical composition, sfc and thermal properties. Eur. J. Lipid Sci. Tech. 2008a, 110, 1014–1024.

Ahmadi, L.; Wright, A.; Marangoni, A.G. Chemical and enzymatic interesterification of tristearin-triolein-rich blends II: microstructure and polymorphism. Eur. J. Lipid Sci. Tech. 2008b, 110, 1025–1034.

Ahmadi, L.; Wright, A.J.; Marangoni, A.G. Structural and mechanical behavior of tristearin/triolein-rich mixtures and the modification achieved by interesterification. Food Biophys. 2009, 4, 64–76.

Ahmadi, L.; Marangoni, A.G. Functionality and physical properties of interesterified high oleic shortening structured with stearic acid. Food Chem. 2009, 117, 668–673

Akoh, C.C.; Min, B.D.; Eds. Food lipids chemistry, nutrition & biotechnology, 2nd ed; Marcel Dekker: New York, 2002.

American Heart Association. http://www.americanheart.org/presenter.jhtml?identifier=3036399 (accessed 2008).

Aro, A.; Jaughiainen, M.; Partanen, R.; Salminen, I.; Mutanen, M. Stearic acid, trans fatty acids, and dairy fat: effects on serum and lipoprotein lipids, apolipoproteins, lipoprotein(a), and lipid transfer proteins in healthy subjects. Am. J. Clin. Nutr. 1997, 65, 1419–1426.

Batte, H.; Wright, A.J.; Idziak, S.H.J.; Marangoni, A.G. Phase behavior, stability and mesomorphism of monostearin-oil-water gels. Food Biophys. 2007a, 2, 29–37.

Batte, H.; Wright, A.J.; Idziak, S.H.J.; Marangoni, A.G. Effect of processing conditions on the structure of Monostearin-oil water gels. Food Res. Int. 2007b, 40, 982–988.

Berger, K.G.; Idris, N.A. Formulation of zero-trans acid shortenings and margarines and other food fats with products of the oil palm. J. Am. Oil Chem. Soc. 2005, 82, 775–782.

Bonanom, A.; Grundy, S.M. Effect of dietary stearic acid on plasma cholesterol and lipoprotein levels. N. Engl. J. Med. 1988, 318, 1244–1248.

Bot, A.; Agterof, W.G.M. Structuring of edible oils by mixtures of γ-oryzanol with β-sitosterol or related phytosterols. J. Am. Oil Chem. Soc. 2006, 83, 513–521.

Bot, A.; Den Adel, R.; Roijers, E.C. Fibrils of γ-Oryzanol and β-Sitosterol in Edible Oil Organogels. J. Am. Oil Chem. Soc. 2008, 85, 1127–1134.

Dassanayake, L.S.K.; Kodali, D.R.; Ueno, S.; Sato, K. Physical properties of rice bran wax in bulk and organogels. J. Am. Oil Chem. Soc. 2009, 86, 1163–1173.

Dijkstra, A.J.; Hamilton, R.J.; Hamm, W. Trans fatty acids, 1st ed. Blackwell Pub.: Oxford, Ames, IW, 2008; 10–42.

Eckel, R.H.; Borra, S.; Lichtenstein, A.H.; Yin-Piazza, S.Y. Understanding the Complexity of Trans Fatty Acid Reduction in the American Diet: American Heart Association Trans Fat Conference 2006: Report of the Trans Fat Conference Planning 2007.

Firestone, D. Physical and Chemical Characteristics of Oils, Fats, and Waxes (2nd Edition). AOCS Press: Online version <http://knovel.com/web/portal/browse/display?_EXT_KNOVEL_DISPLAY_bookid=2836&VerticalID=0> 2006.

Flory, P. Principles of Polymer Chemistry. Cornell University Press: New York, 1953.

Gandolfo, F.G.; Bot, A.; Flöter, E. Structuring edible oils by long-chain FA, fatty alcohols, and their mixtures. J. Am. Oil Chem. Soc. 2004, 81, 1–6.

Gellen, M.J.H.; Tijburg, L.M.B.; Bouma, C.J.; Beynen, A.C. Cholesterol consumption alters hepatic sphingomyelin metabolism in rats. J. Nutr. 1995, 125, 2294–2300.

Ghosh, M.; Bandyopadhyay, S. Studies on the crystal growth of rice bran wax in a hexane medium. J. Am. Oil Chem. Soc. 2005, 82, 229–231.

Ghotra, B.S.; Dyal, S.D.; Narine, S.S. Lipid shortenings: A review. Food Res. Int. 2002, 35, 1015–1048.

Hannam, L. Beware the 'Monday Diet Hangover'. Best Health. Jan–Feb 2010.

Health Canada. http://www.hc-sc.gc.ca/fn-an/nutrition/gras-trans-fats/tfa-age_four-data_quatr-donn_restaurant-eng.php#t1 (accessed 2009).

Hegsted, D.M.; Mcgandy, R.B.; Myers, M.L.; Stare, F.J. Quantitative Effects of Dietary Fat on Serum Cholesterol in Man. Am. J. Clin. Nutr. 1965, 17, 281–295.

Hughes, N.E.; Marangoni, A.G.; Wright, A.J.; Rogers, M.A.; Rush, J.W.E. Potential food applications of edible oil organogels. Trends Food Sci. Tech. 2009, 20, 470–480.

Huschka, B.; Challacombe, C.; Marangoni, A.G.; Seetharaman, K. Effect of a structured shortening and its unstructured components on dough rheology and on starch pasting properties. Cereal Chem. 2010a, Printed online.

Huschka, B.; Challacombe, C.; Marangoni, A.G.; Seetharaman, K. Comparison of oil, shortening, and a structured shortening on wheat dough rheology and on starch pasting properties. Cereal Chem. 2010b, Printed online.

Isomaa, B.; Almgren, P.; Tuomi, T.; Forsen, B.; Lahti, K.; Nissen, M.; Taskinen, M.R.; Groop, L. Cardiovascular morbidity and mortality associated with metabolic syndrome. Diabetes Care 2001, 24, 683.

Keys, A.; Anderson, J.T.; Grande, F. Serum cholesterol response to changes in the diet. IV. Particular saturated fatty acids in the diet. Metabolism 1965, 14, 776–787.

Kobayashi, T.; Shimizugawa, T.; Osakabe, T.; Watanabe, S.; Okuyama, H. A long term feeding of sphingolipids affected the levels of plasma cholesterol and hepatic triacylglycerols but not tissue phospholipids and sphingolipids. Nutr. Res. 1997, 17, 111–114.

Koch, W. Properties and uses of ethylcellulose. Ind. Eng. Chem. 1937, 29 (6) 687–690.

Kris-Etherton, P.M.; Pearson, T.A.; Wan, Y.; Hargrove, R.L.; Moriarty, K.; Fishell, V.; Etherton, T.D. High-monounsaturated fatty acid diets lower both plasma cholesterol and triacylglycerol concentrations. Am. J. Clin. Nutr. 1999, 70, 1009–1015.

Laredo,T; Barbut, S.; Marangoni, A.G. Molecular Interactions of Polymer Oleogelation. Soft Matter 2011, 7, 2734–2743.

Lichtenstein, A.H.; Appel, L.J.; Brands, M.; Carnethon, M.; Daniels, S.; Franch, H.A.; Franklin, B.; Kris-Etherton, P.; Harris, W.S.; Howard, B.; et al. Diet and lifestyle recommendations. Circulation 2006, 114, 82–96.

List G.R.; Mounts T.L.; Orthoefer F.; Neff W.E. Margarine and shortening oils by interesterification of liquid and trisaturated triglycerides. J. Am. Oil Chem. Soc. 1995, 72, 379–382.

Marangoni, A.G.; Idziak, S.H.J.; Vega, C.; Batte, H.; Ollivon, M.; Jantzi, P.S.; Rush, J.W. Oil Microencapsulation Attenuates Acute Elevation of Blood Lipids and Insulin in Humans. Soft Matter 2007, 3, 183–187.

Marangoni, A.G.; Idziak, S.H.J.; Rush, J.W.E. Controlled release of food lipids using monoglyceride gel phases regulates lipid and insulin metabolism in humans. Food Biophys. 2008, 3, 241–245.

Marangoni, A.G. A *trans* fat free future—a commentary. Inform 2007, 18 (4) 281–283.

Martinez, M.A.; Munoz de Benavides, M.; Morales Hernandez, M.E.; and Gallardo Lara, V. Influence of the concentration of a gelling agent and the type of surfactant on the rheological characteristics of oleogels. Il Farmaco 2003, 58, 1289–1294.

Mensink, R.P.; Zock, P.L.; Kester, A.D.M.; Katan, M.B. Effects of dietary fatty acids and carbohydrates on the ratio of serum total to HDL cholesterol and on serum lipids and apolipoproteins: a meta-analysis of 60 controlled trials. Am. J. Clin. Nutr. 2003, 77, 1146 1155.

Mensink, R.P. Effects of stearic acid on plasma lipid and lipoproteins in humans. Lipids 2005, 40, 1201–1205.

Pernetti, M.; van Malssen, K.F.; Flöter, E.; Bot, A. Structuring edible oils by alternatives to crystalline fat. Curr. Opin. Colloid Interface Sci. 2007a, 84, 989–1000.

Pernetti, M.; van Malssen, K.; Kalnin, D.; Flöter, E. Structuring edible oil with lecithin and sorbitan tri-stearate. Food Hydrocolloids 2007b, 21, 855–861.

Purdy, R.H. Oxidative stability of high oleic sunflower and safflower oils. J. Am. Oil Chem. Soc. 1985, 62, 523–525.

Roche, H.M. Fatty acids and the metabolic syndrome. Proc. Nutr. Soc. 2005, 64, 23–29. National Conference of State legislatures. http://www.ncsl.org/programs/health/transfatmenulabelingbills.htm (accessed 2008).

Rogers, M.A.; Smith, A.K.; Wright, A.J.; Marangoni, A.G. A novel cryo-SEM technique for imaging vegetable oil based organogels. J. Am. Oil Chem. Soc. 2007, 84, 899–906.

Rogers, M.A.; Wright, A.J.; Marangoni, A.G. Engineering the oil binding capacity and crystallinity of self-assembled fibrillar networks of 12-hydroxysteric acid in edible oils. Soft Matter 2008a, 4, 1483–1490.

Rogers, M.A.; Wright, A.J.; Marangoni, A.G. Crystalline stability of self-assembled fibrillar networks of 12-hydroxystearic acid in edible oils. Food Res. Int. 2008b, 41, 1026–1034.

Rogers, M.A.; Marangoni, A.G. Non-isothermal nucleation and crystallization of 12-hydroxystearic in vegetable oils. Crystal Growth and Design 2008a, 8, 4596–4601.

Rogers, M.A.; Marangoni, A.G. Crystal engineering of crystalline SAFIN junction zones via enthalpic and entropic forces. J. Physics D 2008b, 41, 215501.

Rogers, M.A.; Wright, A.J.; Marangoni, A.G. Oil Organogels: The Fat of the Future? Soft Matter 2009, In Press.

Rogers, M.A. Novel structuring strategies for unsaturated fats—Meeting the zero-trans, zero-saturated fat challenge: A review. Food Res. Int. 2009, 42, 747–753.

Rogers, M.A.; Bot, A.; Lam, R.S.H.; Pedersen, T.; May, T. Multicomponent Hollow Tubules Formed Using Phytosterol and γ-Oryzanol-Based Compounds: An Understanding of Their Molecular Embrace. J. Phys. Chem. 2010, 114, 8278–8285.

Rush, J.W.E.; Jantzi, P.S.; Dupak, K.; Idziak, S.H.J.; Marangoni, A.G. Effect of food preparation on the structure and metabolic responses to a monostearin-oil-water gel-based spread. Food Res. Int. 2008, 41, 1065–1071.

Rush, J.W.E.; Jantzi, P.S.; Dupak, K.; Idziak, S.H.J.; Marangoni, A.G. Acute metabolic responses to butter, margarine, and a monoglyceride gel-structured spread. Food Res. Int. 2009, 42, 1034–1039.

Schaink, H.M.; van Malssen, K.F.; Morgado-Alves, S.; Kalnin, D.; van der Linden, E. Crystal network for edible oil organogels: possibilities and limitations of the fatty acid & fatty alcohol systems. Food Res. Int. 2007, 40, 1185–1193.

Simopoulos, A.P. Omega-3 fatty acids in health and disease and in growth and development. Am. Soc. Clin. Nutr. 1991, 54, 438–463.

Terech, P.; Weiss, R.G. Low molecular mass gelators of organic liquids and the properties of their gels. Chem. Rev. 1997, 97, 3133–3159.

Toro-Vazquez, J.F.; Morales-Rueda, J.A.; Dibildox-Alvarado, E.; Charo-Alonso, M.; González-Chávez, M.; Alonzo-Macias, M.M. Thermal and textural properties of organogels developed by candelilla wax in safflower oil. J. Am. Oil Chem. Soc. 2007a, 84, 989–1000.

Toro-Vazquez, J.F.; Morales-Rueda, J.A.; Dibildox-Alvarado, E.; Charó-Alonso, M.; Alonzo-Macias, M.; González-Chávez, M.M. Thermal and Textural Properties of Organogels Developed by Candelilla Wax in Safflower Oil. J. Am. Oil Chem. Soc. 2007b, 84, 989–1000.

Wahrburg, U. What are the health effects of fat? Eur. J. Nutr. 2004, 43, Supplement 1, i6–i11.

Weiss, R.G.; Terech, P. Introduction. In molecular gels materials with self-assembled fibrillar networks. Weiss, R.G. Terech, P. Eds; Springer: Dordrecht, Netherlands, 2006; 1–13.

Wright, A.J.; Marangoni, A.G. Formation, structure and rheological properties of ricinelaidic acid-vegetable oil organogels. J. Am. Oil Chem. Soc. 2006, 83, 497–503.

Wright, A.J.; Marangoni, A.G. Time, temperature and concentration dependence of ricinelaidic acid-canola oil organogelation. J. Am. Oil Chem. 2007, 84, 3–9.

Zhang, L.; Hellgren, L.I.; Xu, X. Enzymatic production of ceramide from sphingomyelin. J. Biotech. 2006,123, 93–105.

3

Edible Oil Organogels Based on Self-assembled β-sitosterol + γ-oryzanol Tubules

Arjen Bot and Eckhard Flöter
Unilever Research and Development Vlaardingen, Olivier van Noortlaan 120, NL-3133 AT Vlaardingen, The Netherlands

Introduction

The texture of many food products (e.g., margarine, butter, cookies, etc.) is achieved by including a crystallizing triacylglycerol (TAG) fraction in the lipid phase of the product (de Bruijne and Bot, 1999; Flöter & Bot, 2006). These crystalline TAGs contain considerable amounts of saturated fatty acids (SAFA). It has be demonstrated repeatedly that the intake of saturated fatty acids instead of unsaturated fatty acids raises blood cholesterol, which in turn has been identified as a risk factor for the occurrence of cardio-vascular diseases (Keys et al., 1965; Mensink et al., 2003). It would be desirable, therefore, to reduce the amount of crystallizing TAGs and increase the amount of liquid TAGs, containing predominantly mono-unsaturated and polyunsaturated fatty acids (MUFA and PUFA, respectively). One way to achieve this would be to replace the SAFA-rich crystalline TAGs by an alternative structuring agent. Upfront it is clear that it will not be an easy task to match the functionality of TAG structuring, given the versatility of fats and the practical constraints on food ingredients.

It is best to start with a concise summary of the way that crystalline TAGs structure a lipid phase consisting of liquid TAGs. Fat crystals form according to a nucleation-growth mechanism. The size of the crystals is typically reduced by rapid cooling because low temperature increases the number of nuclei and decreases the growth rate of the crystals. The liquid oil serves as the organic solvent in this network. The melting properties of the network can be tuned by the proper choice of TAG sources (natural, and appropriate physically or chemically modified sources like fractionated or interesterified fats), leading to a network that is suitable for application in foods (de Bruijne & Bot, 1999; Flöter & Bot, 2006). The firmness of the crystalline TAG network increases with smaller crystals and with higher surface-to-volume ratio (i.e., non-sphericity), and the degree of oil exudation decreases with finer pore size of the network.

Alternative structurants can be chosen to mirror this behavior, and this can be achieved in a relatively straightforward way in systems for which structuring is based on the crystallization behavior of fatty acids. On the other hand, one can try to base structuring on fundamentally different modes of molecular assembly. The most interesting candidate is a type of supramolecular ordering that does not result in a simple crystal lattice, but rather in complex structures like ribbons, fibers, and the like. This type of ordering is sometimes referred to as self-assembly, and usually involves chiral molecules (Zinic et al., 2005).

On top of that an alternative food structurant should comply with a number of other restrictions. The ingredient should

- be a food ingredient
- structure an edible oil phase
- allow for some manipulation in melting temperature / amount of structuring
- not interact in an adverse way with any of the other ingredients in a typical food
- have a reasonable price

A number of recent reviews have considered the question to what extent currently known systems satisfy these two sets of requirements (Pernetti et al., 2007; Rogers, 2009; Marangoni, 2009; Wassell et al., 2010). The structuring system based on mixtures of a plant sterol and a plant sterol ester, β-sitosterol and γ-oryzanol, seems to emerge amongst the most interesting candidates. Fig. 3.1shows that these mixtures form a slightly hazy transparent gel in triglyceride oil, whereas the single components do not (Ritter et al., 1997; Bot & Agterof, 2006). The building blocks of this gel

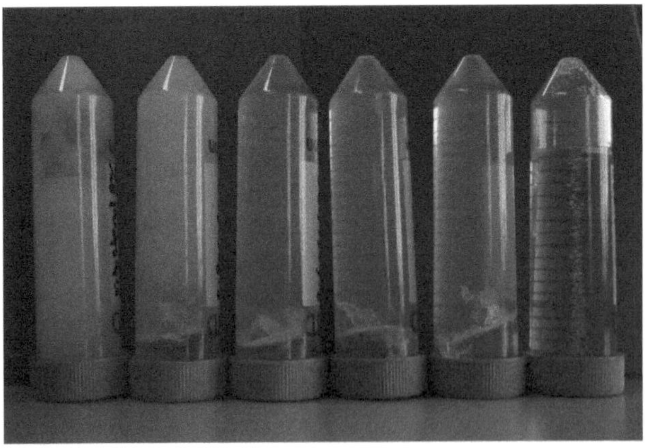

Fig. 3.1. Mixtures of 8% β-sitosterol + γ-oryzanol in sunflower oil at various β-sitosterol:γ-oryzanol ratios (from left to right 100:0; 80:20; 60:40; 40:60; 20:80; 0:100). The binary sterol(ester) systems in sunflower oil are gelled, as can be seen from the fracture pattern introduced near the caps. Dispersions of single sterol(ester)s in sunflower oil are liquid. Source: Bot et al., 2008.

are self-assembled tubules with a diameter of 7.2±0.1 nm and a wall thickness of 0.8±0.2 nm. The structure of the mixture bears little resemblance to that of the pure components.

The present chapter aims to provide a comprehensive review of this system—and of a number of closely related systems. It will start with a description of the single components and continue with the kinetics and structure of the self-assembled structures. This will be followed by a discussion concerning the rheology of the (water-free) organogels. Subsequently, the behavior of this type of structuring in the presence of water will be discussed. Finally, a number of potential applications of the oryzanol + sterol mixture will be presented.

Molecular Structure and Crystal Structure of the Pure Components

Sitosterol and Other Phytosterols

Phytosterols are steroid alcohols that occur in plant cell membranes, having a similar function as cholesterol in mammalian cells. They occur as minor components in a number of vegetable oils, like corn and soybean oil, at typical levels of 0.1–1%. Phytosterols have received a lot of attention in recent years by virtue of their capacity to reduce blood cholesterol in humans by as much as 10% (Katan et al., 2003). This chapter will concern itself mainly with systems with sitosterol in the role of sterol, but the class of materials is much bigger than sitosterol alone. Fig. 3.2 shows a number of plant sterols that are very similar to sitosterol and have the virtue of being relatively easily obtainable commercially: cholestanol, cholesterol, sitosterol, stigmasterol and

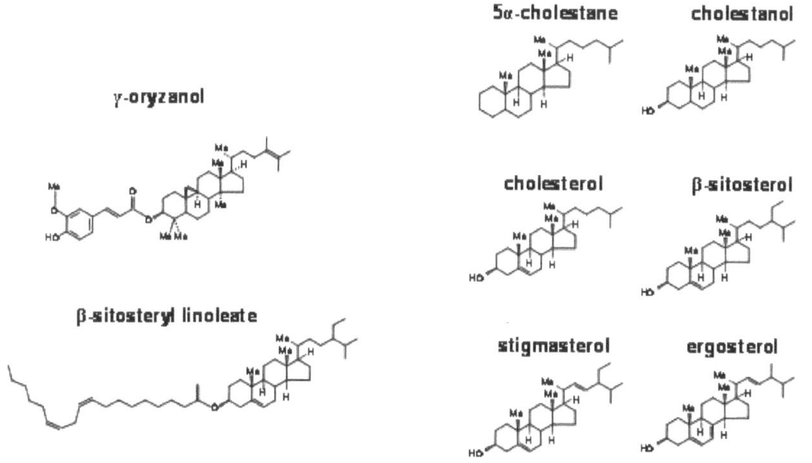

Fig. 3.2. Chemical structures of some commercially available plant sterol esters and plant sterols.

ergosterol (Bot & Agterof, 2006; Bot et al., 2009b). The list of plant sterols is much longer, however.

One can think of these materials as being based on the cholestane molecule, which in itself does not form organogels, and which can be considered to be a "stripped" cholesterol molecule without hydroxyl group or double bond in the steroid ring system (the sterane core). In cholestanol the OH-group is present, but the double bond is still lacking. Sitosterol has a slightly modified alkyl chain relative to cholesterol. Stigmasterol has an additional double bond in the modified alkyl chain. Finally, ergosterol has an extra double bond in the cholesterol ring structure and a small modification in the alkyl chain relative to stigmasterol.

Cholesterol has the interesting property of forming helical ribbons under specific conditions, for example in bile salt (Konikoff et al., 1992). A number of derivatives of cholesterol have been demonstrated to show fibrillar aggregation as well. Amongst these, a few were proven to be gelling agents of non-aqueous phases: cholesterol substituted with amino acid (Li et al., 2006), anthracene (Lin et al., 1989; Terech et al., 1995; 1996), azobenzene (Murata et al., 1994; Sakurai et al., 2001), carbamate (Lu et al., 2000), stilbene (Geiger et al., 1999; Wang et al., 2000), and squaraine groups (Geiger et al., 1999). In addition, dicholesteryl compounds linked by a variety of groups could serve as a gelling agent (Peng et al., 2008). Edible oils were in general not included in the organic fluids that were tested. Given the generally similar behavior of the plant sterols, it is expected that sitosterol and sitosterol derivatives will show the same behavior.

Sitosterol, like cholesterol, arranges in bilayers of approximately 3.8 nm thickness in its crystalline state (Craven, 1986; Argay et al., 1996; Kawachi et al., 2006; den Adel et al., 2010). The order is induced by the stacking of the characteristic sterane core, which leaves some conformational freedom the alkyl chains. The ordering of the anhydrous and hydrated forms is slightly different, driven by the need to maximize the degree of hydrogen bonding between the molecules (Kawachi et al., 2006). Four different temperature ranges can be identified for the stable sitosterol forms: the monohydrate (1 mol water/mol sitosterol) is stable <45°C, the hemi-hydrate (0.5 mol water/mol sitosterol) in the range 45–80°C, the anhydrous form in the range 80–110°C, and dissolution occurs >110°C (Christiansen et al., 2002). The most characteristic X-ray reflection peaks [Small angle and wide angle X-ray scattering (SAXS, WAXS)] resulting from this structure are $d = 2\pi/q_i = 1.76$ and 0.503 nm (monohydrates); d = 1.88 and 0.481 nm (hemi-hydrates); d = 1.76, 0.880 and 0.523 nm (anhydrous). Next to that, all crystal forms typically exhibit Bragg reflections for 1.2 and 0.59 nm. In addition, infrared spectroscopy (IR) reveals the hydration state of sitosterol. Generally, intermolecular hydrogen bonds for these systems appear around ~3450 cm^{-1} (den Adel et al., 2010; Rogers et al., 2010).

Examples of experimental data on pure sitosterol are shown in Fig. 3.3. The most important feature of the X-ray scattering data on pure sitosterol is the sharpness

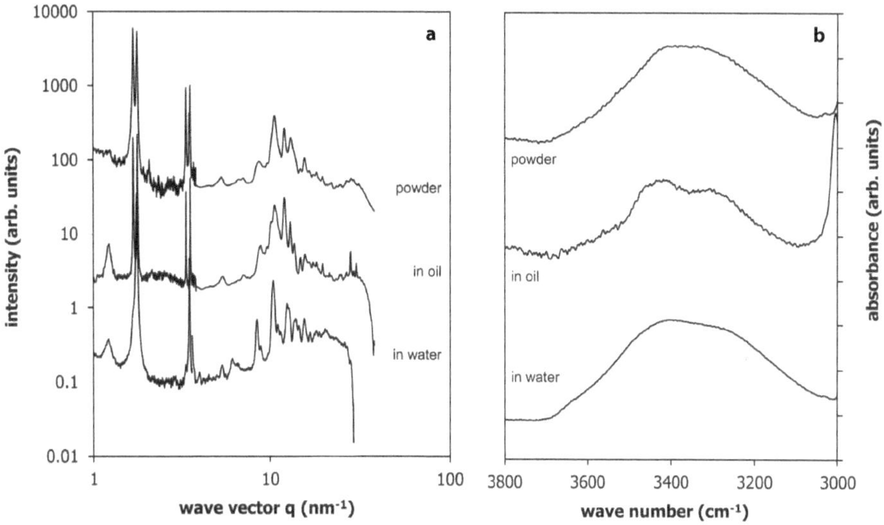

Fig. 3.3. (a) X-ray scattering data and (b) infrared spectra (right) for sitosterol powder, and 16% sitosterol in sunflower oil and water.

of all diffraction peaks and that the sitosterol crystals therefore show long-range order in three dimensions. The peaks at d=3.8 and 1.9 nm (powder and slurry in oil) and d=3.6 and 1.8 nm (slurry in water and in emulsion) confirm the presence of a bilayer structure (den Adel et al., 2010). In addition, the X-ray scattering data in powder and in oil slurry indicate a mixture of the anhydrous and hemi-hydrate forms. The infrared spectra show only a very weak band associated with intermolecular hydrogen bonding, around ~3420–3440 cm^{-1}. The absence of characteristic hemi-hydrate and anhydrate peaks in aqueous slurry and in emulsion suggests the presence of the mono-hydrate form, despite the surprising similarity of the infrared spectrum for the aqueous slurry and that of the powder (den Adel et al., 2010).

Oryzanol and Other Phytosterol Esters

As noted above, sitosterol has the tendency to form three-dimensional crystals in plain oil or water (although it can form more complex structures under specific conditions). In many cases, the crystals can become large enough that they can be perceived in the mouth as sandy when added at the levels required to achieve a blood cholesterol lowering effect. Commercial applications of plant sterols and stanols are therefore usually based on plant sterols/stanol esters because these have higher solubility in oil. The esters are made by interesterifying the sterols with a liquid vegetable oil, e.g., sunflower or canola oil, (and would typically contain sitosteryl linoleate as an important component, see Fig. 3.2). In the digestive tract, the sterol esters are hydrolyzed and the sterols become available to perform their blood cholesterol lowering function. However, natural sterol esters also exist. One of the most interesting plant sterol esters is oryzanol, a minor component of rice bran oil (see Fig. 3.2) (Lerma-Garcia et al., 2009). This chapter will focus mainly on oryzanol as the plant sterol ester.

Cholesteryl esters form monolayers, except if the intermolecular interactions are dominated by the esterified groups, as is the case if that group is a long-chain fatty acid (Craven, 1986). Two types of monolayers may form: (i) type-I monolayers, in which neither the interaction between the steroid ring systems nor the interaction between the esterified acid moiety with the steroid ring system dominates; (ii) type-II monolayers, in which intermolecular interactions between the steroid ring systems dominate (Craven, 1986).

Experimental results for oryzanol are shown in Fig. 3.4. Oryzanol is too soluble in an oil slurry to give distinct peaks, but the pronounced diffraction peaks at d=2.5 nm (in powder) and d=2.6 nm (slurry in water and emulsion) confirm the formation of a monolayer structure (den Adel et al., 2010). Again, it is important to note that the sharp diffraction peaks indicate long-range order in the crystals in three dimensions. Infrared spectroscopy on oryzanol powder shows a weak band, indicating intermolecular hydrogen bonding, except in the case of the aqueous slurry where the spectrum is dominated by water (and no reliable statement on presence or absence of any type of hydrogen bonding can be made).

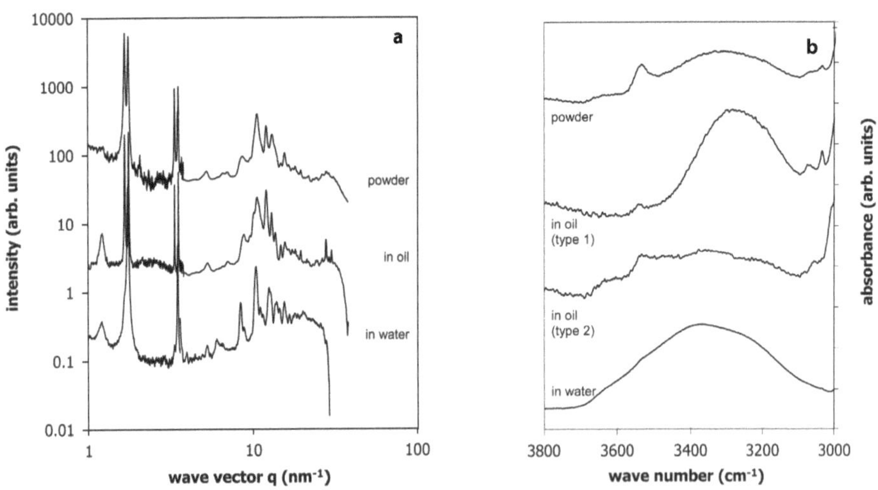

Fig. 3.4. (a) X-ray scattering data and (b) infrared spectra for oryzanol powder and 16% oryzanol in sunflower oil and water. Two types of crystals in oil can be distinguished by infrared spectroscopy.

Mixtures of Plant Sterols and Sterol Esters

The behavior of the individual components does not predict the behavior of the mixture to any extent, as will be shown below. This starts with the observation that the infrared spectrum of a sitosterol + oryzanol slurry in oil contains a novel absorption peak at 3441 cm^{-1}, implying the occurrence of intermolecular hydrogen bonding in these systems (Fig. 3.7) (den Adel et al., 2010). This interaction is most likely related to a potential hydrogen bond between the hydroxyl group of the sitosterol and the carbonyl group of the oryzanol. Preliminary calculations indicate that oryzanol and sitosterol may position their sterane cores over each other to minimize the energy of the complex (Fig. 3.5) (Pernetti et al., 2007). It can be expected that this conformation is sensitive to the presence of water, as water would interfere with hydrogen bond formation in the complex. This arrangement would allow the formation of a column of sterane cores with the ferulic acid group protruding. However, the presence of the hydrogen bond between sitosterol and oryzanol molecule prevents exactly parallel stacking of the sterane cores, as can be seen from Fig. 3.5, and thus the column cannot assemble in a completely straight fashion.

Although the exact size and shape of the complexes is not known, their presence can be identified through light scattering. The onset of assembly is strongly concentration dependent, with 16% total sterol mixtures in sunflower oil aggregating at around 45°C upon cooling and 8–10% total sterol mixtures around room temperature (Sawalha et al., 2011a). These aggregation temperatures are confirmed by DSC and rheology. The concentration dependence of their formation can be interpreted in terms of the Hildebrandt/van't Hoff/Gibbs-Helmholtz equation

$$\left[\left(\frac{\partial \ln x}{\partial 1/T}\right)_{sat,p} = \frac{-\Delta_{melting}}{R}\right] \qquad (1)$$

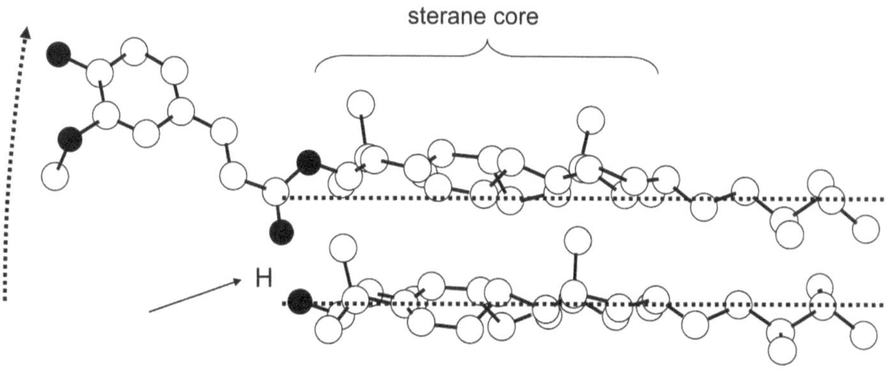

Fig. 3.5. Proposed stacking in the organogel building blocks. The presence of a hydrogen bond (arrow) prevents completely parallel stacking of the molecules, as can be inferred from the different positioning of the each of the molecules relative to the two parallel dotted lines. This leads to a curvature in the stack.

with x the mole fraction of the solute, T the (absolute) temperature, and R the gas constant. The entropy and enthalpy contributions to the aggregation process lead to a binding energy $-G°$ of about 4.5 kJ/mol for a 40:60 sitosterol + oryzanol mixture. Since the enthalpy contribution $\Delta H°$ is negative and the entropy contribution $-T\Delta S°$ is positive, it can be concluded that the aggregation process is enthalpy driven (Sawalha et al., 2011a). This result contrasts with earlier observations on proteinaceous fibril aggregation in an aqueous environment, for which the fibril formation was found to be entropy driven (Kroes-Nijboer et al., 2009). This is not completely unexpected, because the changes in excess entropy of the solvent during fibril formation that occur in a locally tetrahedral ordered fluid like water, will not occur in triglyceride oils.

The kinetics of aggregation depend considerably on the details of the system, as determined from infrared spectroscopy studies on the aggregation process of sitosterol + oryzanol in heavy mineral oil (Rogers et al., 2010). Relatively small variations in the ratio of sitosterol and oryzanol change the activation energy by a factor of two, and small variations in type of sterol (cholestanol or cholesterol instead of sitosterol) change the activation energy by a factor of four. Moreover, small changes in type of sterol change the mechanism of nucleation at a given condition—for the conditions studied, systems with cholesterol seem to show instantaneous nucleation and systems with cholestanol or sitosterol seem to show sporadic nucleation (Rogers et al., 2010). This sensitivity of the aggregation kinetics to experimental conditions is likely to be closely related to the observation that sitosterol + oryzanol systems in sunflower oil tend to supercool relatively easily if the mixture is rich in oryzanol (Bot et al., 2008).

The study by Rogers et al. also shows that the aggregation process is relatively quick—aggregation at temperatures below ~20°C is completed for the most part in 1–2 minutes. This is consistent with the visual observation that the gels develop rapidly once gelling starts (e.g., by mechanical disturbance) (Bot & Agterof, 2006) and observations in SAXS by den Adel et al. (2010) on a gelling system. The low-resolution SAXS data in Fig. 3.6 shows very broad features. This is peculiar, because regular crystals show long-range order, leading to sharp reflections in X-ray scattering data. Such reflections are not observed in the present SAXS data. The resolution of the data is insufficient, however, to interpret the structure more quantitatively.

High-resolution data over a wider wave vector range was obtained at the synchrotron in Grenoble (Fig. 3.7.A), which confirms the broad features for the sitosterol + oryzanol mixture in triglyceride oil with maxima located at 0.99, 1.96 and 2.75 nm^{-1}. The depth of the interference minima indicates a monodisperse, self-assembled structure. Even in the WAXS region of the spectrum, the curve is smooth, suggesting the absence of translational long-range order in any direction in the sitosterol + oryzanol aggregate. Furthermore, none of the characteristic reflections for the pure compounds are present, indicating that monolayer or bilayer structures are not formed. It is interesting to note that this structure occurs in an oil phase only. The

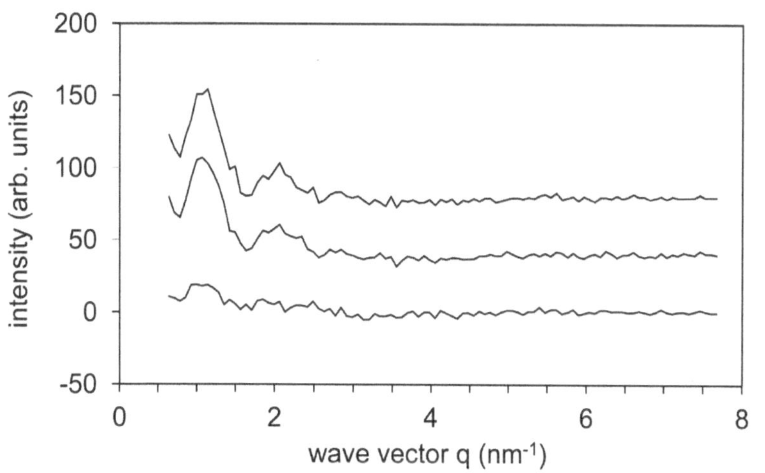

Fig. 3.6. XRD curves for organogels composed of 8% total sterols (40:60 mixture sitosterol:oryzanol) in sunflower oil after (from bottom to top) 0, 20 and 40 min storage time at 5°C after cooling from 100°C at a rate of 10°C/min. Vertical shifts were applied to the curves for clarity.

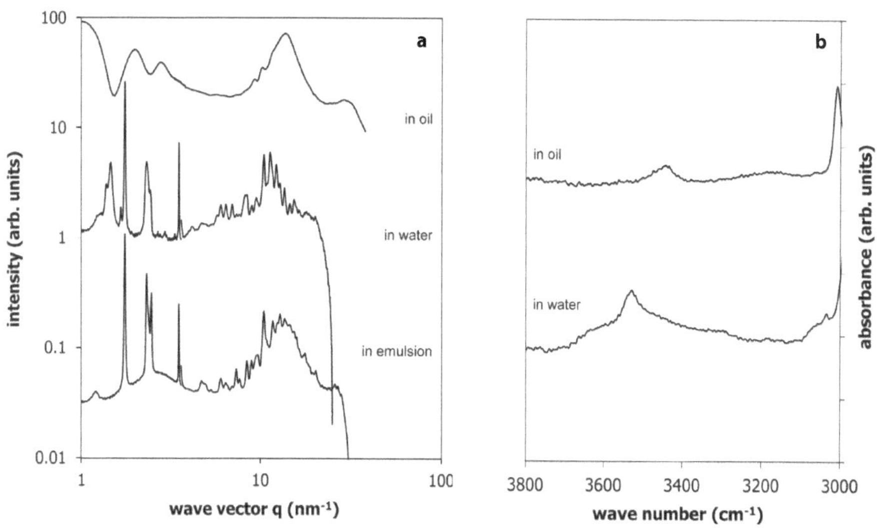

Fig. 3.7. (a) X-ray scattering data and (b) infrared spectra for a 40:60 mixture sitosterol:oryzanol (16% total sterols) in sunflower oil, water, and an water-in-oil emulsions. The small peak at 3441 cm^{-1} in oil is ascribed to intermolecular hydrogen bonding.

sitosterol + oryzanol mixture in water or emulsion does show the sharp crystallographic reflections, and these can be identified as originating from the pure sitosterol and oryzanol crystals (Fig. 3.7.A) (den Adel et al., 2010).

The scattering results above raise questions with regard to the supermolecular organization of sitosterol + oryzanol assemblies, as they seem inconsistent with regular three-dimensional crystals that might have been assumed implicitly. Light or electron microscopy would seem the most obvious techniques to provide insight into the morphology of the aggregates. However, it turns out that microscopy is not the most convenient method to investigate the assemblies. The reason is that separating the aggregates from the oil is difficult because TAG oil cannot be removed easily, e.g., by evaporation. Therefore, other methods should be considered.

One method that can distinguish between the solid and fluid components of the organogel is Atomic Force Microscopy (AFM). Kirkland and Yakubov performed a number of preliminary AFM scans on spin-coated thin films of 16% total sterol (40:60 sitosterol:oryzanol mixture) in sunflower oil (2009). The image in Fig. 3.8 shows bundles of fine linear objects. The dimensions of the finest fibers are difficult to measure, but are estimated to be around 13 nm. This number will be an overestimation of the true figure because it includes a contribution due to the width of the AFM tip. Obviously, the spin-coating process is not identical to the quiescent conditions

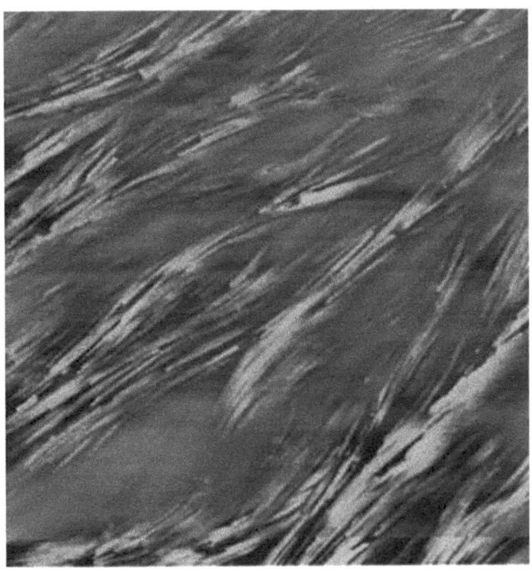

Fig. 3.8. A drop of melted organogel in sunflower oil was added on top of a 90–100°C silanized (hydrophobic) silicon slide, which was mounted on to a PI-KEM P6708D spin coater. After running the spin coater at 3000 rpm for 20s, the slide was stopped and allow to cool. Imaging was done using a Veeco Atomic force microscope in HarmoniX mode. The width of the individual tubules is estimated to be of the order of 10 nm, and the tubules are aggregated in bundles of the order of 100 nm diameter (Kirkland and Yakubov, 2009, unpublished).

used in the experiments above, so one should consider the possibility that some of the larger-scale structures like the bundles could be induced by the sample preparation process. This was not investigated further, however.

The observation of fine linear structures by AFM may help us to interpret the broad X-ray scattering features observed in Fig. 3.7. The AFM image suggests a straightforward interpretation of these structures in terms of solid cylinders. However, solid cylinders imply a wave vector dependence of the maximum intensity of the interference peaks in the SAXS data, which would have to decrease proportional to q^{-4}, whereas the data seems to indicate a decrease proportional to q^{-2}. Such less steep decrease is predicted by a somewhat unphysical model of a tubule with an infinitely thin wall. Tubular structures have been identified previously in cholesterol-derived organogelling agents (Sakurai et al., 2001) and cholesterol in an (aqueous) bile salt environment (Konikoff et al., 1992). A somewhat more complex, but certainly more physical, model can be found by introducing a finite wall thickness w. The scattering intensity as a function of wave vector q for these tubules with diameter r_c and wall thickness w are (Deutch, 1981)

$$I(q) \sim [r_{out}J_1(qr_{out}) - r_{in}J_1(qr_{in})]^2/[q^3 \cdot (r_{out}^2 - r_{in}^2)^2] \tag{2}$$

with r_{out} and r_{in} the outer and inner wall radius, and J_1 the first-order Bessel function (Press et al., 1986). By requiring that the amount of tubule wall material in the cylinder with inner and outer wall at r_{in} and r_c equals the amount in the cylinder with walls at r_c and r_{out}, the values of r_{in} and r_{out} become

$$r_{in} = \sqrt{(r_c^2 - w^2/4)} - w/2 \tag{3}$$

$$r_{out} = \sqrt{(r_c^2 - w^2/4)} + w/2 \tag{4}$$

with $w = r_{out} - r_{in}$. Equations (1–3) can be used to generate the scattering curves for the hollow cylinder model (i.e., w = 0),

$$I(q) \sim J_0^2(qr_c)/q \tag{5}$$

with J_0 the zeroth-order Bessel function, and the solid cylinder model (i.e., $w = r_{out} = \sqrt{2} \cdot r_c$)

$$I(q) \sim J_1^2(qr_{out})/q^3 \tag{6}$$

An example of a fit according to Equation (1) is shown in Fig. 3.9 for a 16% total sterol mixture (40:60 sitosterol:oryzanol mixture) in sunflower oil. It can be seen that the intensity of the first three peaks follows the q^{-2} dependence quite well, but the fourth peak does not appear to have the expected intensity for a hollow tube model. The fit with Equation (1) for tubule diameter $r_c = 7.2 \pm 0.1$ nm and wall thickness $w = 0.8 \pm 0.2$ nm explains the lower intensity as the consequence of a finite wall thickness. Indeed, in this higher wave vector range the scattering intensity is expected to decrease as q^{-4}.

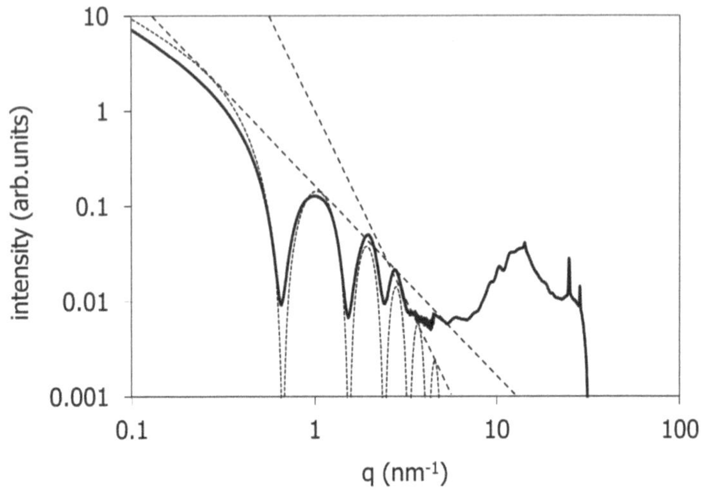

Fig. 3.9. (black curve) SAXS and WAXS curves for 16% total sterols (40:60 sitosterol:oryzanol ratio) in sunflower oil; (dashed curve) A fit for the SAXS range according to Eq. 1 using $2r_c = 7.2$ nm diameter and w = 0.8 nm. The two dashed straight lines represent the q^{-4} and q^{-2} dependence of the intensity of the interference maxima according to the solid cylinder and hollow tubule models.

A wall thickness of 0.8 nm lacks a direct explanation in terms of the main axis of the sterol molecule. As mentioned in Sections 2.1 and 2.2, the long axis of the sterol molecule is about 1.8 nm and of the oryzanol approximately 2.5 nm. However, the length of the long axis of the sterane core is about 0.8 nm. Since the alkyl groups in the sterol molecules are considerably more flexible than the sterane core, these will behave more like the triglycerides than like the rigid steroid ring system. Therefore, the stacked sterane cores are expected to stand out from the surrounding fluid, and to be interpreted as the wall of the tubule in a SAXS experiment.

The details of the tubule structure do not depend on either the total sterol concentration or on the ratio of sitosterol and oryzanol (Bot et al., 2008). However, sitosterol-rich samples (either due to the total concentration or due to the fact that the sterol+sterol ester mixture is rich in sitosterol) do show a tendency towards separate sitosterol crystallisation, as demonstrated by the presence of sharp crystallographic reflections in the SAXS curves. The extent to which this occurs depends somewhat on cooling rate—higher cooling rates tend to suppress the phenomenon.

The nature of the sterol has a somewhat bigger influence on tubule characteristics, although formation was observed with SAXS in all cases (Bot et al., 2009b). The data in Fig. 3.10 indicate that the observed tubule diameter increases with a smaller number of double bonds in the sterane core (ergostanol vs cholesterol, sitosterol, stigmasterol vs cholestanol). Although this increase seems to be related to the structural properties of the tubules, for now, it is impossible to rule out the possibility that small changes in contrast between the triglyceride phase and floppy alkyl chains of the

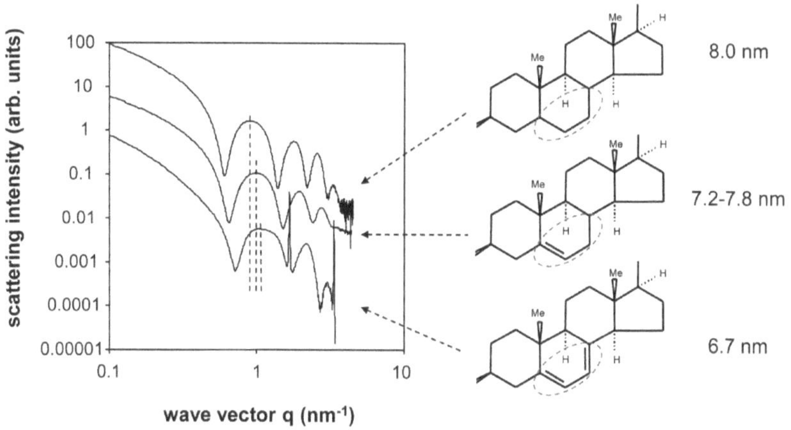

Fig. 3.10. Effect of the number of double bonds in the sterane core on tubules diameter. Two double bonds: 6.7 nm; single double bond: 7.2–7.8 nm; no double bond: 8.0 nm (mixtures of oryzanol with ergostanol, sitosterol, cholestanol, respectively). The sharp peaks in the system with two double bonds are caused by separate crystallization of sterol crystals.

sterol(ester)s play a role here as well. There is no obvious trend in the wall thickness for the various sterols.

The WAXS region of the diffractograms shows a broad peak at ~13–14 nm^{-1}, at slightly smaller values than a typical triglyceride oil peak (Fig. 3.9). The absence of sharp peaks in the WAXS section of the majority of diffractograms indicates a distribution of interatomic distances in the supermolecular assembly. One of the possible causes for this would be a wedged stacking of sitosterol and oryzanol, as suggested above for another reasons. In such a situation, a combination of two phenomena leads to a smooth WAXS spectrum: (i) a distribution of interatomic distances results in a distribution of peaks the WAXS spectrum, and (ii) the absence of long range order leads to a considerable broadening of these peaks) (Bot et al., 2008).

The combined data suggest that the sterol and oryzanol stack to form a supermolecular structure. However, the presence of a hydrogen bond between the sterol and oryzanol molecule prevents parallel stacking as in the crystals of the pure compounds (Pernetti et al., 2007). Instead, wedged stacking occurs, leading to a curved structure, similar to a supermolecular staircase. The staircase winds tightly enough to behave like a tubule, with triglyceride oil both inside and outside the tubule (see Fig. 3.11). The sterane cores of sterol and oryzanol form the wall of the tubules, with the hydrogen bond located at the outside of the tubules. This conformation causes the ferulic acid moieties in the oryzanol to stick out of the exterior wall of the tubules (Bot et al., 2009b). Since the ferulic acid group helps to solubilize the oryzanol in oil compared to sitosterol, it seems likely that they play the same role for the tubule. As a consequence, the interaction between tubules may be reduced, leading to less aggregation of the tubules. An increase in oryzanol may even reduce the tendency of

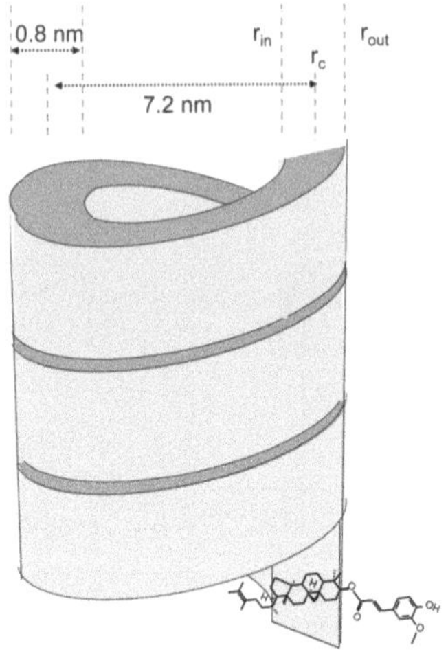

Fig. 3.11. Schematic representation of the proposed helical ribbon structure of the sitosterol+oryzanol tubules with a diameter of 7.2 nm and a wall thickness of 0.8 nm. The hypothetical molecular arrangement in the wall is also shown.

the tubules to aggregate and prevent the formation bundles as observed in Fig. 3.8. Indeed, by increasing the amount of oryzanol, an increase in transparency of the organogels is observed, suggesting a thinner bundle diameter.

The structure proposed in Fig. 3.11 raises some questions concerning the assembly of sitosterol and oryzanol in systems with only minimal oil. On the one hand one would expect the basic complex formation to occur even in the absence of oil, on the other hand the tubular structure would require the presence of some oil to fill the interior of the tubules. Experiments to answer this question are in progress (Sawalha et al., 2011c).

A commercial grade of sitosterol ester, interesterified with the fatty acids of sunflower oil, was tested as an alternative to oryzanol. Such a mixture would typically contain sitosteryl linoleate as an important component (see Fig. 3.2). However, this combination of sterol and sterol ester was found to not form tubules (Bot et al., 2009b), possibly because the long fatty acid chain made the ester too soluble. Although oryzanol is not believed to be unique in its property to form edible oil organogels with sitosterol, this example shows that there is exists a delicate balance for determining whether or not an organogel can be formed. Establishing the "formulation space" for the acid group in the sterol ester in the context of edible oil organogel formation remains one of the open challenges in this field.

Tubules were shown to form in a number of alternative lipid phases, next to triglyceride oils. As a general rule, higher polarity solvents interfere more with tubule formation (Sawalha et al., 2011b). Although this result is interesting and helps to disentangle the subtle balance of effects involved in structuring triglyceride oil, it does not bear directly upon attempts to apply the present structuring agents in edible oils.

Gelling

The tubules can aggregate and form an edible oil organogel with a fat-like consistency and a slippery surface. Most interesting of all, the organogels are transparent/slightly hazy reflecting the fact that the tubules are considerably thinner than the wavelength of visible light (Fig. 3.1). At least 2–4% total sterols are needed to form a gel at 5°C.

The tubules show erratic gelling under quiescent conditions, especially in systems that are rich in oryzanol, because the mixture can remain a metastable liquid state for a considerable length of time after cooling. Under oscillatory shear or other mechanical disturbances the system may gel in a more predictable way. This is illustrated in Fig. 3.12, which displays the development of the storage modulus G′ and the loss modulus G″ of a 6% total sterol organogel (40:60 oryzanol:sitosterol). The storage modulus is often used as a measure for texture development in a gel. When gelling occurs under continuous application of oscillatory shear (Δt = 0s), the modulus increases almost instantaneously and reaches its final value of about 150 kPa in ~200s. However, if first a quiescent period Δt is introduced (Δt = 600, 900 or 1200s), the increase in the modulus is delayed by a period that is slightly longer that this induction period Δt. With longer induction time, the steepness of the initial slope of the modulus decreases and the final value of the modulus of the system decreases. The oscillatory shear amplitude and frequency have less effect, provided the applied stress does not exceed the yield stress or strain of the gel (Bot & Agterof, 2006).

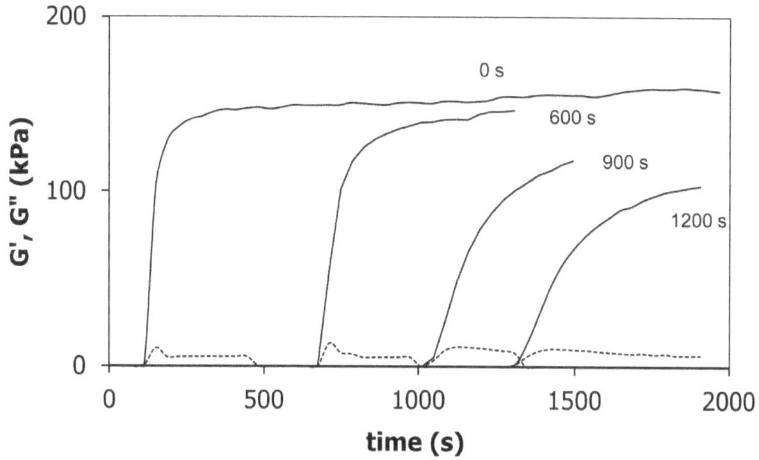

Fig. 3.12. Effect of delay time before an oscillating shear stress profile (amplitude 175 Pa, frequency 1 Hz) is applied to a gelling sterol+oil system in sunflower oil (6 wt% total sterol, 60:40 sitosterol:oryzanol ratio, 10°C) on the storage modulus G′. The numbers in the figure indicate the delay time. Dots represent the loss modulus G″. Source: Bot and Agterof, 2006.

There is no direct evidence to determine the rate-limiting step in the gelling process. On the one hand, the formation of the tubules could be limiting (nucleation and growth). The time scale of minutes of tubule formation derived from X-ray scattering suggests a similar rate of gelling (den Adel et al., 2010), but the time resolution is not sufficient to link the two processes indisputably. The data by Rogers et al. suggest a similar time scale, but for a somewhat different system and show that small differences in composition or concentration affect the tubule formation considerably (Rogers et al., 2010). On the other hand, the rate of aggregation of bundles of tubules into a network may limit gelling. The circumstantial evidence indicates that the aggregation process is the rate-limiting step under quiescent conditions, whereas the nucleation and growth is the limiting step under non-quiescent conditions.

The gelling rate under non-quiescent conditions shows a maximum at 10°C (Fig. 3.13), which leads to the firmest gels (in line with Fig. 3.12) (Bot & Agterof, 2006). Above, the rate of nucleation and growth (Rogers et al., 2010) has been identified as the most likely rate-limiting step. The existence of a maximum as a function of temperature implies that the nucleation and growth show opposite temperature dependence: the formation of the tubules is limited by growth at low temperatures, and by nucleation (and availability of solids) at high temperatures.

The gels have the interesting property that they are much firmer under compression than under shear deformation, as can be established quite easily by physically touching the gel: fingers slip along the surface without too much resistance, but the gel cannot be compressed easily. This can be demonstrated more quantitatively by a number of large deformation stress-strain curves, in compression and in shear. The

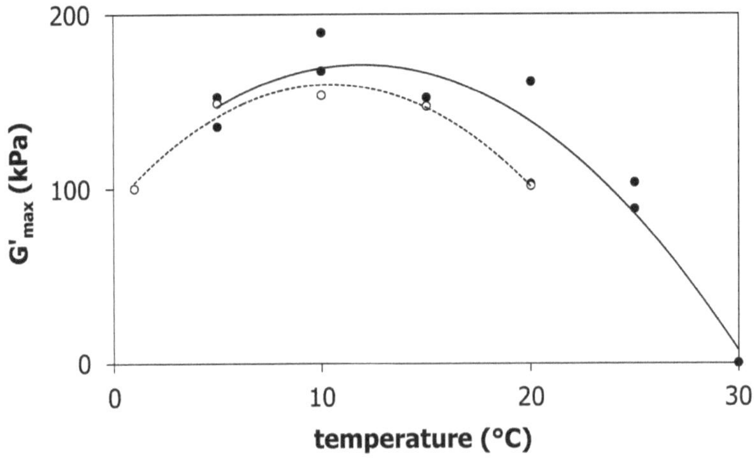

Fig. 3.13. Maximum of the elastic modulus G'_{max} as a function of gelling temperature for 6% total sterol concentration (60:40 sitosterol:oryzanol mixture) in sunflower oil. Measurements were performed at (0•) 150 Pa, (○) 175 Pa (frequency 1 Hz). Parabolic fit guides the eye. Source: Bot and Agterof, 2006.

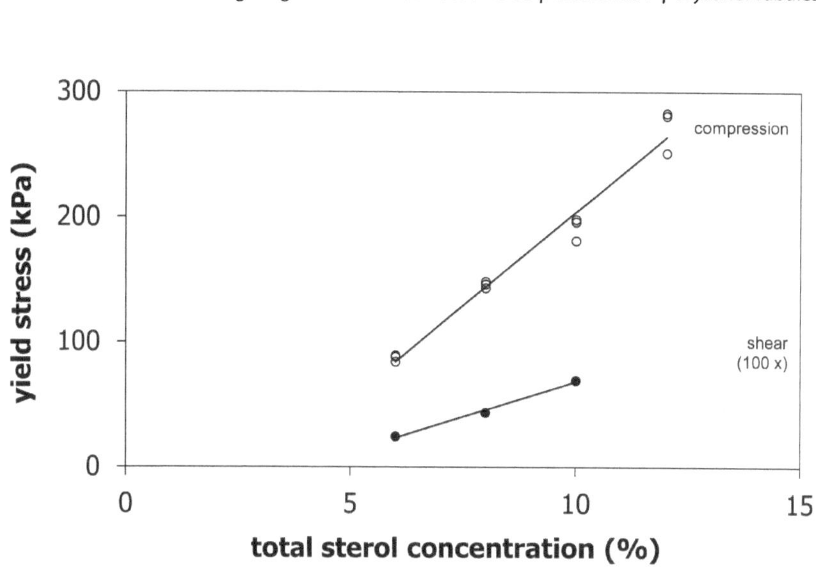

Fig. 3.14. Yield stress of organogels in sunflower oil under (○) compression and (•) shear (40:60 sitosterol:oryzanol mass ratio) as a function of concentration. Yield stress for shear deformation is multiplied by 100.

yield stresses are of the order of 100 kPa in compression and of the order of 0.1 kPa in shear deformation. The gels can be deformed much more in compression than in shear before the gels break (Bot & Agterof, 2006). Fig. 3.14 shows the effect of total sterol concentration. The yield stress increases linearly with total sterol concentration both in compression and shear, which can be interpreted as if network tubules themselves break and not the associations between the tubules. Usually, the cross-links are the weakest spots in an aggregate structure, but long tubules may break even if modest bending forces are applied (Bot & Agterof, 2006).

The ratio of sitosterol and oryzanol and the total amount of sitosterol + oryzanol both affect the strength of the organogel. The optimal composition in terms of maximal organogel firmness is a 1:1 molar ratio of both components (40:60 sitosterol:oryzanol mixture) (Ritter et al., 1997), although Fig. 3.15 shows that the elastic modulus may remain high even at higher oryzanol content (Bot & Agterof, 2006).

The chemical composition of the sterol in the mixture has a modest effect on the rheological properties of the gel (see Fig. 3.16). The critical concentration for gelling may be somewhat higher, as in the case for ergosterol at these concentrations despite the fact that ergosterol does form tubules (Fig. 3.10.). The higher critical concentration may be associated to the fact that ergosterol is amongst the plant sterols with the lowest melting point in triglyceride oil (Bot et al., 2009b). The other sterols—cholestanol, cholesterol, β-sitosterol, stigmasterol—show more or less the same rheological behavior in Fig. 3.16. Gelling of the cholesterol-containing system is relatively slow, again indicating that the system may be close to its critical gelling point due to its lower melting point (Bot et al., 2009b). Generally, these results indicate

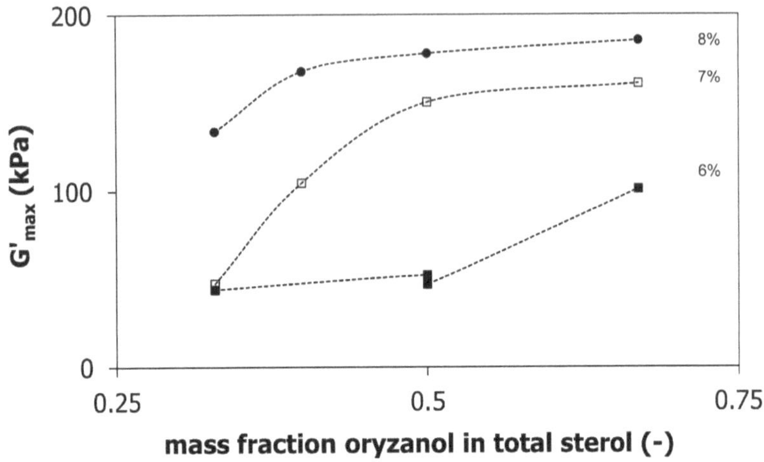

Fig. 3.15. Effect of relative sterol/sterol ester concentration and total concentration on organogel formation at 10°C in sunflower oil: (■) 6 wt%, (□) 7%, (•) 8 wt%. An oscillating shear stress profile (applied deformation during gelling: $\gamma = 0.003$, frequency 1 Hz).

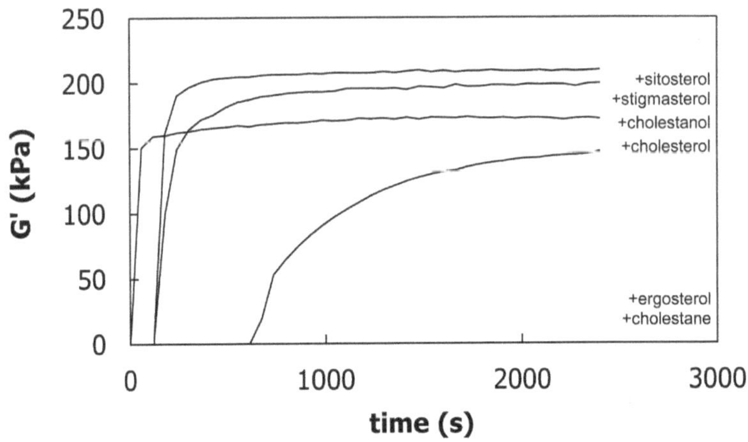

Fig. 3.16. Gelling in sterol+oryzanol systems in sunflower oil (8% total sterols; 40:60 sterol:oryzanol ratio, 10°C) in which the sterol is cholestanol, cholesterol, sitosterol, stigmasterol. No structure is formed in mixtures with cholestane or ergosterol under these conditions (applied oscillatory shear amplitude 175 Pa, frequency 1 Hz).

that gelling is not sensitive to the structure of the alkyl moiety in the plant sterol molecule, but some effect of the structure of the sterane core may be present (fewer double bonds promote quick tubule formation). Finally, the system with 5α-cholestane, the cholestanol-like compound without the hydroxyl group, does not gel at all. This is in line with expectation because 5α-cholestane cannot form intermolecular hydrogen bonds with oryzanol. Similar detailed studies on the nature of the sterol esters have yet to be performed; at present only mixtures containing oryzanol were found to gel oil successfully whereas sitosterol esterified with fatty acids from sunflower oil were not (Bot et al., 2009b), but one has to keep in mind that relatively little effort has been invested until now in finding alternative sterol esters.

Data on the spatial organization of the tubule aggregates in an organogel network have yet to be obtained. It was mentioned above that microscopic techniques are difficult to apply. The transparency of the organogels could make light scattering studies of these organogels feasible (Lui & Sawant, 2001).

Melting

The melting curves of the organogels as studied by DSC feature a pronounced asymmetric peak with a long tail towards lower temperatures. This type of behavior is characteristic for the dissolution of a solid in a solvent. The composition of the solid phase does not affect the temperature at which all solids have dissolved dramatically, at most by 5–10°C (see Fig. 3.17). The highest complete melting temperature is observed for the 60:40 sitosterol:oryzanol mixture. The differences between samples with different sitosterol:oryzanol ratio are small compared to the difference between heating and cooling curves for the same single sample, especially at high oryzanol concentration.

The loss of firmness at higher temperatures can be explained potentially by several mechanisms: (i) a loss of associations between the tubules, (ii) a loss of stiffness of the tubules, (iii) the dissolution of the tubules. However, when the characteristic

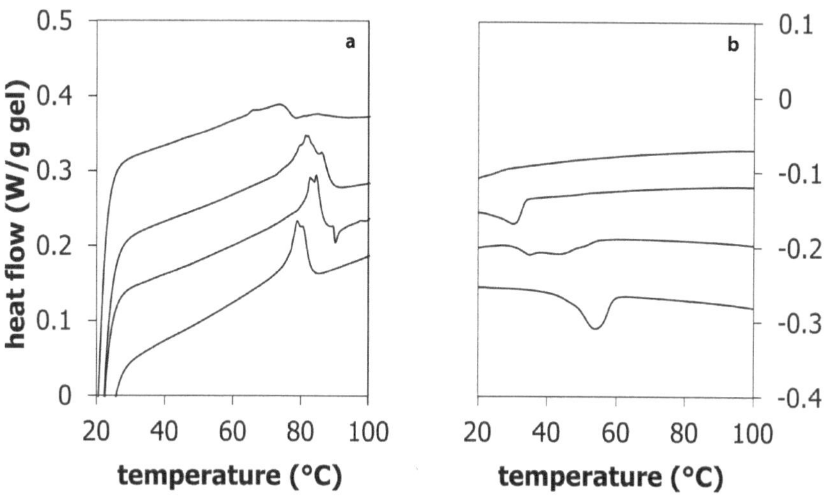

Fig. 3.17. Effect of sitosterol:oryzanol ratio on DSC curves for 16% organogels (heating/cooling rate 10°C/min). Curves were shifted for clarity: (a) heating stage, from top to bottom for 80:20, 60:40, 40:60 and 20:80 sitosterol:oryzanol samples; (b) cooling stage with curves in inverted order (80:20 sitosterol:oryzanol sample is bottom curve). Source: adapted from Bot et al., 2008.

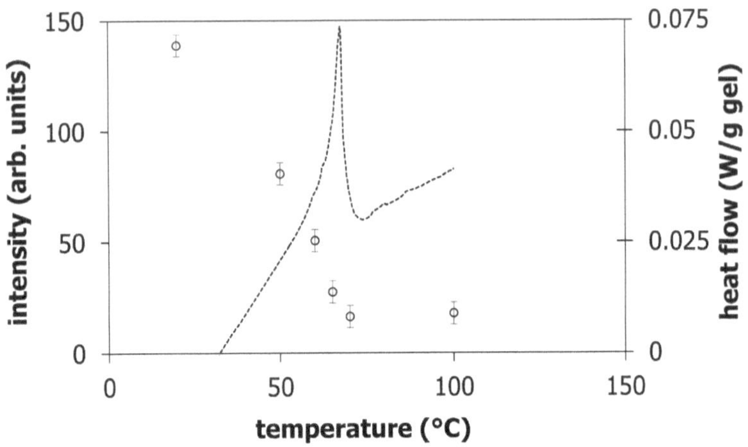

Fig. 3.18. The symbols indicate the temperature dependence of the intensity of the first maximum of the tubule interference pattern near q=1 nm^{-1} for 8% total sterols (40:60 sitosterol:oryzanol mixture) in sunflower oil. The dashed line shows the DSC curve for the same system, heating rate 10°C/min. Source: Bot et al., 2009b.

interference pattern of the tubules is tracked during melting of the gel (Bot et al., 2009b), no trace of any structure above the melting temperature of the gel is found (see Fig. 3.18). This effectively rules out mechanisms (i) and (ii), leading to the conclusion that the gel melts because sitosterol and oryzanol dissolve in the oil.

The progressive dissolution of the sitosterol and oryzanol with increasing temperature can also be inferred from the decrease in elastic modulus with increasing temperature, as is demonstrated in Fig. 3.19. In addition, the figure shows that the temperature at which the modulus vanishes depends strongly on total sterol concentration, as expected.

Assuming that the temperature at which the system is unable to form a network coincides with the temperature at which all solids have dissolved (which is not completely accurate, but a sufficient approximation for the present purpose) and treating the phytosterol mixture as a single component, it is possible to apply the Hildebrandt/van't Hoff/Gibbs-Helmholtz equation to estimate the melting enthalpy for the solid $\Delta H_{melting}$. The result for $\Delta H_{melting}$ is a value of 26±4 kJ/mol (Bot & Agterof, 2006). Obviously, some care must be taken in interpreting this value, but it is interesting to see that it is broadly similar to the values obtained for the melting enthalpy of the pure components and of similar steroid-type organogelling agents of non-edible oil phases (Wade et al., 1986; Terech et al., 1988; Terech, 1997). Similar values would be expected if stacking of the sterane cores is the driving mechanism for the whole group of structurants. This is confirmed by the melting behavior of a range of plant sterol + oryzanol mixtures, which was found to be similar to the sitosterol + oryzanol mixture. The melting temperature of the stigmasterol + oryzanol system was

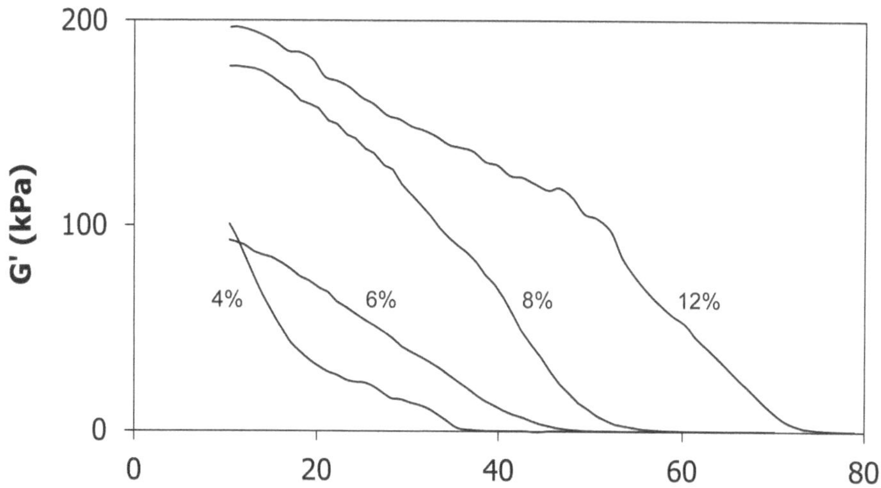

Fig. 3.19. Melting curves for sterol organogels (50:50 sitosterol:oryzanol mass ratio) in sunflower oil during a temperature sweep from 10 to 70°C (applied maximum strain during oscillation is 0.003, frequency 1 Hz). The total sterol concentrations are indicated in the figure. Source: adapted from Bot and Agterof 2006.

found to be ~10°C higher than the sitosterol + oryzanol system, and the cholesterol + oryzanol was found to be ~10°C lower. All other sterol + oryzanol systems (cholestanol, ergosterol) were found to melt at intermediate temperatures (Bot et al., 2009b).

Emulsion Gels Based on Mixtures of Plant Sterols and Sterol Esters

Structuring of the Emulsion on a Molecular Scale

The ultimate goal of non-triglyceride edible oil gelling are emulsion-based applications (Pernetti et al., 2007). These include primarily oil-continuous emulsions (de Bruijne & Bot, 1999), but water-continuous emulsions are of secondary interest as well (Bot et al., 2007). The application of sitosterol and oryzanol in emulsions may introduce a number of issues:

- Hydrate formation by sitosterol and/or oryzanol may interfere with tubule formation.
- Surface activity of sitosterol and/or oryzanol (if any) may cause the structurants to be located preferentially at the interface.
- The lack of solubility of the structurants in the liquid oil may prevent the existence of a dynamic equilibrium between dissolved and precipitated structurant, and thus compromise the dynamic exchange between molecules in the solid and liquid phases that is necessary for curing of the structure, e.g., to re-establish the network after shearing the system.

For the preparation of the emulsions, a distinction should be made between w/o and o/w emulsions. The o/w emulsions can be prepared in a relatively straightforward fashion, as the initial creation of the emulsion can be done more or less independently from the later formation of an organogel network in the lipid phase (Duffy et al., 2009). For a w/o emulsion, which concerns most of the work below, the situation is more complicated because some solids are required to stabilize the emulsion droplets, if no further emulsifiers are added to the formulation. However, too much shear may cause damage to the tubule network that is being built up, and too many solids may render the shearing process inefficient. If the subtle balance between shear and amount of solids is correct, however, a whitish liquid is formed that can be filled in a tub—the loss of transparency is as expected for emulsions containing droplets of different refractive index with sizes beyond a few micrometers. The process is set up such that the emulsion is a thick liquid on filling and firms in the tub during storage. The filling operation is obviously not necessary from a scientific point of view, but quite relevant from an application point of view.

The structurants, more specifically the sitosterol, have a tendency to form hydrates in emulsions (see also Fig. 3.7) (den Adel et al., 2010). The addition of water to a system with the sitosterol + oryzanol structurant (16% total sterols on oil phase, 30% water in the emulsion) does not result in emulsions showing the characteristic broad SAXS interference pattern for tubules, but rather to a pattern with sharp crystallographic reflections that indicate long-range translational symmetry as occurs in crystals. The main reflections can be attributed to the formation of sitosterol monohydrate. The amount of water in the emulsion (in the 10–30% range that was investigated) does not affect the basic aspects of this behavior.

SAXS patterns that bear resemblance to those of the tubules appear for emulsions with higher amounts of organogelling structurants in the oil (32% total sterols on oil phase). The main difference to the organogel tubule pattern is that the first broad peak has become a double peak (see Fig. 3.20). The pattern is sufficiently close to the structures observed previously to surmise that tubule-like structures are formed, but this time with a slightly more complex structure than in the organogel. It does not seem very likely that each of the peaks in the double peak structure reflects the presence of different tubules, as the relatively amplitude of both peaks seems to be rather constant throughout all of these experiments. Nevertheless, it is possible to fit the data, assuming a mixture of two tubules with quite similar widths (diameter 7.0 and 10.0 nm), which makes it likely that the explanation of the structure should be in terms of a more complex wall structure than the tubules in pure oil (Bot et al., 2011). Scattering experiments in which the contrast of tubule, water, and oil can be varied independently—as in small-angle neutron scattering—could shed some further light on the details of the structure of these tubules in emulsions.

Interestingly enough, the absence of the SAXS signature of tubules in Fig. 3.20 for w/o emulsion gels containing a relatively low total sterol concentration does not indicate the absence any fiber-like structures. Fig. 3.21 demonstrates the existence of fibers in a w/o emulsion gel containing 10.5% total sterols and 30% water (left-side

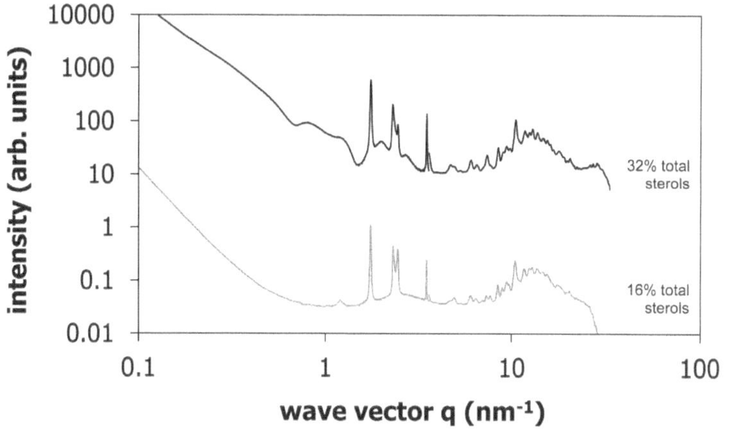

Fig. 3.20. SAXS data for a 40:60 mixture of sitosterol:oryzanol dissolved/dispersed in emulsions with 10% water after 1 week storage at 10°C. From top to bottom: (black) 32% total sterols and (grey) 16% total sterols on oil.

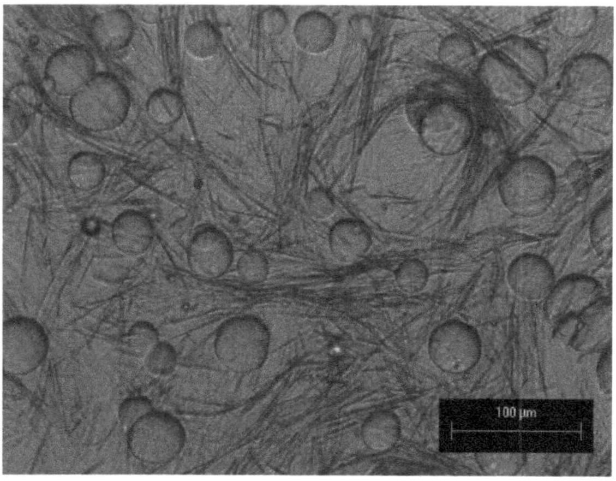

Fig. 3.21. Light microscopy images of an emulsion gel containing 10.5% total sterols on oil phase (40:60 sitosterol:oryzanol mixture) and 30% water. Image width 475 μm.

panel). The fibers trap the emulsion droplets. The fibers are clearly wider than 1 μm (because they can be observed easily with light microscopy) and their typical length is around 100 μm or more. The SAXS and WAXS data demonstrate that these fibrils must be more or less regular crystals, as can be inferred from the presence of sharp crystallographic reflections. These properties match with those of the sitosterol monohydrate crystals (Christiansen et al., 2002; von Bonsdorff-Nikander et al., 2003).

These observations are in line with results for o/w emulsions, in which fibers also formed that could be imaged by light microscopy (Duffy et al., 2009). These fibers should be identified as sitosterol monohydrate fibres too. Under specific conditions the thin "organogel" tubules are formed too, especially when the emulsion is produced by mixing a concentrated organogel with the water phase and the sample is imaged by electron microscopy immediately after production.

Because the sitosterol monohydrate fibers are much thicker that the organogel tubules, their surface to volume ratio is smaller, which implies less efficient structuring per amount of structurant. It is indeed observed that the firmness of w/o emulsion gels after storage is less compared to the expected firmness of an organogel with the same total sterol concentration (Bot et al., 2009a; Sawalha et al., 2011b). Raising the total sterol concentration leads to an increase in firmness of the emulsion gel, as expected (Bot et al., 2009a). The firmness of the emulsion gel is more sensitive to the ratio of sitosterol and oryzanol than in the organogel—the optimum being a 1:1 molar ratio in the emulsion gel. The firmness of the emulsion gels shows surprisingly little sensitivity to amount of dispersed water phase. Only close to the inversion point of the emulsion, a minor increase in firmness is observed (Bot et al., 2009a). Addition of gelling agents to the water phase has not been tested extensively, but is expected to increase firmness. Although addition of low amounts of salt to the aqueous phase (<5%) has been shown to have little effect on emulsion firmness, higher amounts considerably increase the firmness (Sawalha et al., 2011b).

The tubules in w/o emulsion gels are generally not found to be stable during storage. The transformation can be monitored by following the increasing intensity of the sharp crystallographic reflections in the SAXS/WAXS diffractograms with increasing storage time, and the diminishing intensity of the "tubule" interference pattern in SAXS under simultaneously shifting of elements of the pattern to smaller wave vectors (i.e., longer length scales) as demonstrated in Fig. 3.22. In systems with a high fiber-to-surface ratio aging of the emulsion is associated with a loss of sphericity of the droplets of the dispersed water phase (Bot et al., 2011). Studies are underway to identify factors that may delay the formation of hydrates (Sawalha et al., 2011b).

For emulsion gels based on the sitosterol+oryzanol structuring system, the most important phenomenon during heating is the mono-hydrate to hemi-hydrate transition for sitosterol at ~45°C (Fig. 3.23). The tubules are unstable during storage below this temperature, but show more stability at higher temperatures. The amount of water in the emulsion does not play an important role, under the assumption that emulsions contain at least 10% water under practical conditions. There is no obvious change in behavior at the hemi-hydrate to anhydrate transition at ~80°C, although any effects due to the type of hydrate on the kinetics of destabilization have yet to be investigated. In systems without hydrates, the melting behavior of the emulsion gels seems to resemble that of the pure organogel (Bot et al., 2011).

Application of a heating-cooling temperature cycle to the emulsion causes the sitosterol monohydrate fibers to dissolve and reform. The structures obtained during

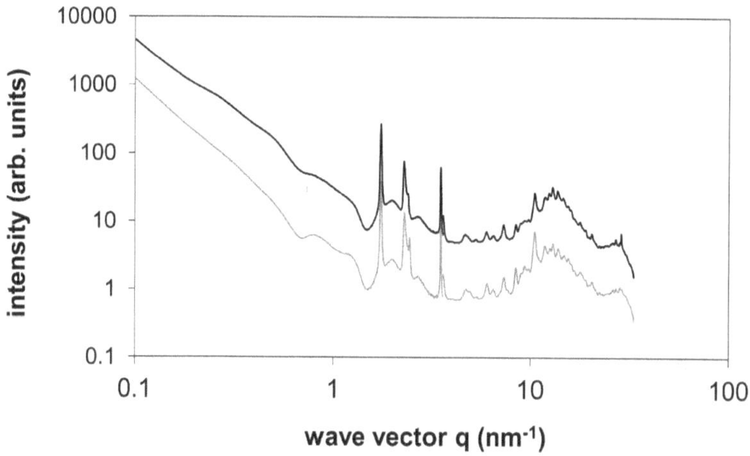

Fig. 3.22. Effect of aging on X-ray diffractograms of a 32% total sterol on oil phase w/o emulsion containing 10% water at 10°C. From top to bottom: (black) after 27 weeks; (grey) after 1 week.

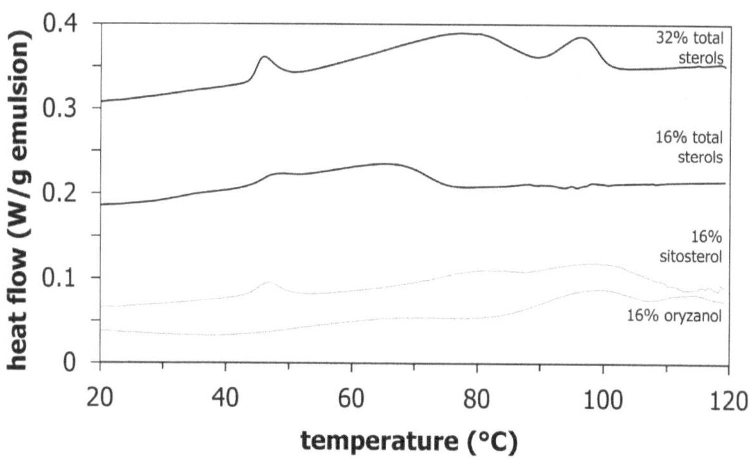

Fig. 3.23. DSC data for sitosterol, oryzanol and mixed sitosterol+oryzanol in an emulsions after 1 week storage at 10°C. From top to bottom: (black) curves for 32% and 16% total sterols on oil in emulsions containing 10% water, respectively; (grey) curves for 16% sitosterol and oryzanol in an emulsion containing 30% water, respectively (scanning rate was 10°C/min).

reformation of the emulsion tend to show more branching than the original fibers. Most surprisingly, is the nucleation of water droplets at the tip of the fibers at the last stage of cooling (T<50°C) (Fig. 3.24). The details of this process have yet to be well studied (Bot et al., 2011).

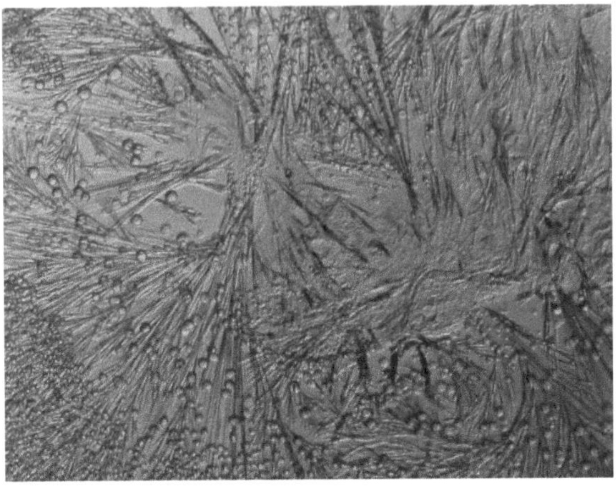

Fig. 3.24. Droplet formation in an emulsion gel containing 10.5% total sterols on oil phase (40:60 sitosterol:oryzanol mixture) and 30% water after heating the emulsion gel to 90°C and subsequent cooling to 5°C (heating/cooling rate of 10°C/min). Image width is 1190 μm.

Potential Applications

Until now, study of the sitosterol + oryzanol system has lead to many surprising and interesting insights into the physical chemistry of this complex system. However, in practice this has yet to lead to many applications.

The most obvious potential application is low-SAFA structuring of edible oils (Pernetti et al., 2007; Rogers, 2009; Marangoni, 2009; Wassell et al., 2010). This type of structuring was already considered at an early phase of the study of the system (Ritter et al., 1997). In principle, the system can be applied in this way, but in the presence of water the effectiveness of structuring by the sitosterol + oryzanol mixture becomes less (Bot et al., 2009a). To retain the effectiveness of the system, additional measures need to be taken to prevent the transformation of tubules in sitosterol mono-hydrates. At present, therefore, generic application of the sitosterol + oryzanol system in foods is not yet expected, although application in niche applications would still be possible. Such examples could be water-free applications, or specific applications with minimal amounts of water. Furthermore, it should not be taken for granted that application of these ingredients at the present levels will be possible without any further regulatory approval.

Sitosterol and oryzanol mixtures have also been investigated as structurants of the oil phase in an oil-in-water emulsion. This system shows the same tendency to form sitosterol monohydrate as other water-containing emulsions, but this has fewer consequences for the texture of the emulsion than in the case of firm w/o emulsions. However, at high total sterol concentration and short storage time, the emulsions are shown to contain tubules. An interesting property of the emulsions is that they seem to reduce the rate of lipolysis, leading to higher storage temperature and longer storage time. This suggests that the presence of sitosterol monohydrates results in slower lipid digestion, as shown by a reduction in the release of free fatty acids during hydrolysis (Duffy et al., 2009; Diks et al., 2007). The exact mechanism of action is not completely understood, but could find an explanation in the reduced accessibility of the oil within the tubules to the lipolytic enzymes.

Finally, the sitosterol + oryzanol system has been studied as a structurant of the non-aqueous phase in non-food applications—in particular in anti-perspirant sticks (Franklin et al., 2000). The composition of oil and aqueous phase in these sticks is different compared to the systems described in Section 3, leading to a much higher effectiveness of the tubule-based structuring in this system. The impact of applying alternative aqueous and lipid phases has yet to be investigated in detail, but studies have begun looking into these effects (Sawalha et al., 2011b).

Outlook

This chapter reviewed the results that have been obtained over the last few years on the self-assembling mixtures of sitosterol and oryzanol in edible oil and in emulsions. The physical chemistry of the system is found to be extremely interesting. First of all, the building blocks of the network are fundamentally different from those in a regular TAG network: the components tend to form tubules with a diameter of ~7 nm and a wall thickness of ~1 nm, which can aggregate to form a firm network that results in a so-called organogel. The system shows great sensitivity—relatively small changes in the environment result in major differences in the behavior of the system, as can be seen for example from the effect of water on the tubule formation in emulsions. Further work is needed to explain the detailed behavior of this system.

The first aspect that deserves further attention concerns the details of the intermolecular assembly of the tubules. Despite the fact that the overall structure of the tubules is understood, the exact docking of the molecules has been assessed only in a rather qualitative sense. More detailed computer simulations may shed more light on this, and especially on the conformation of the solvent molecules in the tubules and the interaction between solvent molecules and the tubules. From the experimental side, a number of sterol molecules have been identified that show the ability to self-assemble with oryzanol (Bot & Agterof, 2006; Bot et al., 2009b). On the other hand, alternatives to oryzanol have not yet been identified. A synthetic experimental program may be required to establish the rules that determine which sterolesters have the capability to self-assemble with sitosterol-type molecules in edible triglyceride oil. Furthermore, experimental work on the effect of changes in aqueous phase and organic phase have been only superficially touched (Ritter et al., 1997) but are currently undergoing more detailed investigations (Sawalha et al., 2011b). The main aim of such work would be to establish the rules under which the tubule-based organogels can form in terms of composition of aqueous and organic phase, and of temperature and presence of water.

Unfortunately, the rich physico-chemical behavior also seems to be a hurdle in straightforward application of the structuring system as an alternative to structuring based on crystalline triglycerides. In relevant model systems, like w/o emulsions, the presence of water reduces the efficiency of structuring compared to structuring in (mostly) water-free systems. Since the sterols and sterolesters are expensive ingredients compared to crystallising triglycerides, this is a rather important drawback. This implies for the moment that application of the system is most likely limited to relatively niche applications, unless further progress in the understanding in the mechanism of the tubule formation in triglyceride phases leads to a solution for reducing the sensitivity of the system to the presence of water.

Acknowledgments

The authors would like to thank their collaborators who helped in collecting the data described in this chapter over the years: Ruud den Adel, Wim Agterof, Christiaan Beindorff, Alain Belibi, Han Blonk, Magali Cazade, Guus Duchateau, Nicola Duffy, Patricia Heussen, Mark Kirkland, Volkmar Müller, Christos Regkos, Eli Roijers, Rob van de Sande, Yvonne Veldhuizen, Tim Voordouw, Gleb Yakubov (all Unilever), Mike Rogers (University of Saskatchewan), Hassan Sawalha, Paul Venema, Erik van der Linden (Wageningen University).

References

den Adel, R.; Heussen, P.C.M.; Bot, A. Effect of water on self-assembled tubules in β-sitosterol + γ-oryzanol-based organogels. J. Phys. Conf. Ser. 2010, 247, 012025.

Argay, G.; Kálmán, A.; Vladimirov, S.; Zivanov-Stakic, D.; Ribár, B. Crystal structure of stigmast-5-en-3β-ol monohydrate, $C_{29}H_{52}O_2$. Z. Kristallographie 1996, 211, 725–727.

von Bonsdorff-Nikander, A.; Karjalainen, M.A.; Rantanen, J.T.; Christiansen, L.I.; Yliruusi, J.K. Physical stability of a microcrystalline β-sitosterol suspension in oil. Eur. J. Pharm. Sci. 2003, 19, 173–179.

Bot, A.; Agterof, W.G.M. Structuring of edible oils by mixtures of γ-oryzanol with β-sitosterol or related phytosterols. J. Am. Oil Chem. Soc. 2006, 83, 513–521.

Bot, A.; Kleinherenbrink, F.A.M.; Mellema, M.; Magnani, C.K. Cream cheese as an acidified protein-stabilised emulsion gel. Handbook of food products manufacturing; Hui, Y.H., et al., Eds; Wiley: New York, 2007; 649–670.

Bot, A.; den Adel, R.; Roijers, E.C. Fibrils of γ-oryzanol + β-sitosterol in edible oil organogels. J. Am. Oil Chem. Soc. 2008, 85, 1127–1134.

Bot, A.; Veldhuizen, Y.S.J.; den Adel, R.; Roijers, E.C. Non-TAG structuring of edible oils and emulsions. Food Hydrocolloids 2009a, 23, 1184–1189.

Bot, A.; den Adel, R.; Roijers, E.C.; Regkos, C. Effect of sterol type on structure of tubules in sterol + γ-oryzanol-based organogels. Food Biophysics 2009b, 4, 266–272.

Bot, A.; den Adel, R.; Regkos, C.; Sawalha, H.; Venema, P.; Flöter, E. Structuring in β-sitosterol + γ-oryzanol-based emulsion gels during various stages of a temperature cycle. Food Hydrocolloids 2011, 25, 639–646.

de Bruijne, D.W.; Bot, A. Fabricated fat-based foods. Food Texture: Measurement and Perception; Rosenthal, A.J., Ed; Aspen: Gaithersburg, MD, 1999; 185–227.

Christiansen, L.I.; Rantanen, J.T.; von Bonsdorff, A.K.; Karjalainen, M.A.; Yliruusi, J.K. A novel method of producing a microcrystalline β-sitosterol suspension in oil. Eur. J. Pharm. Sci. 2002, 15, 261–269.

Craven, B.M. The physical chemistry of lipids—from alkanes to phospholipds. Handbook of lipid research, volume 4; Hanahan, D.J. Small D.M., Eds; Plenum Press: New York, 1986; 149–182.

Deutch, J.M. Light-scattering from hollow finite cylinders. Macromolecules 1981, 14, 1826–1827.

Diks, R.M.M.; Duchateau, G.S.M.J.E.; Duffy, N; Peters, H.P.F. Food products, ingredients therefor and useof those food products and ingredients. Patent application WO2007/054197.

Duffy, N.; Blonk, J.C.G.; Beindorff, C.M.; Cazade, M.; Bot, A.; Duchateau, G.S.M.J.E. Organogel based emulsion systems, micro-structural features and impact on in-vitro digestion. J. Am. Oil Chem. Soc. 2009, 86, 733–741.

Flöter, E.; Bot, A. Developing products with modified fats. Improving the fat content of foods, Williams, C., Buttriss, J., Eds; Woodhead: Cambridge, 2006; chapter 17, 411–427.

Franklin, K.R.; Grainger, L.; Kowalski, A.J. Antiperspirant compositions. Patent application WO 2000/061096 (2000).

Geiger, C.; Stanescu, M.; Chen, L.H.; Whitten, D.G. Organogels resulting from competing self-assembly units in the gelator: structure, dynamics, and photophysical behavior of gels formed from cholesterol-stilbene and cholesterol-squaraine gelators. Langmuir 1999, 15, 2241–2245.

Lerma-Garcia, M.J.; Herrero-Martinez, J.M.; Simó-Alfonso, E.F.; Mendonca, C.R.B.; Ramis-Ramos, G. Composition, industrial processing and applications of rice bran gamma-oryzanol, Food Chem. 2009, 115, 389–404.

Li, Y.G., Liu, K.Q.; Liu, J.; Peng, J.X.; Feng, X.L.; Fang, Y. Amino acid derivatives of cholesterol as "latent" organogelators with hydrogen chloride as a protonation reagent. Langmuir 2006, 22, 7016–7020.

Lin, Y.C.; Kachar, B.; Weiss, R.G. Novel family of gelators of organic fluids and the structure of their gels. J. Am. Chem. Soc. 1989, 111, 5542–5551.

Liu, X.Y.; Sawant, P.D. Formation kinetics of fractal nanofibre networks in organogels, Appl. Phys. Lett. 2001, 79, 3518–3520.

Lu, L.; Cocker, T.M.; Bachman, R.E.; Weiss, R.G. Gelation of organic liquids by some 5α-cholestan-3β-yl N-(2-aryl)carbamates and 3β-cholesteryl 4-(2-anthrylamino)butanoates. How important are H-bonding interactions in the gel and neat assemblies of aza aromatic-linker-steroid gelators? Langmuir 2000, 16, 20–34.

Katan, M.B.; Grundy, S.M.; Jones, P.; Law, M.; Miettinen, T.; Paoletti, R.; et al. Efficacy and safety of plant stanols and sterols in the management of blood cholesterol concentrations. Mayo Clinic Proc. 2003, 78, 965–978.

Kawachi, H.; Tanaka, R.; Hirano, M.; Igarashi, K.; Ooshima, H. Crystallization of beta-sitosterol using a water-immiscible solvent hexane. J. Chem. Eng. Jpn. 2006, 39, 869–875.

Keys, A.; Anderson, J.T.; Grande, F. Serum cholesterol response to changes in the diet, IV. Particular saturated fatty acids in the diet. Metabolism 1965, 14, 776–787.

Konikoff, F.M.; Chung, D.S.; Donovan, J.M.; Small, D.M.; Carey, M.C. Filamentous, helical and tubular microstructures during cholesterol crystallization from bile: evidence that biliary cholesterol does not nucleate classic monohydrate plates. J. Clin. Invest. 1992, 90, 1155–1160.

Kroes-Nijboer, A.; Venema, P.; Bouman, J.; van der Linden, E. The critical aggregation concentration of β-lactoglobulin-based fibril formation. Food Biophysics 2009, 4, 59–63.

Marangoni, A.G. Novel strategies for nanostructuring liquid oils into functional fats. Proceedings of the 5th International Symposium on Food Rheology and Structure, Fischer, P., Pollard, M., Windhab, E., Eds.; ETH Zürich: Zürich, 2009; 38–44.

Mensink, R.P.; Zock, P.L.; Kester, A.D.M.; Katan, M.B. Effects of dietary fatty acids and carbohydrates on the ratio of serum total to HDL cholesterol and on serum lipids and apoproteins: a meta analysis of 60 controlled trials. Am. J. Clin. Nutr. 2003, 77, 1146–1155.

Murata, K.; Aoki, M.; Suzuki, T.; Harada, T.; Kawabata, H.; Komori, T.; Ohseto, F.; Ueda, K.; Shinkai, S. Thermal and light control of the sol-gel phase transition in cholesterol-based organic gels. Novel helical aggregation modes as detected by circular dichroism and electron microscopic observation. J. Am. Chem. Soc. 1994, 116, 6664–6667.

Peng, J.; Lin, K.; Liu, J.; Zhang, Q.; Feng, X.; Fang, Y. New dicholesteryl-based gelators: chirality and spacer length effect. Langmuir 2008, 24, 2992–3000.

Pernetti, M.; van Malssen, K.F.; Flöter, E.; Bot, A. Structuring of edible oils by alternatives to crystalline fat. Curr. Opin. Colloid Interface Sci. 2007, 12, 221–231.

Press, W.H.; Flannery, B.P.; Teukolsky, S.A.; Vetterling, W.T. Numerical Recipes: The art of scientific computing, Chapter 6; Cambridge University Press: Cambridge, UK, 1986.

Ritter, H.; van de Sande, R.L.K.M.; Müller, V.K. Liquid fatty component containing composition. Patent application WO 97/42830 (1997).

Rogers, M.A. Novel structuring strategies for unsaturated fats—Meeting the zero-trans, zero-saturated fat challenge: A review. Food Res. Int. 2009, 42, 747–753.

Rogers, M.A.; Bot, A.; Lam, R.S.H.; Pedersen, T.; May, T. Multi-component hollow tubules formed using phytosterol and γ-oryzanol based compounds: an understanding of their molecular embrace. J. Phys. Chem. A 2010, 114, 8278–8285.

Sakurai, K.; Ono, Y.; Jung, H.J.; Okamoto, S.; Sakurai, S.; Shinkai, S. Synchrotron small angle X-ray scattering from organogels. Part 1. Changes in molecular assemblies of cholesterol gelators during gel-sol transition. J. Chem. Soc.—Perkin Trans. II 2001, 108–112.

Sawalha, H.; Venema, P.; Bot, A.; Flöter, E.; van der Linden, E. The influence of concentration and temperature on the formation of γ-oryzanol + β-sitosterol tubules in edible oil organogels. Food Biophysics 2011a, 6, 20–25.

Sawalha, H.; den Adel, R.; Venema, P.; Bot, A.; Flöter, E.; van der Linden, E. Organogel-emulsions with mixtures of β-sitosterol + γ-oryzanol: Influence of water activity and type of oil phase. 2011b. In preparation.

Sawalha, H.; Venema, P.; Flöter, E.; Bot, A.; den Adel, R.; van der Linden, E. Phase diagram of γ-oryzanol + β-sitosterol binary mixtures and ternary mixtures in edible oil organogel. 2011c. In preparation.

Terech, P.; Ostuni, E.; Weiss, R.G. Structural study of cholesteryl anthraquinone-2-carboxylate (CAQ) physical organogels by neutron and X-ray small angle scattering. J. Phys. Chem. 1996, 100, 3759–3766.

Terech, P.; Wade, R.H. The relationship between a dried and native steroid gel, J. Colloid Interface Sci. 1988, 125, 542–551.

Terech, P.; Furman, I.; Weiss, R.G.; Bouas-Laurent, H.; Desvergne, J.P.; Ramasseul, R. Gels from small molecules in organic solvents: Structural features of a family of steroid and anthryl-based organogelators, Faraday Discuss. 1995, 101, 345–358.

Terech P. Low-molecular weight organogelators. Specialist Surfactants, Robb, I.D., Ed., Blackie Academic and Professional: Glasgow, 1997; chapter 8, 208–268.

Wade, R.H.; Terech, P.; Hewat, E.A.; Ramasseul, R.; Volino, F. The network structure of a steroid cyclohexane physical gel, J. Colloid Interface Sci. 1986, 114, 442–451.

Wang, R.; Geiger, C.; Chen, L.; Swanson, B.; Whitten, D.G. Direct observation of sol–gel conversion: the role of the solvent in organogel formation. J. Am. Chem. Soc. 2000, 122, 2399–2400.

Wassell, P.; Bonwick, G.; Smith, C.J.; Almiron-Roig, E.; Young, N.W.G. Towards a multidisciplinary approach to structuring in reduced saturated fat-based systems—a review. Int. J. Food Sci. Technol. 2010, 45, 642–655.

Žinic, M.; Vogtle, F.; Fages, F. In Low molecular mass gelators, Topics in Current Chemistry, volume 256; Fages, F., Ed.; Springer: Berlin, 2005; 39–76.

4

Vegetable Oil-based Ricinelaidic Acid Organogels—Phase Behavior, Microstructure, and Rheology

Amanda J. Wright[1] and Alejandro G. Marangoni[2]

[1]Department of Human Health and Nutritional Sciences, College of Biological Sciences, University of Guelph, Guelph, Ontario, Canada and[2] Department of Food Science, Ontario Agricultural College, University of Guelph, Guelph, Ontario, Canada

Introduction

Organogels constitute a diverse group of gelled materials in which a network of self-assembled molecules forms a thermally reversible gel upon cooling, immobilizing a non-aqueous liquid. These viscoelastic structures form through non-covalent associations of gelator molecules at low concentrations (i.e., often <1%). The aggregated molecules form superstructures, often long fibers, which entangle or form pseudo-crystalline regions, immobilizing the liquid largely by surface tension and forming a gel of variable consistency (Terech & Weiss, 1997; Van Esch & Feringa, 2000). The networks formed are reminiscent of macromolecular gels and polymers. However, their thermoreversibility and nucleation involving one-dimensional crystal growth distinguish organogels from materials like cross-linked polystyrene (Mukkamala & Weiss, 1996).

Traditional applications of organogels include gelation of flammable solvents, aerogel formation, and media for nonaqueous reactions (Abdallah & Weiss, 2000; Van Esch & Feringa, 2000). However, there is considerable interest in the potential uses of organogels across various industries. For example, the ability of these materials to self assemble on the nano-scale and to solubilize lipophilic guest molecules makes them useful in purification and separation processes, as media for non-aqueous synthesis or reactions, and in drug delivery. In addition, the sharp temperature and moisture sensitivities of many organogels make them useful in sensing and controlled release scenarios. The wide-ranging uses of organogels include environmental clean-up, safe transport of flammable liquids, aerogel formation, ingredients in lubricants and coatings, and pharmaceutical and food technologies uses (Abdallah & Weiss, 2000; Terech & Weiss, 1997; Van Esch & Feringa, 2000). In the food industry, attempts are being made to apply the principles of organogelation (Bot & Agterof,

2006; Gandolfo et al., 2004; Hughes et al., 2009; Rogers et al., 2009) to structure oils without the need for high levels of saturated or *trans* fatty acids, which have recognized implications for cardiovascular disease risk. There is also active research in using organogels to minimize oil migration in composite food products and for enhanced stability and delivery of lipophilic bioactive molecules. However, it remains a significant challenge to identify appropriate food-grade organogelator molecules.

Organogelator molecules have relatively low molecular weights compared to the molecules that form the gel networks in typical hydrogels. Some organogelators include derivatives and metal salts of fatty acids (Uzu, 1975; Terech et al., 1994) and steroids (Mukkamala & Weiss, 1996), amino-acid type molecules (Hanabusa et al., 1997; Okabe et al., 2004), carbohydrate amphiphiles (Friggeri et al., 2002), and organometallic compounds (Terech et al., 1987; Terech, et al., 1992). These molecules are amphiphilic in nature and self-assemble in hydrophobic liquids when an appropriate balance between solubility and aggregation forces exists (Terech & Weiss, 1997). Aggregation occurs primarily through dipolar interactions, specific intermolecular hydrogen bonding, and metal co-ordination bonding (Terech & Weiss, 1997). The nature of both the solvent and gelator influence organogel formation and morphology (Terech et al., 2000). In fact, a fine balance between aggregation and dissolution forces is critical in the formation of an organogel and the identification of organogelator-solvent combinations capable of forming these gels remains a challenge and an area of considerable activity (Hanabusa et al., 1997; Terech & Weiss, 1997). There is still much to learn in identifying suitable gelator-solvent combinations (Terech & Weiss, 1997).

Hydroxystearic acid (hydroxy octadecanoic acid, HSA) is a well known organogelator molecule. Studies with its isomers (Eloundou et al., 2005; Marton et al., 1941; Tamura et al., 1994; Terech et al., 1994, Terech et al., 2000) and other fatty acids (Daniel & Rajasekharan, 2003) have led to a greater understanding of the relationships between molecular structure and organogel formation and structure. 12-hydroxy octadecanoic acid (12-HSA) in particular is commonly used in cosmetics as well as lubricating greases and coatings (Terech et al., 2000; Marton et al., 1941). The hydroxylated fatty acid is formed through the hydrogenation of castor oil, a material rich in ricinoleic acid (12-hydroxy, 9-octadecenoic acid). In organic solvents, 12-HSA forms long, rigid fibers through hydrogen bonding, and these fibers then are linked together in crystalline monoclinic domains (Terech & Weiss, 1997).

HSA organogel research has served as a model for further exploration of organogels (Terech & Weiss, 1997). For example, in our quest to identify new organogelator molecules, we turned to 12-hydroxy, 9-*trans*-octadecenoic acid (ricinelaidic acid, REA). REA is a derivative of 12-HSA and has a similar molecular structure. Given the *trans* configuration of the double bond on carbon 9, it was hypothesized that the molecule would physically align in a similar manner as the saturated molecule, 12-HSA. Our lab was the first to report on the ability of REA to act as an organogelator. (Wright & Marangoni, 2006; Wright & Marangoni, 2007). This chapter summarizes our current understanding of REA gelation of vegetable oils, including the influence of temperature and concentration on gel formation, microstructure, and stability.

Experimental Procedures

Materials and Sample Preparation

Ricinelaidic acid (REA, >99 % purity, Nu-Chek Prep, Elysian, MN) was blended with refined, bleached, and deodorized canola oil (CO, Bunge, Toronto, Canada), unrefined sesame oil (Sunfresh Ltd., Toronto, Canada), and diacylglycerol oil (ENOVA, ADM Kao LLC, Decatur, IL) at 0.1–5.0 wt% levels. The samples were heated to 80°C, mixed vigorously, and held at 80°C for 10 minutes prior to subsequent experiments. The melting temperature of REA was determined by differential scanning calorimetry (DSC) using a Q-1000 calorimeter (TA Instruments Water-LLC, New Castle, DE); 8–10 mg of sample was hermetically sealed in an aluminum pan, crystallized at 5°C for 24 hours, and then heated at a rate of 5°C/minute. The peak melting temperature was determined using the software provided with the instrument.

Gelation Behavior

Phase diagrams were constructed using the aforementioned blends of REA and CO, unrefined sesame oil, and diacylglycerol oil. Vials containing 1 mL of the blends were held at temperatures between –10 and 50°C. After 24 hours, the vials were inverted and the self-standing ability of the samples was assessed visually. Depending on appearance, samples were described as "liquid", "thickened liquid", "clear gel", "translucent-opaque gel", or "fat-like".

Gelation times (t_{gel}) were determined for the REA-CO samples by recording the time when the material was self-standing in an inverted vial. For the 0.5, 1.0, and 5.0% REA-CO samples at 5°C, gelation was also monitored by small deformation rheology using an AR 2000 rheometer (TA instruments, Mississauga, ON, Canada). Oscillatory stress sweeps (0.15% strain and 1 Hz) were performed periodically (i.e., every 14 minutes) after loading the samples at 80°C into a temperature controlled aluminum recessed end concentric cylinder (1 mm spacing between cylinders and 2000 µm gap).

Gel Microstructure and Rheological Properties

Gel microstructure was visualized by bright field and polarized light microscopy using an Olympus BH microscope (Olympus, Tokyo, Japan). Small samples (~15 mg) of the gels were placed on glass microscope slides and gently covered with glass cover-slips. Slide temperature was maintained using a Linkham LTS350 cold stage (Linkam Scientific Instruments Inc., Surrey, England). Digital images of the partially and fully polarized specimens were acquired using a Sony XC75 CCD camera and LG-3 capture board (Scion Corporation, Frederick, USA).

Gel microstructure was also studied using a Rigaku MultiFlex X-Ray Diffractometer (RigakuMSC Inc., The Woodlands, TX). Glass holders with sample wells (0.5 mm depression) were loaded with hot sample and held for 1, 7, 14, and 28 days at 5, 15, 20, and 25°C. Sample temperature during analysis was maintained by placing a Peltier plate at the desired temperature directly beneath the sample holder. Angular scans was performed at 0.5 degree per minute from 1 to 30° 2θ using a Cu source X-ray tube at 40 kV and 44 mA. Background subtractions, to remove liquid oil scattering, and peak detection/labeling were performed using the MDI Jade 6.5 software.

Characteristic small deformation dynamic rheological parameters of the gels were determined after 1, 7, 14, and 28 days at temperatures between 5 and 35°C. Three mL of each sample was placed in a 4 mL vial at the desired temperature and, after the specified time, loaded onto the temperature-controlled flat plate of the AR2000 rheometer (TA instruments, New Castle, DE) and storage moduli in the linear viscoelastic region (G'_{LVR}) were determined. A 6 cm, 2 degree acrylic cone with truncation gap of 66 μm was used. Strain sweeps were performed from 6.0×10^{-3} to 2.0% at a frequency of 1 Hz. Frequency sweeps from 0.1 to 10 Hz were also performed after 24 h, using a controlled strain of 0.1%.

Results and Discussion

Phase Behavior and Gelation

The 24 hour phase diagrams for REA with canola oil (REA-CO) and sesame oil (REA-sesame) are shown in Fig. 4.1. In the case of the REA-CO system (1A), four phases were identified, according to the visual appearance of the samples. These were fat-like (i.e., the material appeared more polycrystalline than gel-like), nontransparent gel (i.e., gels ranged from slightly translucent to completely opaque), clear gel (i.e., self-standing gel was completely transparent) and liquid (i.e., no thickening or gelation was observed).

The minimum REA concentration at which gelation occurred was 0.5%. A clear gel formed at this concentration at 5°C, but not at 10°C. According to Fig. 4.1.A, the concentration and temperature range at which a clear gel formed was narrow. By 1%, the gels were slightly cloudy. With increasing REA content, increased opacity was observed. By 3.5%, the samples were gelled, but opaque. Gel appearance depended more on concentration than temperature, i.e., within each concentration, sample opacity was similar across the temperature range where setting occurred. Larger changes in the appearance of the samples were observed with increasing temperature. Similarly, there was no transition region in the upper temperature range for gelation. For example, a clear gel was not observed for the 2% sample around 25°C, nor was an opaque-gel observed for the 4% sample around 40°C.

The peak melting temperature of pure REA is 50.3 ± 0.1°C, while the maximum gelation temperature was 35°C with 5% REA. Maximum gelation temperatures

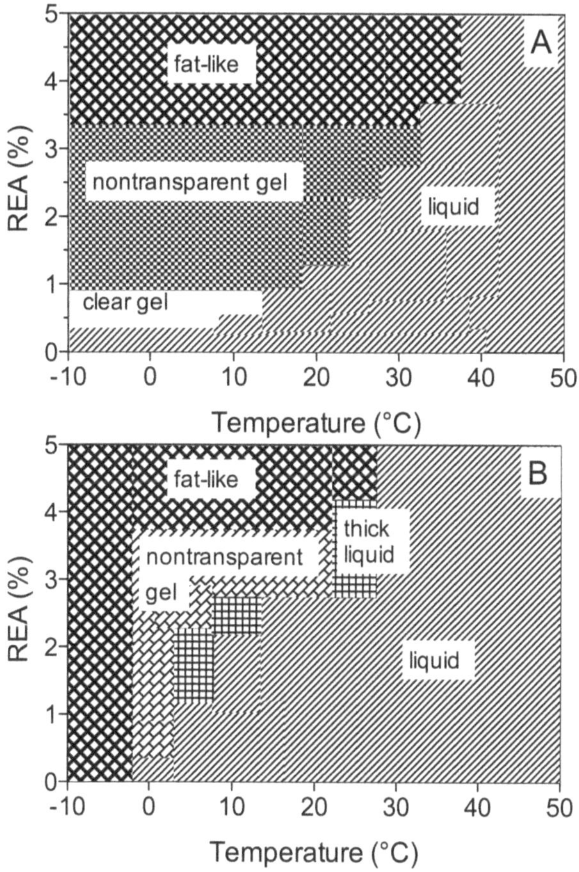

Fig. 4.1. Phase diagram for REA-canola oil (A) and REA-sesame oil (B). Fat-like (i.e., the material appeared more crystalline than gelled), Nontransparent gel (i.e., gels ranged from slightly translucent to completely opaque), Clear gel (i.e., self-standing gel was completely transparent, Liquid (i.e., no thickening or gelation observed), and Thick liquid (i.e., liquid was visibly thickened, but self-standing gel was not observed).

generally do not correspond to the melting temperatures of the neat gelators (Terech & Weiss, 1997). According to the Van't Hoff equation, some of the REA will be solubilized in the oil phase, thereby lowering the melting temperature of the gels versus pure REA. Even above the solubility limit, gelation will be dependent on having a threshold REA concentration, which results in enough organized molecules to form a space-filling network capable of entrapping canola oil. Also, based on the similarities of REA to HSA, hydrogen bonding is expected to play a critical role in the aggregation of REA molecules into the long network strands. Hydrogen bonding between hydroxyl groups causes HSA aggregation in organic solvents (Terech et al., 1994). Since the strength of hydrogen bonding decreases with increasing temperature, this may also partially limit REA aggregation at higher temperatures.

Solvent nature, polarity, and the presence of co-surfactants can influence the aggregation of organogelator molecules (Terech & Weiss, 1997). To investigate the influence of solvent on REA gelation, phase diagrams were constructed using unrefined sesame oil and diacylglycerol oil. The phase diagram for REA with the sesame oil is shown in Fig. 4.1.B. Comparing Figs. 4.1.A and 4.1.B shows that the upper limits for REA gelation were lower for sesame oil than with canola oil. Unrefined sesame oil contains relatively higher levels of free fatty acids, sterols, and tocopherols than RBD canola oil (AOCS, 1996). Each of these compounds contains hydroxyl groups with the potential to participate in hydrogen bonding and therefore impede REA self-assembly. In addition, sterols and tocopherols contain large ring structures that may sterically interfere with the ability of REA to aggregate. A clear gel was never observed with sesame oil. However, a fifth type of phase was apparent, i.e., some samples were visibly thickened, but not gelled (thick liquid). This behavior represented a transition between the formation of a non-transparent gel and a liquid sample. No such intermediary phases were observed with the canola oil samples.

The phase behavior was also studied for REA with diacylglycerol (DAG) oil. However, no thickening or gelation was observed after 24 hours (data not shown). DAG oil is prepared from canola and soybean oils. It differs from natural seed oils, including canola oil and sesame oil, in that it is made up of diacylglycerol molecules (as the name implies), rather than triacylglycerols. In a DAG molecule there is a free hydroxyl group on the glycerol backbone, which is otherwise esterified to a fatty acid in a triacylglycerol. This high number of hydroxyl groups in the DAG oil was correlated with a lack of REA gelation. This lends further evidence to the fact that hydrogen bonding is critical to REA gelation.

Gelation times for the REA-CO blends are shown in Table 4.A. The concentration dependence of REA gelation is evident. Even after several months at 5°C, no gelation was observed below 0.5% REA. At 0.5% a gel was observed after 20.5 hours. With 1% REA, t_{gel} decreased dramatically to roughly 11 minutes. With increasing concentration, an exponential decrease in t_{gel} was observed at 5°C. Within each temperature studied, concentration had a significant influence on t_{gel} ($P<0.05$). However, for each concentration above 1%, significantly higher values of t_{gel} were not observed until above 20°C. For concentrations of 2% and above, t_{gel} was independent of temperature, up to 20°C. Above 20°C, a dramatic increase in t_{gel} was observed for the gels, indicating the existence of a different gelation regime.

Gelation behavior at 5°C was also monitored by small deformation rheology. Fig. 4.2 shows the changes in the storage modulus (G′) during gelation of the 0.5, 1.0, and 5.0 % REA blends at 5°C. With 1.0 and 5.0 % REA, gelation occurred rapidly at 5°C. The samples quickly reached a maximum in the storage modulus G′, ~24,000 Pa and ~65,000 Pa, respectively). Such sharp sol-to-gel transitions are characteristic of "strong" organogels, which demonstrate solid-like viscoelastic behavior (Terech & Weiss, 1997). In contrast, the most dilute sample (0.5%) gelled gradually

Table 4.A. Gelation Times (t_{gel}, s)* for 0.5, 1.0, 2.0, 3.0, 4.0 and 5.0 % REA-canola Oil Blends Between 5 and 40°C.

Temperature (°C)	0.5%	1.0%	2.0%	3.0%	4.0%	5.0%
5	7,4025 ± 4512	660 ± 91 [a]	446 ± 12 [a]	341 ± 6 [a]	290 ± 38 [a]	207 ± 59 [a]
10	ng	886 ± 80 [b]	458 ± 49 [a]	326 ± 41 [a]	306 ± 36 [a]	265 ± 35 [a]
15	ng	3359 ± 993 [c]	353 ± 5 [a]	227 ± 4 [b]	187 ± 6 [a]	145 ± 8 [a]
20	ng	ng	517 ± 24 [a]	342 ± 4 [a]	224 ± 27 [a]	190 ± 6 [a]
25	ng	ng	1819 ± 28 [b]	1693± 46 [c]	1136 ± 240 [b]	795 ± 75 [b]
30	ng	ng	ng	Ng	2959 ± 475 [c]	1638 ± 125 [c]
35	ng	ng	ng	Ng	ng	ng
40	ng	ng	ng	Ng	ng	ng

*Time when material was self-standing in inverted vial

Mean ± standard deviation (n=6)

ng indicates no gelation after 24 hours

Within each column, different superscript (a–c) indicates significant difference ($P < 0.05$)

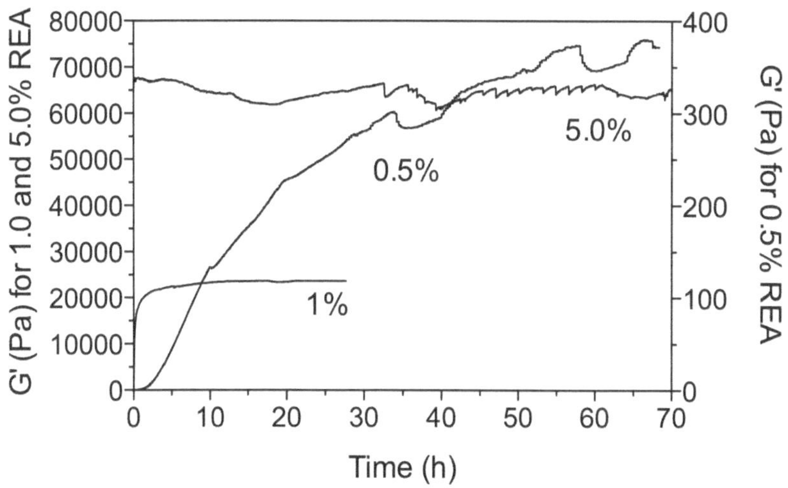

Fig. 4.2. Changes in storage modulus (G′, Pa) of 0.5, 1.0 and 5.0% REA-canola oil blends during gelation at 5°C (0.15% strain, 1 Hz).

in a manner more typical of "weak" gel-like systems, which possess liquid-like viscoelastic properties (Terech & Weiss, 1997). An increase in G′ was observed throughout the duration of the experiment, reaching a G′ of ~370 Pa after 68 hours.

24-Hour Gel Rheology and Scaling Relationships

Fig. 4.3 shows the storage modulus in the linear viscoelastic region (G'_{LVR}, Pa) for samples of the 0.5, 1.0, 2.5, 4.0 and 5.0% REA-CO gels after 24 hours at 5, 10, 20 and 30°C.

REA concentration had a significant influence on gel elasticity at each temperature studied (Figs. 4.3.A–4.3.D, $P<0.05$). In contrast, Figs. 4.3.E–4.3.H show that storage temperature was only significantly correlated with gel elasticity at the

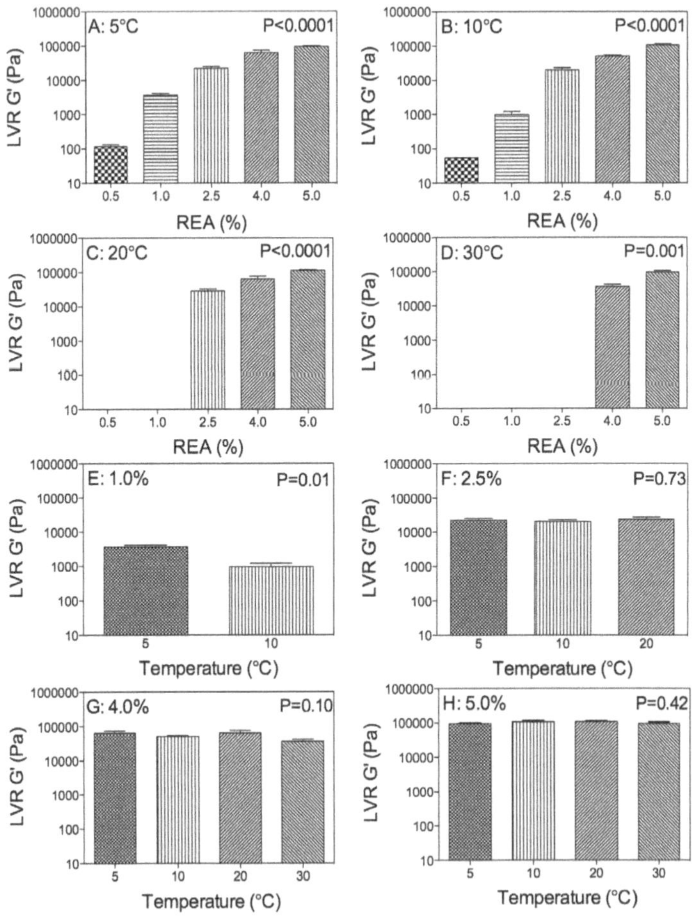

Fig. 4.3. Influence of REA concentration (0.5, 1.0, 2.5, 4.0 and 5.0%) and gelation temperature (5, 10, 20 and 30°C) on the 24 hour linear viscoelastic region storage modulus (G'_{LVR}, Pa) for REA-CO gels.

1.0% REA level (P=0.01). Above 1.0% REA, there was no significant effect of temperature on G'_{LVR} for the gels that formed (P>0.05). This trend correlates with similar trends in the REA-CO phase diagram (Fig. 4.1). Within each region where gelation occurred, only 1 type of phase was observed across all temperatures. Therefore, both gel appearance and strength depended more on concentration than on temperature. Comparing t_{gel} for the 2–5% REA-CO gels, indicates significant changes only above 20°C. However, based on the REA concentrations in Fig. 4.3.G and 4.3.H, there is no evidence from G'_{LVR} that a change in gelation mechanism around 20°C resulted in a change in network elasticity. To further explore this and to identify differences in gel microstructure across the temperature range studied, the scaling relationship between G'_{LVR} and REA concentration are shown in Fig. 4.4 for 5, 10, 15, 20 and 25°C.

For HSA organogels, the elastic modulus scales with concentration in a power-law fashion. According to Fig. 4.3, this scaling behavior was observed for the REA-CO gels at each temperature, suggesting that, regardless of concentration, within each temperature, the gel structure is similar. However, comparing the scaling exponents in Fig. 4.4, similar values were observed at 5, 10, 15 and 20°C, but not at 25°C (2.69, 3.12, 2.36, and 3.16 vs 8.18, P<0.05). The values observed between 5 and 20°C are within the range of those reported for HSA-solvent gels in the literatures (i.e., 1.89–3.87) (Eloundou et al., 2005; Terech et al., 2000). The change in scaling exponent observed at 25°C coincides with differences in t_{gel}, i.e., increases in gelation time for the 2, 3, 4 and 5% gels above 20°C. However, this change is primarily the result of a low G'_{LVR} for only the 2% sample at 25°C (Fig. 4.4.D). Differences in gelation time and the scaling exponent at 25°C may suggest differences in gelation

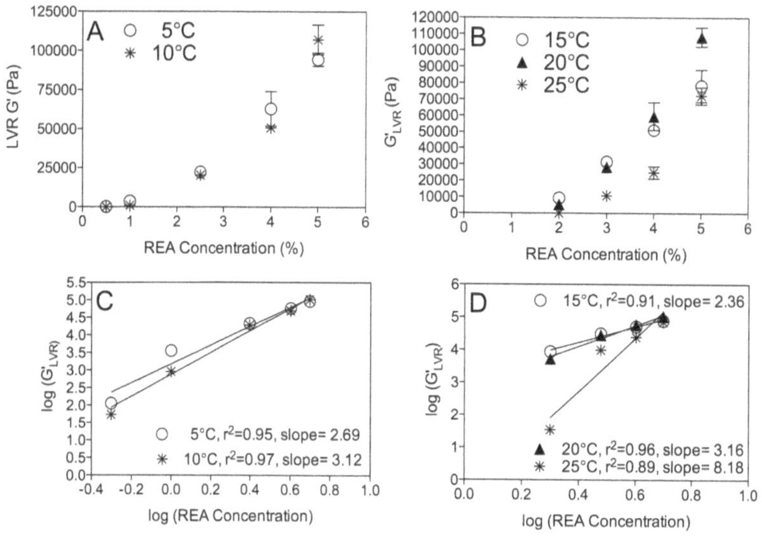

Fig. 4.4. Scaling relationships between G'_{LVR} and REA concentration (wt%) for REA-CO gels formed at 5, 10, 15, 20 and 25°C.

mechanism and structure compared with the gels at the lower temperatures, particularly at lower concentrations. Any potential differences in gelation mechanism did not translate into differences in G'_{LVR} for the 4 or 5% REA gels (i.e., Figs. 4.3.G and 4.3.H) between 5 and 30°C.

Gel Microstructure

When neat REA crystallizes, it forms large spherulitic microstructures made up of thick, needle-like structures (Fig. 4.5.A). A significantly different network is observed for the CO gels. During gelation, the REA molecules aggregate into long, thin, fibers that intertwine and cluster, immobilizing the liquid oil. The polarized light micrographs of the REA gels in Figs. 4.5.B and 4.6 show the networks of long birefringent REA fibers characteristic of organogels (Terech & Weiss, 1997). According to Fig. 4.6, some of the fibers exceed ~800 μm in length. Also Fig. 4.6.A nicely shows the edges between two clusters of fibers.

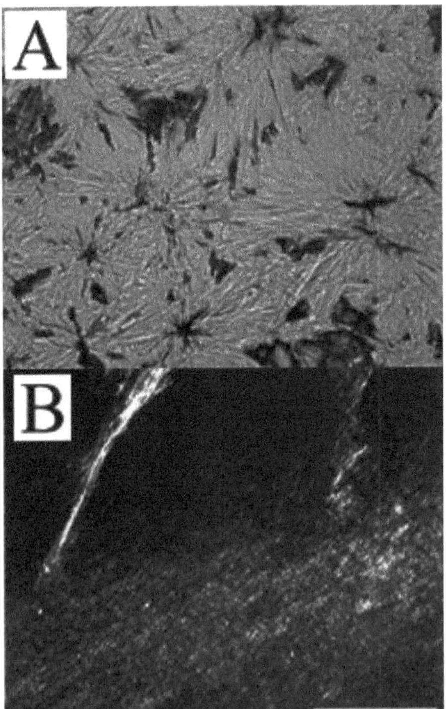

Fig. 4.5. Partially polarized light micrographs of neat REA (A) and 3% REA-canola oil (B) after 1 day at 25°C. Magnification bar represents 100 μm.

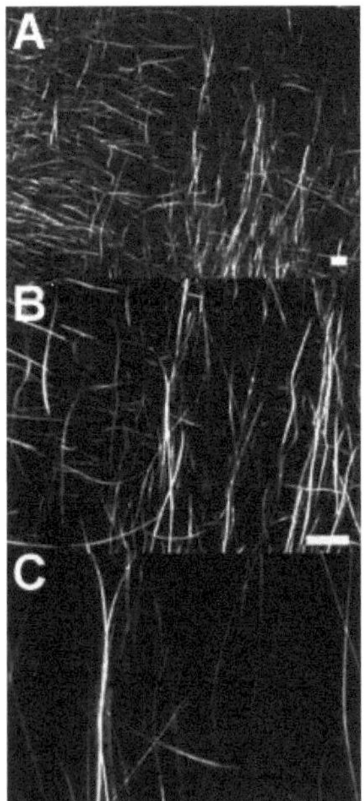

Fig. 4.6. Polarized light micrographs of 1.5% REA-CO gel at 4, 10, and 40× magnification after 24 hours at 20°C. Magnification bars represent 100 μm.

Gel microstructure was systematically studied for the REA-CO samples using partially polarized light microscopy. Fig. 4.7 shows the 0.5, 1.0 and 5.0% gels at 5°C. Images of the gels at 2 and 5% REA, but at higher temperatures (15, 20, and 25°C) are shown in Fig. 4.8.

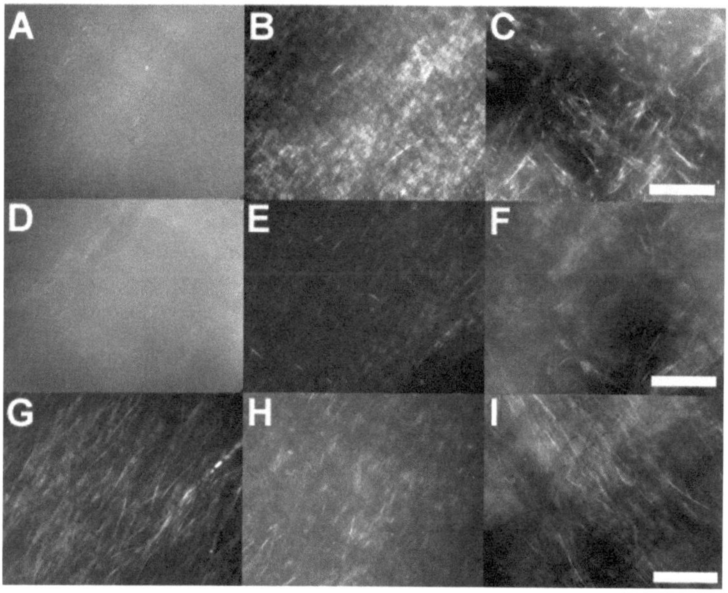

Fig. 4.7. Partially polarized light micrographs for 0.5% (A, D, G), 1.0% (B, E, H) and 5.0% (C, F, I) REA-canola oil gels after 1 (A, B, C), 7 (D, E, F) and 28 (G, H, I) days at 5°C. Magnification bars represents 100 μm.

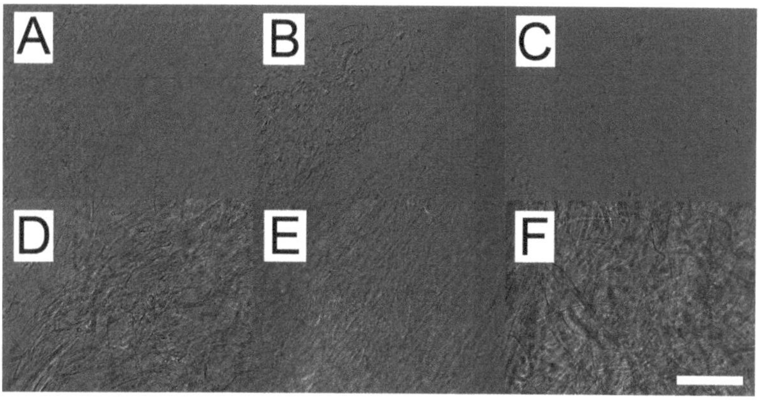

Fig. 4.8. Partially polarized light micrographs of 2 (A, B, C) and 5% (D, E, F) REA-canola oil gels after 1 day at 15 (A, D), 20 (B, E), and 25°C (C, F). Magnification bar represents 100 μm.

The influence of REA concentration on gel microstructure is apparent after 24 hours in Fig. 4.7 A (0.5%), B (1%) and C (5%). At 0.5% REA, the networks were difficult to visualize. Only when the samples were flooded with light and partially polarized could any structure be seen. At higher concentrations, more of the individual strands were apparent and clusters of the strands were evident. According to Fig. 4.8, the nature of the REA fibers was influenced by REA concentration. The fibers in the 5% gels (Figs. 4.8.D–4.8.F) were longer and thicker than those at 2% (Figs. 4.8.A–4.8.C). Also, smaller and circular structures (~5–10 μm) were observed, on occasion, in the gels examined in this study (i.e., in the 2, 3, 4 and 5% gels at 15, 20 and 25°C). However, this type of structure is most apparent in the 2% REA-CO gels (Figs. 4.8.A–4.8.C).

There were no apparent differences in microstructure between 15, 20 and 25°C, in the case of the 3, 4 (data not shown) and 5% REA-CO gels. However, for the 2% REA-CO gels in Figs. 4.8.A–4.8.C, temperature did influence gel microstructure. At 25°C (Fig. 4.8.C), the fibers were shorter and more difficult to visualize than at 15 and 20°C (Figs. 4.8.A and 4.8.B). A similar microstructure was observed for the clear 0.5% REA-CO gel at 5°C (Fig. 4.7.A). According to Fig. 4.1, 5°C was the upper temperature limit of gelation for the 0.5% REA-CO blend and 25°C was the upper temperature limit of gelation for the 2% gel. Of the gels formed at 15, 20 and 25°C, the 2% sample at 25°C was also the least opaque of the gels.

XRD

Powder X-ray diffraction (XRD) was used to explore the structure of the REA gels in both the small and wide angle regions. The XRD patterns for neat REA and the 5% REA-CO gel after 24 hours at 5°C are shown in Fig. 4.9. The peaks identified in

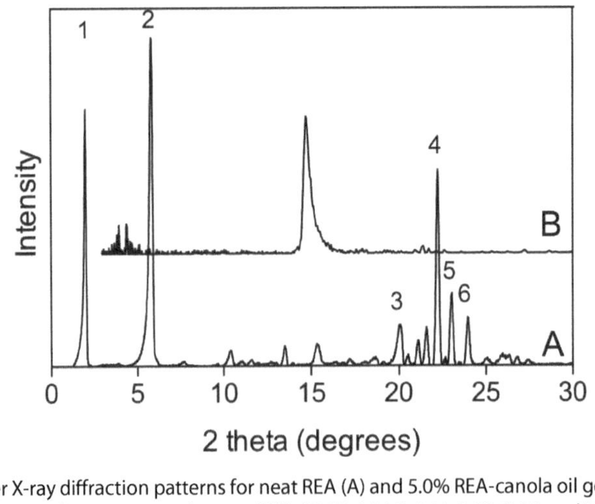

Fig. 4.9. Powder X-ray diffraction patterns for neat REA (A) and 5.0% REA-canola oil gel (B) after 24 hours at 5°C. Numbers 1-6 refer to spacings for neat REA: 43.28, 15.16, 4.41, 3.99, 3.86, and 3.71 Å, respectively.

Fig. 4.9 correspond to those for neat REA, and spacings for the REA-CO samples are shown in Table 4.B.

The patterns for the neat REA and REA organogels are similar, suggesting a relationship between the molecular packing in the neat and gel phases. Both neat REA and the gel had small angle reflections, which roughly correspond to the 001 and 003 planes. This is similar to neat HSA, which shows characteristically intense (001) and (003) reflections that shift to different extents depending on the solvent. HSA organogels consist of randomly distributed long fibers, interconnected through junction zones, i.e., monoclinic crystalline domains. The fibers themselves are made up of fatty acid dimers formed through hydrogen bonding between the hydroxyl groups at C12 and carboxyl groups that are aligned head to head (Terech et al., 1994). The small angle reflections observed for the REA gels suggest the presence of dimers (i.e., ~ peak at 41 Å) and a structure similar as HSA gels. Crystallinity is evidenced

Table 4.B. Spacings (Å) Detected by Powder X-ray Diffraction for 0.5, 1.0 and 5.0% REA-Canola Oil Blends Stored at 5°C for 1, 7, and 28 Days.

Days	0.5%	1.0%	5.0%
1	nd	nd	40.5 ± 0.0
			14.8 ± 0.0
			4.4 ± 0.0
			4.0 ± 0.0
			3.9 ± 0.0
			3.8*
7	40.8 ± 1.3	41.1 ± 0.8	42.7 ± 0.2
	14.78*	14.9 ± 0.1	15.0 ± 0.0
			4.4 ± 0.0
			4.0 ± 0.0
			3.9 ± 0.0
28	41.1 ± 1.0	40.7 ± 1.3	41.7 ± 0.8
	14.9 ± 0.0	14.9 ± 0.1	14.9 ± 0.1
		4.4*	4.4 ± 0.0
		4.0*	4.0 ± 0.0
			3.9 ± 0.0

nd: no peak detected

* Indicates spacing was only detected in one replicate

Spacings (Å) for neat REA: 43.28 ± 0.001, 15.16 ± 0.01, 4.41 ± 0.00001, 3.99 ± 0.00001, 3.86 ± 0.001, 3.71 ± 0.0001 by the presence of the short Bragg spacings. Neither small nor wide angle reflections were observed for the 0.5 and 1.0% REA-CO gels after 24 hours at 5°C. This is likely related to equipment sensitivity issues at low solids, as gelation was observed and structure was evident by microscopy.

There was a shift towards slightly higher peak positions in the gel phase versus neat REA (for example, 43.28 Å for neat REA versus 40.54 Å for the gel). Shifts and peak broadening can occur when liquid molecules are included in the network strands (Terech & Weiss, 1997). For example, HSA aerogel structures (formed by the evaporation of the organic solvents from organogels) have slightly, but consistently higher values of d-spacings than neat HSA (Terech et al., 1994). In our case, a shift towards smaller spacings in the REA gels was observed, indicating a different and more closely packed arrangement of the REA dimers within the strands than in the neat crystalline state. The properties of the solvent strongly influence the shape and dimensions of organogelator assemblies (Abdallah & Weiss, 2000). These results point to differences between gelator aggregation in organic solvents versus vegetable oil. REA also exhibits polymorphism (data not shown). Differences in REA polymorphic behavior between the neat and gel phases may also explain the differences observed.

Powder XRD was also used to study differences in the REA-CO gels formed at higher temperatures, i.e., 15, 20 and 25°C. In this case, peaks corresponding to the (001) and (003) reflections and Bragg spacings were again observed for the gels (data not shown). However, there was more variability associated with these determinations and no trends with respect to peak position and concentration or temperature. Also, in several samples, a double peak around the (003) position was observed, indicating the existence of multiple structures within the gel network, a characteristic absent from the gels at 5°C.

Frequency Sweeps

To study the nature of the REA-CO gels formed at different temperatures after 24 h, frequency sweeps were performed. These results are shown in Fig. 4.10.

In all cases, the initial storage modulus (G′) was greater than the loss modulus (G″). The 0.5% gel behaved differently than the more concentrated samples. Fig. 4.10.A shows that G′ was frequency dependent and a cross-over point was observed for the 0.5% gel at ~ 4 Hz. A strong frequency dependence of G′, tan(δ) >1, and a cross-over point where G″ exceeds G′ are characteristics of entangled polymeric solutions (Tosh et al., 2004).

A slight positive slope for G′ is consistent with the presence of a weak gel. This was observed for the 5.0% sample at 5, 20 and 30°C (Figs. 4.10.D–4.10.F). At 5°C, the 1.0 and 3.0% gels showed a relatively higher frequency dependence, although tan(δ) was less than one and no cross-over points were observed for these gels (Figs. 4.10.B and 4.10.C). These 5°C results suggest that, while the 0.5% gel consists of entangled fibers, the 1.0, 3.0, and 5.0% gels are stabilized by junction zones between the fibers. Similarly, the 5% samples at 20 (Fig. 4.10.E) and 30°C (Fig. 4.10.F) had behavior consistent with the presence of a weak gel stabilized by junction zones. Each

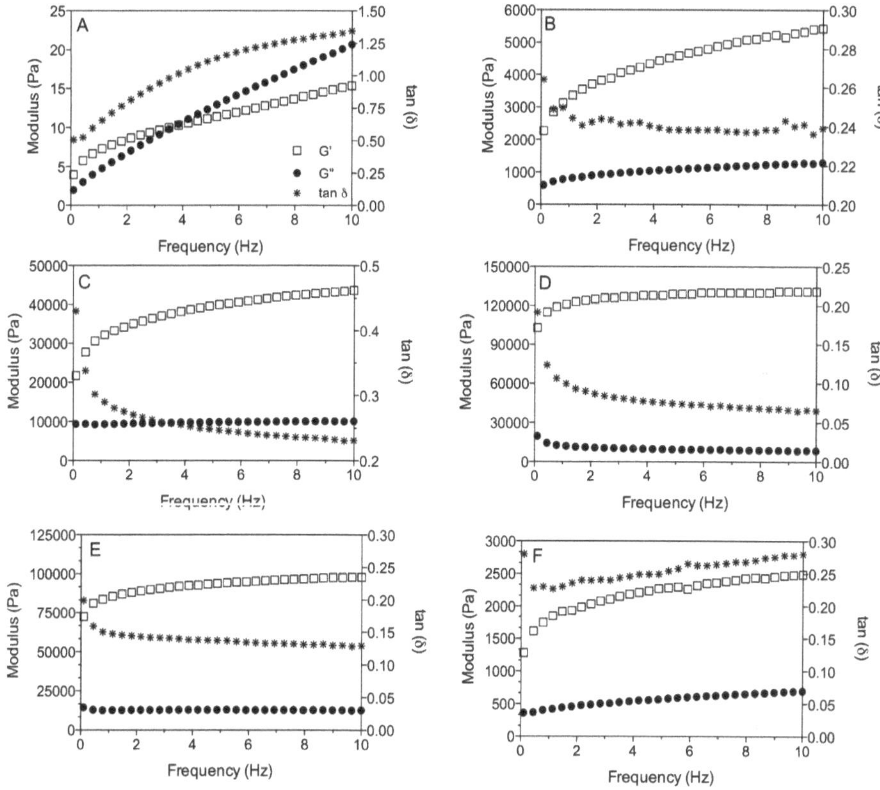

Fig. 4.10. Storage (G′) and loss (G″) moduli and tan (δ) during frequency sweeps (0.1 to 10 Hz) for 0.5 (A), 1.0 (B), 3.0 (C) and 5.0% (D) REA-canola oil gels stored at 5°C and 5% REA-canola oil gels stored at 20 (E) and 35°C (F) for 24 hours.

sample demonstrated a small frequency dependence of G′ and G″ and had tan delta values < 1. Recall that 35°C was around the upper temperature range for gelation with 5% REA. According to Fig. 4.10, G′ and G″ were significantly less at 35°C (Fig. 4.10.F) versus 5 and 20°C (Fig. 4.10.E) ($P<0.05$). However, the differences in t_{gel} above and below 20°C did not correlate with large differences in the gel's frequency response for the 5% gels. In terms of visual appearance, the gels at 5, 20 and 35°C were also similar (data not shown).

The fact that entangled fibers versus junction zone stabilization are suggested for the 0.5% versus 1, 3, and 5% gels, respectively, is consistent with the XRD and microscopy data. Little, if any, birefringence is observed for the 0.5% gels after 24 hours (Fig. 4.7.A) and no diffraction peaks were observed in the wide-angle region (refer to Table 4.B). At higher concentrations, both the light microscopy and XRD analysis indicate the presence of crystallinity, which is expected in the junction zones between REA fibers (i.e., the microcrystalline domains).

REA-CO Gel Stability with Time

After 7 days at 5°C, changes were observed in the opacity of the 0.5 and 1.0% gels, i.e., the 0.5% gel changed from a clear to a cloudy gel after 7 days and the 1.0% sample changed from a cloudy to a more opaque gel. Visible changes were not observed in the 0.25 and 5.0% samples during 28 days storage, i.e., the 0.25% sample remained liquid and the 5.0% appeared "fat-like" throughout. The most significant change in gel microstructure was observed for the 0.5% sample between 7 and 28 days (Fig. 4.7.D versus 4.7.G). After 28 days, an increase in the number and thickness of fibers were evident in the 0.5% sample. The 0.5, 1.0 and 5.0% gels had similar networks as visualized by microscopy after 28 days, although the gels differed substantially in terms of their visual appearance; the 0.5, 1.0 and 5.0% gels appeared as a cloudy gel, an opaque gel, and fat-like, respectively.

Changes in gel elasticity and microstructure at 5°C in the region of the apparent transition temperature of 20°C (i.e., around room temperature) were determined. Fig. 4.11 compares the G'_{LVR} for the gels after 1, 7 and 28 days at 5, 15, 20, and 25°C.

Significant increases in G' were observed for the 1.0 and 5.0% samples in time ($P<0.05$), although not for the 0.5% gels ($P=0.16$) (Fig. 4.11.A). These results indicate continued setting and/or restructuring of the gels. The observed increases in gel opacity and changes in microstructure of the 0.5% sample also indicate the dynamic nature of the gels. Similarly, Table 4.B shows that changes were observed

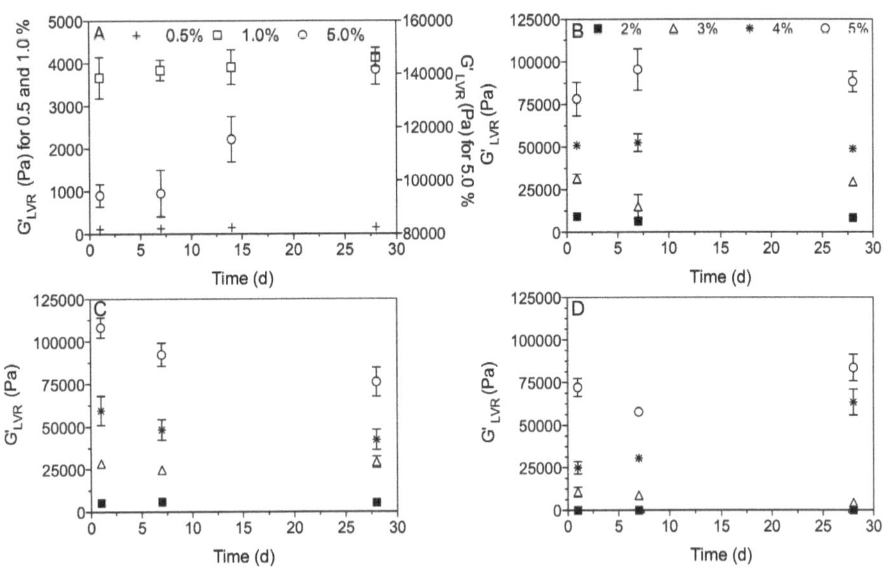

Fig. 4.11. Storage modulus in the linear viscoelastic region (G'_{LVR}) for the 0.5, 1 and 5% REA-CO gels at 5°C (A) and the 2, 3, 4, and 5% REA-CO gels at 15 (B), 20 (C), and 25°C (D) after 1, 7, and 28 days.

in the scattering properties of the gels during storage. After 1 day, no scattering was observed for the 0.5 and 1.0% gels. However, after 7 days, peaks in the small-angle region were detected, suggesting the presence of more REA dimers. In addition, wider angle reflections were detected in the 1.0% gel after 28 days.

Although more order was observed for the dilute gels, there were no significant trends in the XRD patterns for the 5.0% gels during storage at 5°C or higher (P>0.05). The positions of the ~41 and ~15 Å peaks neither increased nor decreased over time for the 5.0% gel at 5°C (Table 4.B). Similarly, the interplanar spacings were not significantly different after 28 days (P>0.05). The 2 and 5% REA-CO gels at 15, 20, and 25°C appeared to increase in opacity over time (data not shown), although there also was no evidence of microstructural changes either from microscopy or by XRD (data not shown). Overall, these results indicate continual reorganization of the REA gels at 5°C, leading to an increase in the microcrystalline domains, which may be likened to retrogradation in polysaccharide gels. Evidence for continual structuring at the higher temperatures was less conclusive.

Conclusion

The hydroxylated fatty acid ricinelaidic acid (12-hydroxy-9-*trans*-octadecenoic acid, REA) is an organogelator molecule, capable of gelling triacylglycerol-based vegetable oils, depending on temperature, concentration and oil purity. Gelation occurred with as little as 0.5% (w/w) REA. Reduced gelation tendency was correlated with the presence of potential hydrogen-bonding moieties in the oils. REA concentration also had a significant influence on gelation kinetics and gel rheology. At 5°C, the 0.5% canola oil gel behaved like a weak, viscoelastic network composed of entangled strands. Between 1.0 and 5.0% REA, solid-like, viscoelastic gels were formed at 5°C. Between 5 and 30°C, REA concentration had a significant influence on gel elasticity (P<0.05), whereas temperature had a relatively lesser influence on gel rheology. Differences were observed in the scaling exponent of G'_{LVR} with concentrations above 20°C. These were correlated with significant differences in gelation time, also at 20°C. Values for gelation time indicated a change in behavior below 2% REA and above 20°C.

The 5% REA gels at 5, 20 and 35°C displayed similar microstructures and behaved like weak gels stabilized by junction zones. Most of the gels studied (i.e., the 2, 3, 4, and 5% gels at 15, 20 and 25°C) consisted of long, thin fibrous REA strands. Polarized light microscopy revealed that the gels were formed through the interactions of long, thin and birefringent fibers. However, at 25°C, the 2% gel was characterized by more transient and circular entities. Structural analysis using XRD indicated the presence of repeating REA dimers and increasing order with REA concentration and gel storage time. Increases in gel opacity, birefringence, XRD scattering, and fiber clustering were observed for the dilute gels during storage at 5°C. However, with 28 days of storage at higher temperatures, there were no apparent changes detected in the 2 and 5% gel microstructure by microscopy or XRD, despite increases in gels opacity.

References

Abdallah, A.J.; Weiss, R.G. Organogels and Low Molecular Mass Organic Gelators, Adv. Mater. 2000, 12, 1237–1247.

American Oil Chemists' Society. Official methods and recommended practices, Section 1: Physical and chemical characteristics of oils, fats and waxes; American Oil Chemists' Society: Champaign, IL, 1996; 44.

Bot, A.; Agterof, W.G.M. Structuring of Edible Oils by Mixtures of γ-oryzanol with β-Sitosterol or Related Phytosterols. J. Am. Oil Chem. Soc. 2006, 83, 513–521.

Daniel, J.; Rajasekharan, R. Organogelation of Plant Oils and Hydrocarbons by Long-Chain Saturated FA, Fatty Alcohols, Wax Esters, and Dicarboxylic Acids. J. Am. Oil Chem. Soc. 2003, 80, 417–421.

Eloundou, J.; Girard-Reydet, E.; Gerard, J.; Pascault, J. Calorimetric and rheological studies of 12-hydroxystearic acid/diglycidyl ether of bisphenol A blends. Polym. Bull. 2005, 53, 367–375.

Friggeri, A.; Gronwald, O.; van Bommel, K.; Shinkai, S.; Reinhoudt, D.N. Charge-Transfer Phenomena in Novel, Dual-Component Sugar-Based Organogels. J. Am. Chem. Soc. 2002, 124, 10754–10758.

Gandolfo, F.G.; Arjen, B.; Flöter, E. Structuring of Edible Oils by Long-Chain FA, Fatty Alcohols, and Their Mixtures. J. Am. Oil Chem. Soc. 2004, 81, 1–6.

Hanabusa, K.; Tanaka, R.; Suzuki, M.; Kimura, M.; Shirai, H. Excellent Gelators for Organic Fluids: Simple Bolaform Amides Derived from Amino Acids. Adv. Mater. 1997, 9, 1095–1097.

Hughes, N.E.; Marangoni, A.G.; Wright, A.J.; Rogers, M.A.; Rush, J.E. Potential Food Applications of Edible Oil Organogels. Trends Food Sci. Technol. 2009, 20, 470–480.

Marton, L.; McBaln, J.W.; Vold, R.D. An Electron Microscope Study of Curd Fibers of Sodium Laurate. Rational Design of Low Molecular Mass Organogelators: Toward a Library of Functional N-acyl-1, ω-amino acid derivative. J. Am. Chem. Soc. 1941, 63, 1990–1993.

Mukkamala R.; Weiss, R.G. Physical Gelation of Organic Fluids by Anthraquinone-Steroid Based Molecules. Structural Features Influencing the Properties of Gels. Langmuir 1996, 12, 1474–1482.

Okabe, S.; Ando, K.; Hanabusa, K; Shibayama, M. Dynamic Light Scattering and Small-Angle Neutron Scattering Studies on Organogels Formed with a Gelator. J. Polym. Sci. B– Polym. Phys. 2004, 42, 1841–1848.

Rogers, M.A.; Wright, A.J.; Marangoni, A.G. Oil Organogels: The Fat of the Future. Soft Matter 2009, 5, 1594–1596.

Tamura, T.; Suetake, T.; Ohkubo, T.; Ohbu, K. Effect of alkali metal ions on gel formation in the 12-hydroxystearic acid/soybean oil system. J. Am. Oil Chem. Soc. 1994, 71, 857–861.

Terech, P.; Chachaty, C.; Gaillard, J.; Godquin-Giroud, A.M. Electron Paramagnetic Resonance Study of the Physical Gelation of a Copper Complex in Cyclohexane. J. Phys. (Paris) 1987, 48, 663.

Terech, P.; Schaffhauser, V.; Maldivi, P.; Guenet, J.M. Living Polymers in Organic Solvents. Langmuir 1992, 8, 2104–2106.

Terech, P.; Rodriguez, V.; Barnes, J.D.; McKenna, G.B. Organogels and Aerogels of Racemic and Chiral 12-Hydroxyoctadecenoic Acid. Langmuir 1994, 10, 3406–3418.

Terech, P.; Weiss, R.G. Low molecular mass gelators of organic liquids and the properties of their gels. Chem. Rev. 1997, 97, 3133–3159.

Terech, P.; Pasquier, D.; Bordas, V.; Rossat, C. Rheological Properties and Structural Correlations in Molecular Organogels. Langmuir 2000, 16, 4485–4494.

Tosh, S.M.; Wood, P.J.; Wang, Q.; Weisz, J.; Structural Characteristics and Rheological Properties of Partially Hydrolyzed Oat b-Glucan: The Effects of Molecular Weight and Hydrolysis Method. Carb. Polym. 2004, 55, 425–436.

Uzu, Y.; Phase Behavior and Colloidal Structure of Lithium Stearate in Hydrocarbons. J. Jpn. Oil Chem. Soc. 1975, 24, 261.

Van Esch, J.H.; Feringa, B.L. New Functional Materials Based on Self-Assembling Organogels: From Serendipity Towards Design. Angew Chem. Int. Ed. 2000, 39, 22636–22666.

Wright, A.J.; Marangoni, A.G. Formation, structure and rheological properties of ricinelaidic acid-vegetable oil organogels. J. Am. Oil Chem. Soc. 2006, 83, 1–7.

Wright. A.J.; Marangoni, A.G. Time, temperature and concentration dependence of ricinelaidic acid-canola oil organogelation. J. Am. Oil Chem. Soc. 2007, 43, 3–9.

5

Hydroxystearic Acid Oleogels

Michael A. Rogers[1] and Alejandro G. Marangoni[2]

[1]School of Environmental and Biological Sciences, Department of Food Science; Rutgers University; The State University of New Jersey, New Brunswick, New Jersey and [2]Department of Food Science, University of Guelph, Guelph, Ontario, Canada

Production of 12-Hydroxystearic Acid

12-hydroxystearic acid (12HSA) is obtained via hydrogenation of castor oil, resulting in complete saturation of the hydrocarbon chain (Maskaev et al., 1971). Hydrogenation is then followed by saponification, and free fatty acids are separated by the addition of mineral acid. If the conditions of hydrogenation are not optimized, dehydration and reduction reactions of the hydroxyl group lead to the formation of stearic acid (Maskaev et al., 1971). The processing conditions must be tailored to avoid the production of ricinoleic acid. Optimal processing conditions to produce 12HSA occur at 100°C, 2% Raney metallic catalyst (68.8% nickel), 12.5 atm, for 300 min (Maskaev et al., 1971). Using these parameters, 76.1% of the end products are 12HSA, 6% ricinoleic acid, 12.4% stearic acid, 3.2% oleic acid, 2.1% plasmatic and arachic acid, and 0.2% linoleic acid (Maskaev et al., 1971).

12-Hydroxystearic Acid Crystallization

Crystal Structure

The crystal structure of 12HSA, which is an 18 carbon fatty acid with a hydroxyl group at position 12, was elucidated by Kuwahara et al. (1996) (Fig. 5.1). 12HSA stacks via a hydrogen bond sequence between the 12 hydroxyl groups, resulting in a zig-zag pattern at a distance of 2.87Å (Kuwahara et al., 1996). The zig-zag pattern arises due to an all *trans* configuration of the molecule and the angle at the C12—O3—O3 is 117° (Kuwahara et al., 1996). Furthermore, the angles of C4—C5—C6—C7 and C15—C16—C17—C18 are 180°, while the C10—C11—C12—C13 has an angle of 173° (Kuwahara et al., 1996). The deviation from 180° for 12HSA was not observed in stearic acid and is thought to be partly responsible for the twisted helical structure observed in 12HSA fibers (Kuwahara et al., 1996). When 12HSA dimerized between the carboxylic acid groups, the distance between the adjacent O2—O3 atoms was 2.62Å (Kuwahara et al., 1996).

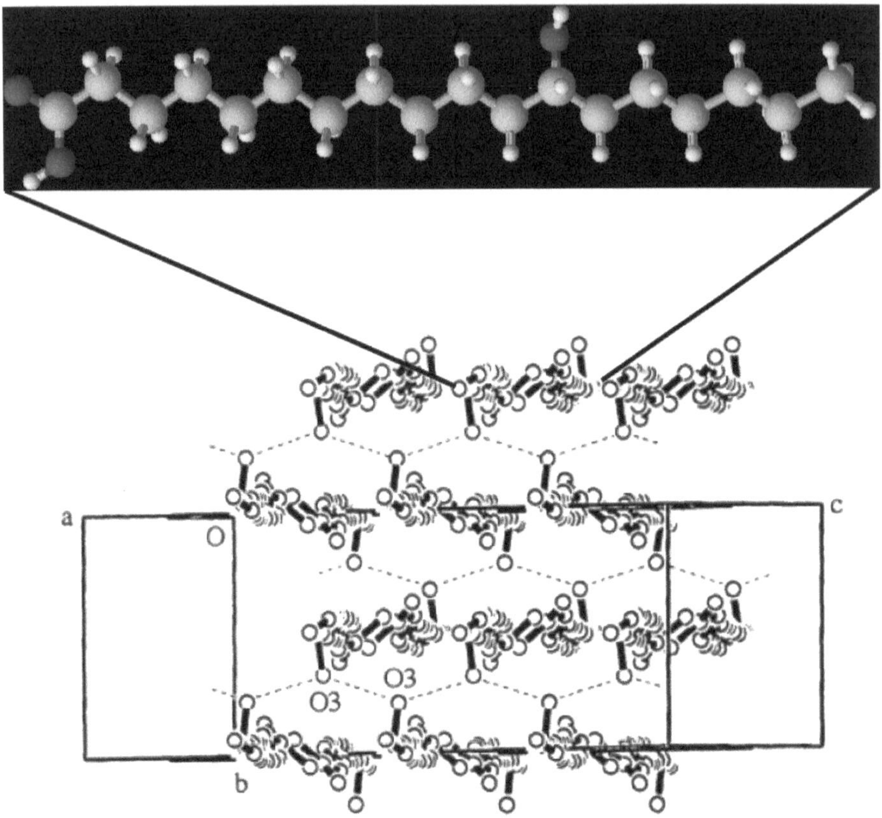

Fig. 5.1. Crystal conformation of DL 12HSA (Adapted from Kuwahara et al., 1996).

Nucleation Behavior

The nucleation behavior affects numerous structural features including fiber size, fiber morphology, number of crystal aggregates, and the spatial distribution of crystalline mass (Marangoni & McGauley, 2003). Once the organogelator-solvent melt is cooled below the melting point of the gelator, the sol becomes supersaturated and causes the gelator to assemble via nucleation events (Weiss & Terech, 2007). In industrial practices, crystallization typically occurs under non-isothermal cooling conditions (i.e., the temperature changes as the material crystallizes).

To quantify the nucleation process, the number of crystals (N) may be counted on polarized light micrographs and plotted as a function of time (t) for each cooling rate (Figs. 5.2.A and 5.2.B). The first derivative of this curve corresponds to the rate of nucleation (Figs. 5.2.C and 5.2.D). Further analysis of this data utilizes the peak nucleation rate (J_p) calculated from the first derivative of the nucleation rate curve

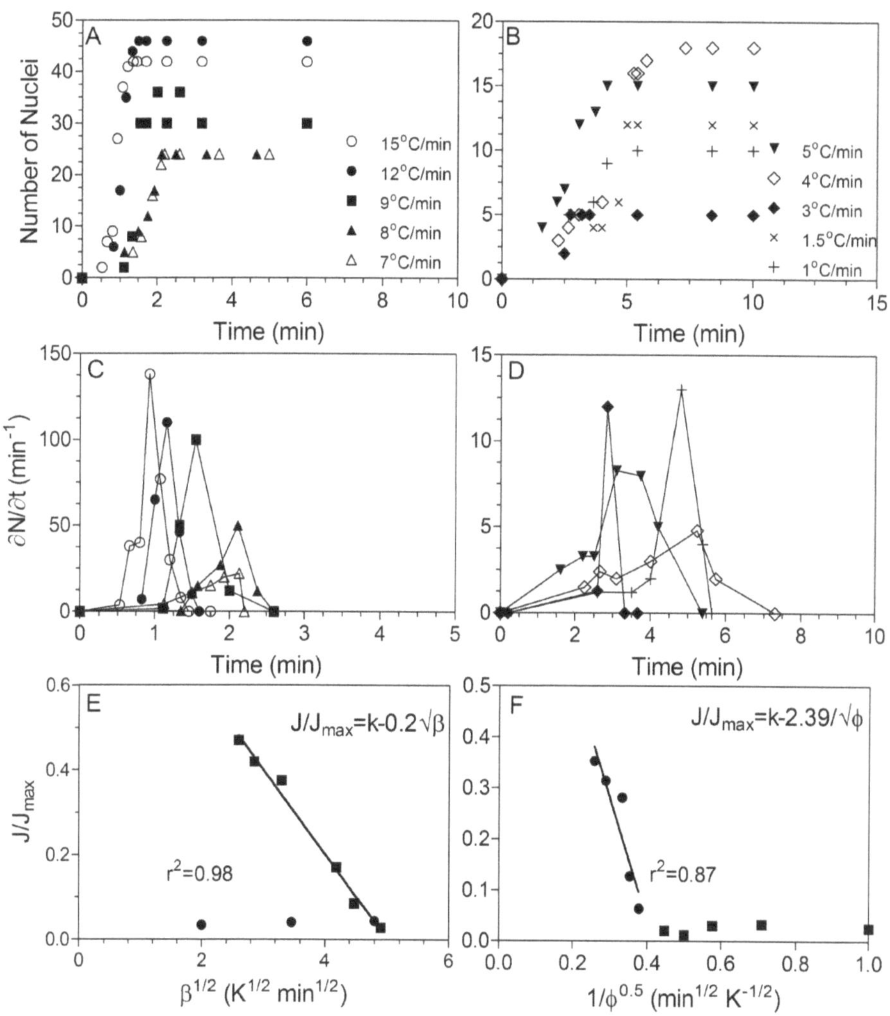

Fig. 5.2. Number of nuclei counted on brightfield micrographs as the sample was cooled non-isothermally (A,B), the rate of nucleation determined by the change in the number of nuclei per unit time (C, D), and the J_{max}, the maximum nucleation rate, which is a fitted parameter that is a function of cooling rate (Source: Rogers & Marangoni, 2008).

corresponding to the maximum value of the first derivative. A simple inverse relationship between the rate of nucleation and the cooling rate was found for 12HSA in most solvents (see Figs. 5.2.E and 5.2.F) (Rogers & Marangoni, 2008). However, for certain solvents that are capable of interacting with the growing fiber, an exponential relationship is found leading to the development of spherulitc crystals (Rogers & Marangoni, 2008; 2009).

Two dissimilar nucleation regimes with differing sensitivities to cooling rate were observed between J_p and cooling rate and J_p and the time-temperature (β) parameter that corresponds to the changing thermodynamic driving force as the sample cools (see Figs. 5.2.E and 5.2.F) (Rogers & Marangoni, 2008; 2009). Cooling rates less than 5–7°C/min did not effect the nucleation rate. However, cooling rates greater than 5–7°C/min showed dependence between the rate of nucleation and cooling rate (Fig. 5.2.F). Similar trends were observed for the supercooling time exposure parameter for rapidly cooled samples and slowly cooled samples (Fig. 5.2.E). This implies that fast cooling rates lead to a time-dependent thermodynamic driving force and at slow cooling rates the nucleation behavior was a function of mass transfer (Rogers & Marangoni, 2008; 2009).

Using a statistical approach the activation energy was calculated from the slopes of Figs. 5.2.E and 5.2.F (Rogers & Marangoni, 2008; 2009). For 12HSA (MW=300g/mol) the energy of activation for nucleation was determined in various oils: 2.2 kJ/mol for 12HSA/methyl oleate, 5.4 kJ/mol for 12HSA/mineral oil, 10.6 kJ/mol for 12HSA/triolein, 15.8 kJ/mol for 12HSA/glycerol, and 12.1 kJ/mol for 12HSA/canola oil (Rogers & Marangoni, 2009; 2008). Originally, it was believed that the static relative permittivity (i.e., solvent polarity) would greatly influence the activation energy of nucleation. Since the rates of phase separation and nucleation are a function of the difference between the solvent and gelator polarity, the quantifiable activation energy was thought to be a function of the static relative permittivity. However, a relationship between the polarity and the activation energy of nucleation was not found (Rogers & Marangoni, 2009). Methyl oleate and triolein both have static relative permittivities of 3.2, however, their activation energies are 2.2 kJ/mol and 10.6 kJ/mol, respectively (Rogers & Marangoni, 2009). Although it is logical to assume that polarity influences the activation energy, the ability of the solvent to interact with the gelator is of significance regarding the nucleation parameters, and thus effects the crystal morphology, number of crystals, and the spatial distribution of mass.

Crystal Growth

The nucleation process requires highly specific interactions to induce preferential one-dimensional growth. Following nucleation, fibers grow one-dimensionally (Fig. 5.3) and the rate of branching, interpenetration, and entanglement depends on the degree of supersaturation or undercooling (Wang et al., 2006). The ability for SAFiNs to self-assemble into rod-like structures requires a balance between opposing parameters, such as solubility and those parameters that control epitaxial growth, to grow into axially symmetric elongated aggregates (Weiss & Terech, 2007; Suzuki et al., 2003).

Fiber growth rate was determined by monitoring the length of each fiber on the polarized light micrographs as a function of time under different cooling rates (Fig. 5.3). The data was fitted to a modified Avrami equation characterizing the crystal growth kinetics (Figs. 5.4.A and 5.4.B) (Rogers & Marangoni, 2009; Lam, et al., 2010; Lam & Rogers, 2010). The modified Avrami model has the form:

$$Y = Y_{max}(1 - e^{-k_{app}(x - x_0)^n})$$

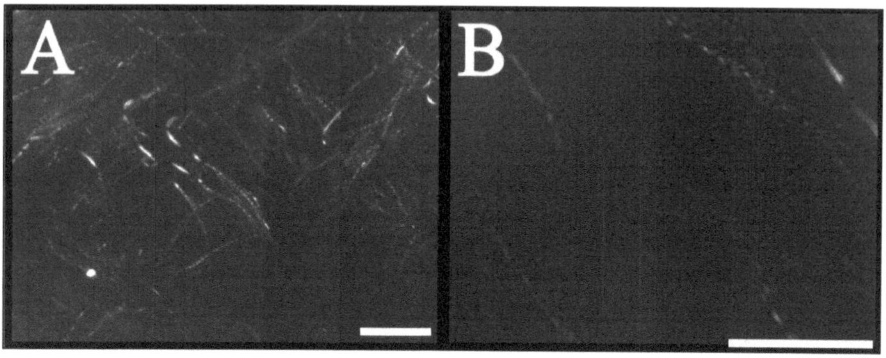

Fig. 5.3. Two-dimensional cross polarized light microscopy of 2 wt % HSA/Canola oil; under 10× magnification (A) and 40× magnification (B). Magnification Bar = 100 um. Rogers et al., 2007.

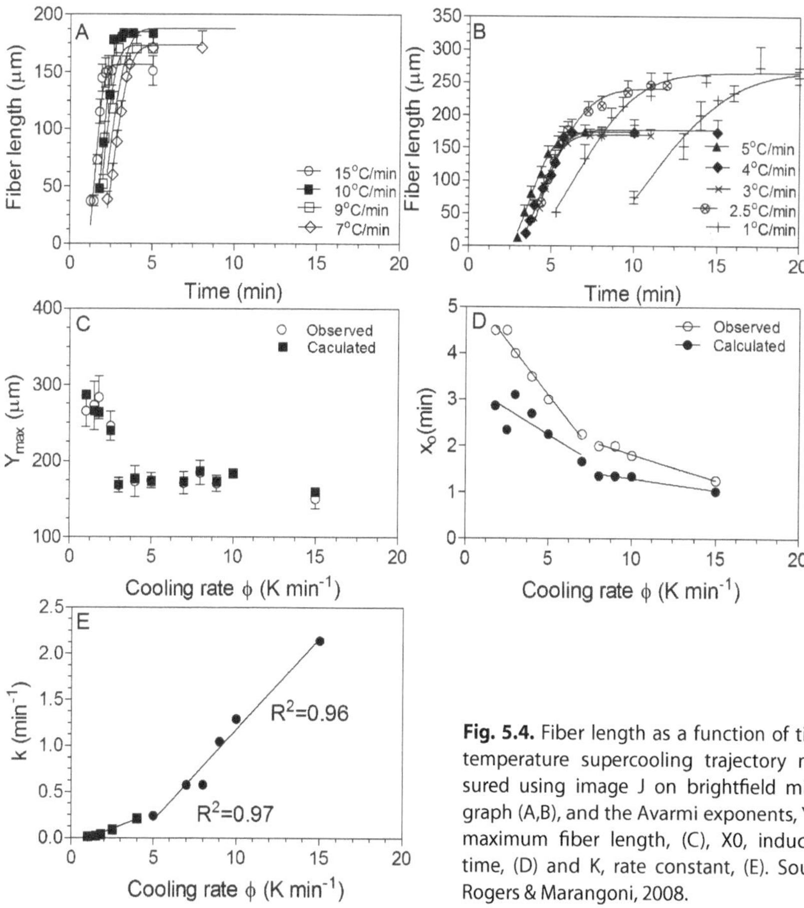

Fig. 5.4. Fiber length as a function of time/temperature supercooling trajectory measured using image J on brightfield micrograph (A,B), and the Avarmi exponents, Y_{max}, maximum fiber length, (C), X0, induction time, (D) and K, rate constant, (E). Source: Rogers & Marangoni, 2008.

where Y is the fiber length, Y_{max} is the maximum length of the fiber, k_{app} is the apparent rate constant of fiber elongation, x is the time, x_o is the induction time, and n is the dimensionality of growth.

The Avrami index (n) is related to the mechanism of nucleation and the dimensionality of crystal growth. Hence, n is typically an integer between 1 and 4 for instantaneous or sporadic nucleation, and corresponds to 1, 2, or 3-dimensional growth (Table 5.A). Upon fitting the modified Avrami model to the experimental data (Figs. 5.4.A and 5.4.B), the constants for Y_{max}, x_o and k_{app} were determined (Figs. 5.4.C–5.4.E). The Avrami exponent was 1 when corresponding to instantaneous nucleation and one-dimensional growth (Lam & Rogers, 2010). The calculated and experimentally determined Y_{max} corresponded well. Thus, the Avrami model is a good characterization tool for non-isothermal SAFiN growth (Lam & Rogers, 2010). A drastic difference in the final fibers lengths was observed depending on the cooling rate (Lam & Rogers, 2010). Cooling rates faster than 5–7°C/min produced short fibers, suggesting a different mode of assembly for 12HSA, regardless of the solvent employed (Rogers & Marangoni, 2008; 2009; Lam & Rogers, 2010). The Avrami constant, K, followed the same trends as the peak nucleation rates (Figs. 5.2.E and 5.2.F). For fast cooling rates, the time-dependent thermodynamic driving force leads to greater rates of both crystal growth and nucleation. This regime of nucleation would be similar to that described by Wang et al.'s nucleation-growth-noncrystallographic branching-growth mechanism (2006). Samples cooled faster and slower than 5–7°C/min indicate significant differences in the morphology of the SAFiN network (Marangoni & Rogers, 2009). Slow cooling rates develop few nuclei, and the fibers have little branching and are several hundred microns in length. This type of system does not effectively entrain liquid oil and exhibits syneresis (Rogers et al., 2008). Conversely, under fast cooling rates, nucleation is a function

Table 5.A. Possible Forms of the Avrami Model for the Different Modes of Nucleation and Their Associated Dimensionality of Crystal Growth (Lam & Rogers, 2010).

Mode of Nucleation	Dimensionality of Crystal Growth	Avrami Equation	N
Instantaneous	1-dimensional	$\frac{Ys}{Y_{max}} = 1 - e^{Kt^1}$	1
Sporadic	1-dimensional	$\frac{Ys}{Y_{max}} = 1 - e^{Kt^2}$	2
Instantaneous	2-dimensional	$\frac{Ys}{Y_{max}} = 1 - e^{Kt^2}$	2
Sporadic	2-dimensional	$\frac{Ys}{Y_{max}} = 1 - e^{Kt^3}$	3
Instantaneous	3-dimensional	$\frac{Ys}{Y_{max}} = 1 - e^{Kt^3}$	3
Sporadic	3-dimensional	$\frac{Ys}{Y_{max}} = 1 - e^{Kt^4}$	4

of the supercooling-time exposure, leading to the creation of highly branched networks effective at entraining oil (Rogers et al., 2008). Therefore, to engineer the oil binding capacity and hardness/elasticity of an organogel, external fields, such as heat and mass transfer, may be modified. The external fields not only dictate the number of nuclei present, but also define the degree of branching and fiber length of the SAFiN network. Future work should define those critical ranges of nucleation behavior.

Crystallographic Mismatches and Transient Junction Zones

Crystallographic mismatches, or crystal imperfections, allow for the development of branched fibers. Recent work by Liu has provided support for the nucleation-growth-crystallographic-mismatch-branching (CMB) model (Liu & Sawant, 2001; 2002). The CMB model suggests that at low degrees of undercooling, following nucleation, one-dimensional fibers grow with little branching, interpenetration, or entanglement (Wang et al., 2006). Low bulk supersaturation cannot overcome the high crystallographic mismatch nucleation energy barrier (ΔG^*). This favors one-dimensional fiber growth with large correlation lengths (χ) (Fig. 5.5) (Wang et al., 2006; Lam & Rogers, 2010). However, under isothermal cooling profiles, when the crystallization temperature decreases, there is an increase in the supersaturation. The increased supersaturation results in the crystallographic mismatch energy barrier being overcome, resulting in increased fiber tip branching (Wang et al., 2006; Lam & Rogers, 2010). This has also been confirmed using a rheological fractal dimension (D_f) whereby cooling rates less than 5°C/min have a fractal value close to one and at fast cooling rates the fractal value is approximately 1.2, which indicates a highly branched

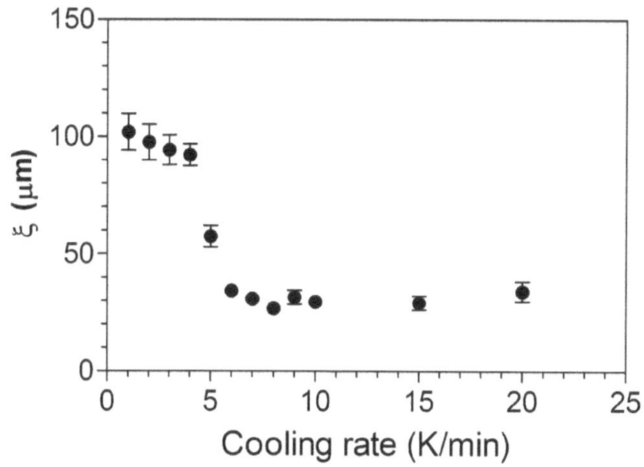

Fig. 5.5. Correlation length of 2.5 wt % 12HSA fibers in mineral oil measured from the micrographs obtained at 30°C after they had been cooled from 90°C at varying cooling rates (Lam & Rogers, 2010).

fiber (Lam et al., 2009). The highly branched fibers [i.e., short correlation length (χ)] coincide with smaller pore sizes (Wang et al., 2006).

12HSA fiber formation occurs when two monomers dimerize between their carboxylic acid groups, and longitudinal growth occurs via hydrogen bonding between the hydroxyl groups at carbon 12. Spectral features associated with hydrogen bonding (3200 cm^{-1}) and COOH—COOH dimerization (1700 cm^{-1}) were monitored using Fourier transform infra-red spectroscopy (FTIR) (Figs. 5.6.A and 5.6.B). More specifically, when 12HSA dimerizes there is a shift in the spectral feature from 1710 cm^{-1} (COOH monomer) to 1700 cm^{-1} (COOH dimer) (Rogers & Marangoni, 2008; 2009; Lam & Rogers, 2010). Similarly, hydrogen bonding was monitored by the area of the peak at position 3200 cm^{-1} (Figs. 5.6.A and 5.6.B) (Rogers & Marangoni, 2008; 2009; Lam & Rogers, 2010). The area under the peaks at 1700 cm^{-1} and 3200 cm^{-1} were measured and plotted as a function of time while undergoing non-isothermal cooling (Figs. 5.6.C and 5.6.D). To assess if any changes in the rate of evolution between these two peaks were observed, the modified Avrami model was applied (Rogers & Marangoni, 2009). Upon fitting the data to the Avrami model, regardless of the cooling rate, similar intensities of the peak area were measured (Fig. 5.6.E). An exponential decrease in the induction time was observed (Fig. 5.6.G) (Lam & Rogers, 2009). The Avrami exponent was found to be 1 for all samples, indicating that crystals undergoes instantaneous nucleation and linear one-dimensional growth, which corresponds to Fig. 5.4 using PLM (Fig. 5.6.H). However, examining the apparent rate constant of hydrogen bond formation between the 12HSA carboxylic acid groups and between the hydroxyl groups, a new insight into self-assembly and into crystallographic mismatches was observed (Fig. 5.6.F) (Lam et al., 2009). Previously, branched fibers were thought to arise from molecules being incorporated into the lattice in an imperfect manner. This would suggest that there is a significant decrease in the rate of hydrogen bonding above 5°C/min, which corresponds to the significant change in fiber length (Fig. 5.5). However, the rate of hydroxyl hydrogen bonding changes linearly with cooling rate and does not plateau as the cooling rate is increased, which was observed by a linear increase in the rate of hydroxyl interactions (Lam et al., 2009). However, it was observed at cooling rates greater than 5–7°C/min that the rate of dimerization between carboxylic acid groups levelled off (Fig. 5.6.F). This suggests that the crystallographic mismatch for 12HSA is a result of the inability of the 12HSA monomers to effectively dimerize at rapid cooling rates. When incorporated into the crystal lattice, monomers prevent the uninterrupted longitudinal growth observed at slower cooling rates.

The elastic properties of 12HSA organogels result from the formation of both permanent and transient junction zones within these soft materials. The permanent junction zones arise from crystallographic mismatches while the transient junction zones are a consequence of fiber entanglement. The non-covalent interactions responsible for the occurrence of transient junction zones in 12HSA arise from hydroxyl-hydroxyl hydrogen bonding. This may result from the preferential arrangement of

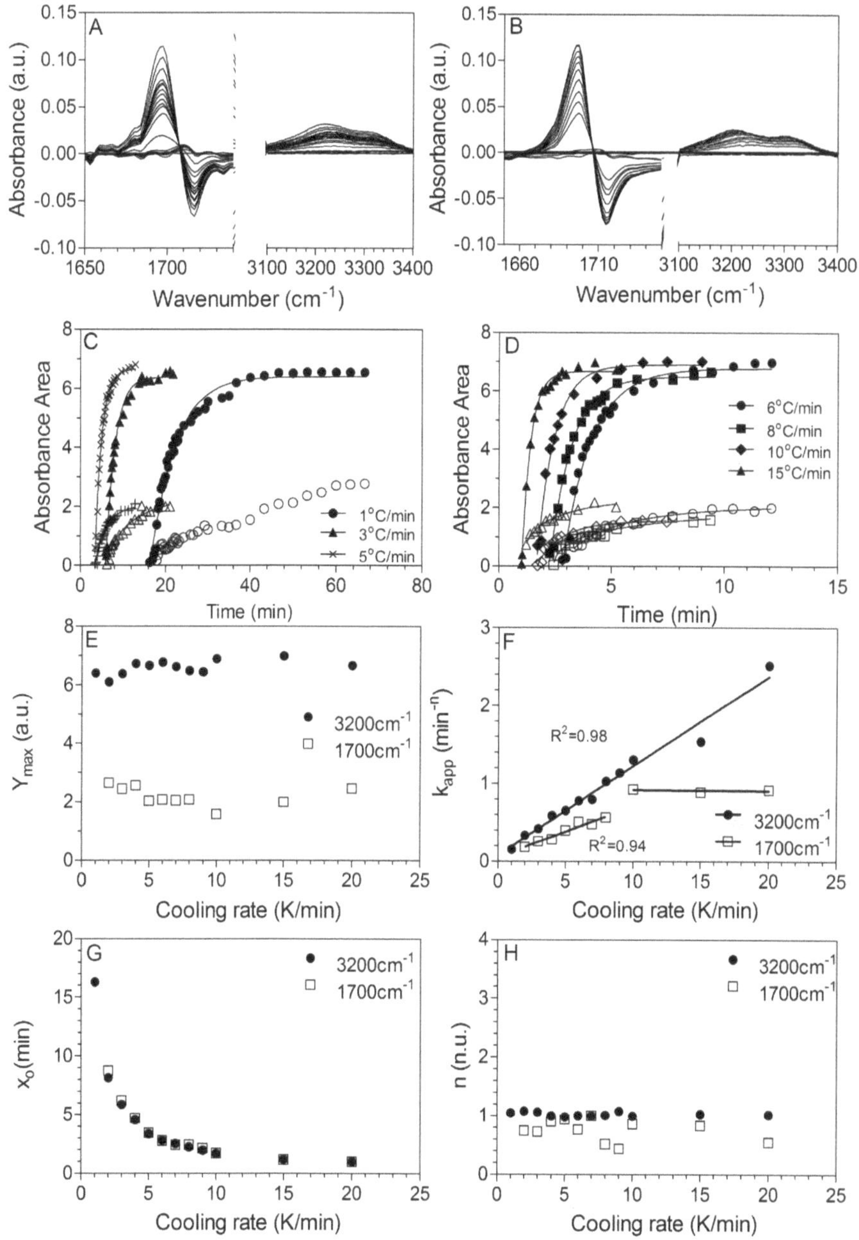

Fig. 5.6. Differential FTIR spectra of 12HSA assembly during slow cooling at 1 °C/min and was collected from 90°C to 30°C over a 60 min period (A) and fast cooling at 10°C/min from 90°C to 30°C over a 6 min period (B) non-isothermal cooling profiles. Integration of the FTIR spectra (C and D) to determine the area under the spectral features of interest (filled symbols represent absorbance at 3200 cm^{-1} and open symbols represent the absorbance at 1700 cm^{-1}). Avrami constants were determined by fitting the integrated area versus time to obtain maximum signal (E), apparent rate constant (F), induction time (G), and dimensionality of growth (H) (Source: Lam & Rogers, 2010).

carboxylic acid groups to dimerize within the fiber, causing a reduction of the fiber-solvent interfacial tension (Roger et al., 2009). Also, synchrotron infrared spectromicroscopy indicates that the crystallization process of SAFiNs efficiently displaces solvent from the interface of the growing crystals, contributing to the epitaxial growth into axially symmetric elongated aggregates (Rogers et al., 2009). The periodicity in the density of hydrogen bonding resolves the supramolecular chirality of 12HSA fibers (Fig. 5.7).

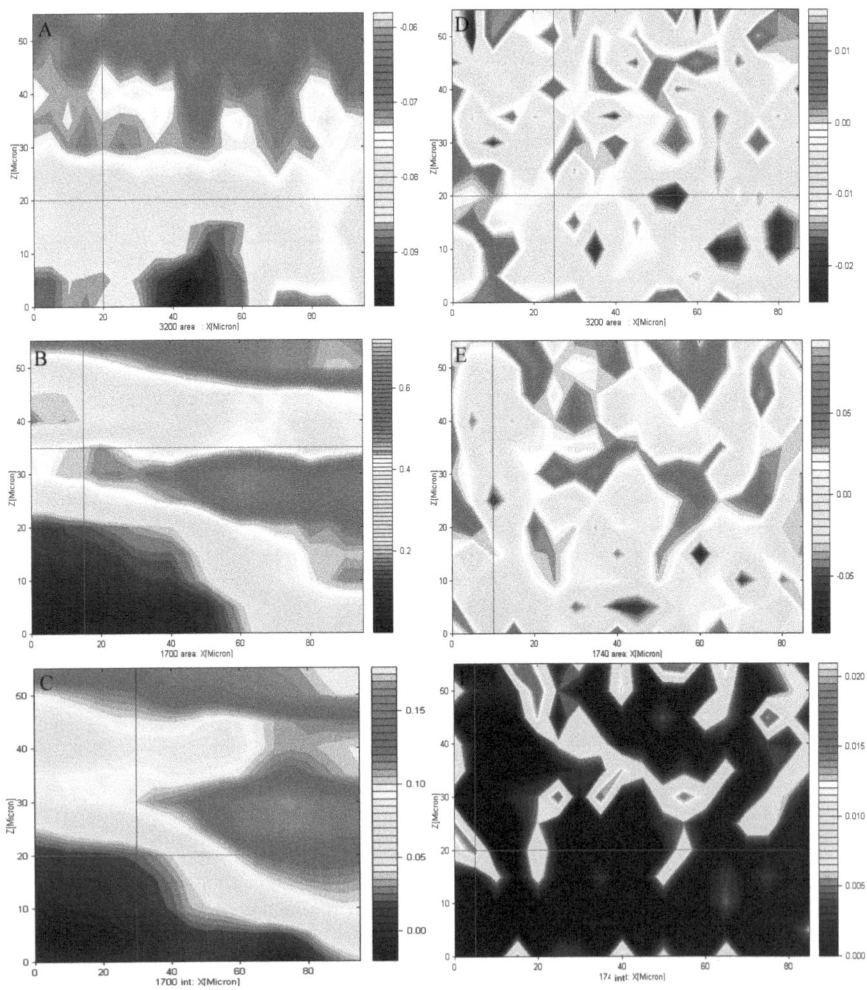

Fig. 5.7. Fourier transform infrared maps of the relative change in hydroxyl hydrogen bonding area (A,D), carboxylic acid dimerization area (B), carboxylic acid dimerization intensity (C), ester band area (E), and intensity (F) for mineral oil (A—C) and methyl oleate (D—F) (Source: Rogers et al., 2009).

For 12HSA in mineral oil the area of the peak at 3200 cm^{-1}, corresponding to the hydroxyl-hydroxyl interactions between molecules (Fig. 5.7.A), and the area (Fig. 5.7.B) and intensity (Fig. 5.7.C) of the dimerization of carboxylic acid groups on 12HSA at 1700 cm^{-1}, were mapped. The two-dimensional maps for 12HSA/mineral oil revealed two distinct fibers with an approximate width of ~8–10μm (Rogers et al., 2009). The width corresponds well with previously obtained cryo-SEM images (Rogers et al., 2007). Figs. 5.7.A–5.7.C present the same field-of-view and are superimposable. Strands of 12HSA have a helical nature due to the supramolecular arrangement of fibers arising from the chirality at position 12 on 12HSA (Dumas et al., 2004; Tachibaba & Kambara, 1968; Uzu & Sugiura, 1975; Fuhrhop et al., 1987; Terech et al., 1994). Previously, 12HSA strands in mineral oil where shown to be approximately 100 μm long and 3–8 μm wide (Rogers et al., 2009). 12HSA in methyl oleate have shorter fibers ranging in length from 5 to 75 μm and are approximately the same width as observed in mineral oil (Rogers et al., 2009). The supramolecular helical configuration instilled from the chiral carbon may be observed via the hydrogen-bonding periodicity within the fibers as illustrated by reddish-pink zones containing concentrated dimerization in the crystalline structure (see Fig. 5.7.B).

The area (Fig. 5.7.B) and intensity (Fig. 5.7.C) of the carboxylic acid interactions were used to visualize the fibers. Figs. 5.7.B and 5.7.C indicate a reduction in the amount of COOH—COOH interactions between fibers. This suggests that upon formation of the fiber the COOH groups are effectively excluded from the fiber-solvent interface, thereby reducing the interfacial tension. Fig. 5.7.A indicates that the fiber non-permanent junction zones interact via hydrogen bonding of the hydroxyl groups while preferentially excluding the carboxylic acid groups from the surface of the fiber (Rogers et al., 2009). As a result, the hydroxyl groups at position 12 are exposed to the surface. The idea that nanofibers interact to form the supramolecular network via van der Waals interactions alone, in certain cases, may be over simplified.

12HSA in methyl oleate produced few long fibers, resulting in small, densely packed crystals (Fig. 5.7.D). Both Figs. 5.7.D and 5.7.E show numerous 5–10 μm crystals. The areas elevated in hydroxyl-hydroxyl interactions (Fig. 5.2.D reddish nuclei) correspond to the areas depleted of the ester band for methyl oleate (Figs. 5.7.E and 5.7.F blue solvent), suggesting that the development of 12HSA fibers is extremely efficient at displacing the solvent from the growing fiber surface and as a result develops densely packed crystals. However, numerous other crystallization processes, as in the case of triacylglyceride (TAGs), often entrain liquid TAGs during the formation of large diffusion-limited clusters.

Oil Binding Capacity

Organogels have tremendous potential for topical applications in drug delivery (Giorgano et al., 1998). The rate of delivery is governed by the pore-size distribution within a gel, therefore understanding the fiber nanostructure is of utmost importance (Kantaria et al., 1999). The rate of release is governed by the rate of syneresis, which is the expulsion of oil from a matrix due to shrinkage. This results in the solvent (oil) being expelled from the organogel matrix. Thus, to modify the rate of release, the fiber branching/interactions must be adjustable. To prevent syneresis in a gel, the interfacial tension between the solvent and SAFiN must exceed or balance the gravitational force. It has been postulated that if the interfacial area between the oil and 12HSA network is larger, a corresponding higher interfacial tension will occur. If the interfacial tension exceeds the gravitational forces, the network will retain oil. However, if the interfacial area is smaller than the average capillary dimensions, with an even distribution of mass, the network will entrain the oil and prevent syneresis. Several authors have observed differences in the oil binding capacity of organogels (Markovic et al., 2004; Tamura & Ichikawa, 1997).

Samples of 12HSA in canola oil were prepared by isothermally cooling them to 5°C and 30°C. They were removed from the incubators after 24 hrs and were equilibrated at 20°C for 1 hr before being placed on Whatman filter papers (Whatman, Number 5, 15cm diameter) for 1 hr (Rogers et al., 2008). The distance that the oil front moved on the Whatman filter paper was measured and plotted as a function of time. Fig. 5.8.A indicates that the oil mobility is significantly greater for the sample stored at 30°C than the sample stored at 5°C. The ratio of the rates of syneresis (D) at the two temperatures studied, $D_{30°C}/D_{5°C}$, was 1.35±0.02 (Rogers et al., 2008).

Samples were subjected to T_2 relaxation measurements on a Bruker mq20 Series NMR Analyzer (Bruker, Milton, ON, CA). A Carr-Purcell-Meiboom-Gill (CPMG) pulse sequence was used to determine the spin-spin relaxation times for the liquid relaxations. The data underwent a Laplace transformation and was solved using a constrained regularization method for data inversion (Provencher, 1982a; 1982b). Three main distributions of liquid relaxations are distinguished in Fig. 5.8B as having T_2 in the range of 50 to 70 ms, 80 to 130 ms, and 200 to 500 ms, respectively. The

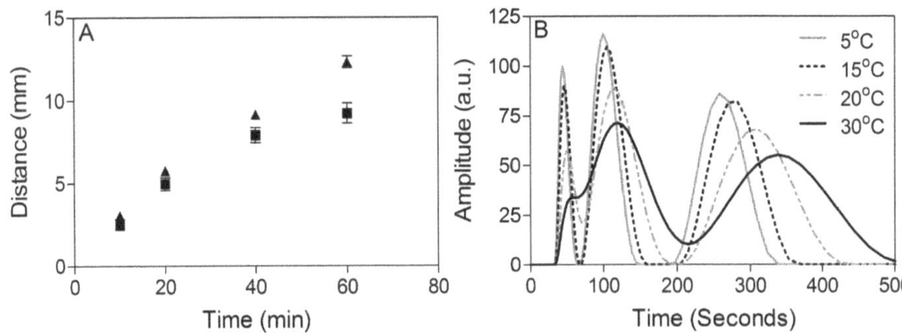

Fig. 5.8. Oil syneresis from a 2 cm diameter disc of .5% HSA/canola oil on a Whatman 5 filter paper (A) and Pulse NMR *T2* relaxation distributions 2.5% HSA/canola oil stored at different temperatures (Adapted from Rogers et al., 2009).

50 to 70 ms relaxation was attributed to the immobilized or entrained oil (less mobile oil) and the 200 to 500 ms relaxation to the more mobile liquid oil. With increasing crystallization temperature, the immobilized or entrained oil at 50 to 70 ms becomes part of the 80 to 130 ms relaxation and the 200 to 500 ms relaxation shifts to higher relaxations. The absence of the highly immobilized oil at 30°C suggests that the SAFiN network is not entraining the liquid oil efficiently.

T_2 relaxations in porous systems have been related to both the time required for the molecule to diffuse to the surface of the pore (Strange et al., 1993), as well as to levels of confinement in a pore (Baba, 1999). Regardless of this assumption, if there is a change in the level of confinement or a change in the pore size, it should affect the T_2 relaxation time. If T_2 is directly related to the pore size, then the $T_{2(30°C)}/T_{2(5°C)}$ ratio should be numerically similar to $D_{30°C}/D_{5°C}$ (1.35(0.02). The ratios of the three T_2 relaxations at 30°C (51ms, 113ms, and 326ms) and 5°C (39ms, 90ms, and 250ms) were determined to be 1.30±0.03 (average). This suggests that the ratios of the T_2 relaxations and the rates of syneresis are similar.

The 12HSA matrix is comprised of capillaries where the solvent interacts with the surface of a fiber. Thus the force entraining the liquid oil will be a function of the radius of the pore (r) and the contact angle (α). The Laplace pressure (P_c) that arises from these interactions is expressed as:

$$p_c = \frac{2T_s \cos \alpha}{r}$$

For flow to occur, the capillary pressure must be overcome. Poiseuille's equation describes the pressure required to cause a material to flow with viscosity η though a capillary of radius r, length l, and rate υ as:

$$P_f = \frac{\upsilon 8 \eta l}{r^2}$$

For syneresis to occur in a capillary, a minimum value of P_f must be equal to or exceed P_c, and the length will be replaced with a variable diffusion distance χ_0 (Vesely, 2001). If P_c is equal to P_f, then the minimum force required for flow to occur is overcome and the following expression for the flow velocity can be obtained:

$$\upsilon = \frac{T_s r \cos\alpha}{4\eta\chi_0}$$

It is assumed that the interfacial tension, contact angle, and diffusion distance remains constant for samples stored at 30°C and 5°C since they are essentially the same system. Therefore, to determine the relative differences between the rates of syneresis our expression simplifies to:

$$\frac{\upsilon 30°C}{\upsilon 5°C} = \frac{r30°C}{r5°C}\frac{\eta 5°C}{\eta 30°C}$$

The complex viscosity was determined using small deformation rheology, while the pore radius was determined by microscopy at different magnifications (Fig. 5.9, Table 5.B). Using the pore radius and complex viscosity (17 kPa at 30°C and 8kPa at 5°C), a relative change in the flow rate may then be determined. The ratio was found to be 1.35±0.14. Analysis of variance was used to compare the ratio of the flow rate determined using radius and viscosity to the T_2 relaxations and the oil mobility. It was found that there was no significant difference in the ratio of mobility regardless of the

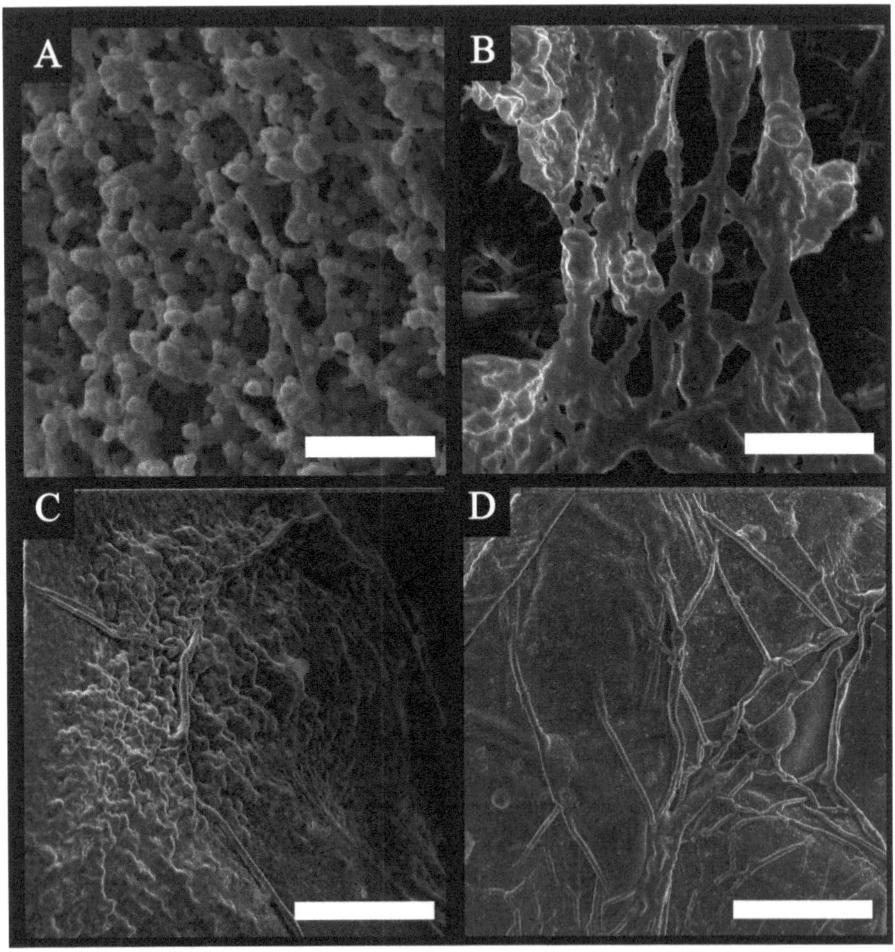

Fig. 5.9. Cryogenic scanning electron micrographs of 2% 12HSA/canola oil organogels stored for 24 hours. After storing for 24 hours at 5°C (A,C) and 30°C (B,D) the samples were treated with 80/20 (v:v) hexane/acetone. Error bars for A,B = 30μm and C,D=3μm (Source: Rogers et al., 2009).

Table 5.B. The Length of 12HSA Strands and Size of Pores in Canola Oil were Determined Using Different Magnifications of the Xerogel and the Osmium Tetroxide/Isobutanol Treated Samples (*Source:* Rogers et al., 2009).

	5°C			30°C		
	300X	1000X	10 000X	300X	1000X	10 000X
Strand Length (μm)	13.4±7.5	8.0±2.6	1.0±0.3	19.2±14.7	12.2±4.3	2.8±1.2
Pore Diameter (μm)	9.9±2.0	9.4±2.8	1.7±0.7	26.9±20.0	29.0±15.0	4.4±2.0

method used. This confirms that changes in oil syneresis may be directly measured using both the T_2 relaxation times as well as the ratio of the viscosity to pore radius.

Applications of 12-Hydroxystearic Acid

The first study addressing the thickening action of 12HSA occurred in 1972 when it was added to peanut butter (Elliger et al., 1972). In this study, low levels (0.5 to 1.0%) of 12HSA were added to peanut butter in an attempt to prevent the syneresis of oil. Industrially it is used to increase the recovery of oil from oil wells; trihydroxystearin is also used as an industrial lubricant. This investigation found that other hydroxystearates (9, 10-dihydroxystearic acid and 12-ketostearic acid) did not form gel-like structures; these molecules crystallized and settled out (Elliger et al., 1972).

Recently, 12HSA has been investigated as a delivery system for pharmaceuticals and nutraceuticals (Hughes, et al., 2009). The rate of release of biologically active compounds from food matrices and their bioavailablity are crucial in the development of functional foods. For functional foods, it is often desirable for the bioactive to have limited solubility and not be released immediately after consumption (Felt et al., 1998). Instead, it is desirable to control the rate of release and hence the rate at which it enters the bloodstream. 12HSA organogels have been shown to slow the rate of beta carotene micellarisation using an in-vivo model for digestion (Hughes et al., 2009). The rate of micellellarisation was at its maximum from liquid oil after 30 min, however, using a 12HSA organogel the process took 60 min to reach the maximum release rate (Hughes et al., 2009).

Oral administration of ibuprofen in organogels in vivo also showed significant encapsulation (Iwanaga et al., 2010). Ibuprofen in an aqueous suspension or soybean oil resulted in rapid increase in the ibuprofen in the blood serum (Fig. 5.10) (Iwanaga et al., 2010). However, when the ibuprofen was administered in a 12HSA organogel, the absorption rate decreased considerably while the bioavailibity increased significantly (Iwanaga et al., 2010). More complex delivery systems are being developed where a colloidal dispersion of gelled oil nanoparticles have been made using 12HSA as the gelling compound (Kirilov et al., 2009).

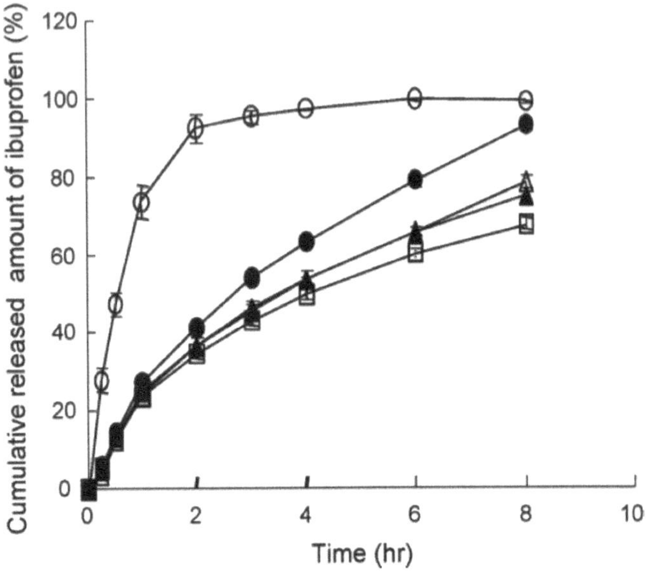

Fig. 5.10. Release profiles of ibuprofen from organogel including (○) 0%, (•) 2%, (△) 3%, (▲) 5%, and (□) 10% of 12-HSA in simulated intestinal fluid. Each value represents the mean±SE of three experiments (Taken from Iwanage et al., 2010).

References

Baba, M.; Gardette, J.-I.; Lacoste, J. Crosslinking on ageing of elastomers II. Comparison of solvent freezing point depression and conventional crosslinking evaluation. Polym. Degrad. Stab. 1999, 65, 415.

Dumas, P.; Jamin, N.; Teillaud, J.L.; Miller, L.M.; Beccard, B. Imaging capabilities of synchrotron infrared microspectroscopy. Faraday Discuss. 2004, 126, 289–302.

Elliger, C.A.; Guadagni, D.G; Dunlap, C.E. Thickening action of hydroxystearates in peanut butter. J. Am. Oil Chem. Soc. 1972, 49, 536–537.

Fuhrhop, J.-H.; Schnieder, P.; Rosenberg, J.; Boekema, E. The Chiral Bilayer Effect Stabilizes Micellar Fibers. J. Am. Chem. Soc. 1987, 109, 3387–3390.

Giorgano, J.; Daleo, C.; Stacks, S.M.; Topical ondansetron attenuates nociceptive and inflammatory effects of intradermal capsaicin in humans. Eur. J. Pharmacol. 1998, 354, R13–141.

Hughes, N.E.; Marangoni, A.G.; Wright, A.J.; Rogers, M.A.; Rush. J.W. Potential Food Applications of Edible Oil Organogels. Trends in Food Sci. 2009, 20, 470–480.

Iwanaga, K.; Sumizawa, T.; Miyazaki, M.; Kakemi, M. Characterization of organogel as a novel oral controlled release formulation for lipophilic compounds. Int. J. Pharm. 2010, 3888, 123–128.

Kantaria, S.; Rees, G.D.; Lawrence, M.J. Gelatin-Stabilised microemulsion-based organogels.: Rheology and application in iontophoretic transdermal drug delivery J. Control. Rel. 1999, 60, 355–365.

Kirilov, P.; Gauffre, F.; Franceschi-Messant, S.; Perez, E.; Rico-Lattes, I. Rheological Characterization of a New Type of Colloidal Dispersion Based on Nanoparticles of Gelled Oil. J. Phys. Chem. B. 2009, 113, 11101–11108.

Kuwahara, T.; Nagase, H.; Endo, T.; Ueda, H.; Nakagaki, M. Crystal Structure of DL-12-Hydroxystearic acid. Chem. Lett. 1996, 25, 435–436.

Lam, R.; Pederson, T.; Quaroni, L.; Rogers, M.A. A Molecular Insight Into The Nature of Crystallographic Mismatches in Self-Assembled Fibrillar Networks Under Non-Isothermal Crystallization Conditions. Soft Matter 2010, 6, 404–408.

Lam, R.; Rogers, M.A. Experimental Validation of the Modified Avrami Model for Non-Isothermal Crystallization Condition. CrystEngComm. 2011. DOI: 10.1039/C0CE00532C.

Liu, X.Y.; Sawant, P.D. Mechanism of the Formation of Self-Organized Microstructures in Soft Functional Materials. Ad. Mater. 2002, 14, 421–426.

Liu, X.Y.; Sawant, P.D. Formation Kinetics of Fractal Nanofiber Networks in Organogels. App. Phys. Lett. 2001, 79, 3518–5320.

Marangoni, A.G.; McGauley, S.E. Relationship between Crystallization Behavior and Structure in Cocoa Butter. Cryst. Grow. Des. 2003, 3, 95–102.

Markovic, N.; Ginic-Markovic, M.; Dutta, M.K. Benzene physical and chemical organogels: Effect of network scaffolding on the thermodynamic behavior of entrapped solvent molecules. J. App. Polym. Sci. 2004, 94, 1253.

Maskaev, A.K.; Man'Kovskaya, N.K.; Lend'el I.V.; Fedorvskii, V.T.; Simurova E.I.; Terent'eva, V.N.; Preparation of 12-hydroxystearic acid, the raw material for plastic greases. Chem. Techn. Fuels Oil. 1971, 7, 109–112.

Provencher, S.W. A Constrained Regularization Method for Inverting Data Represented by Linear Algebraic or Integral Equations. Comp. Phys. Comm. 1982a, 27, 213–217.

Provencher, S.W.; Contin: A General Purpose Constrained Regularization Program for Inverting Noisy Linear Algebraic and Integral Equations. Comp. Phys. Comm. 1982b, 27, 229–242.

Rogers, M.A.; Pederson, T.; Quaroni, L. Hydrogen Bonding Density of Supramolecular Structures for Self-Assembled Fibrillar Networks Probed Using Synchrotron Infrared Spectromicroscopy. Cryst. Grow. Des. 2009, 9, 3621–3625.

Rogers, M.A.; Marangoni, A.G. Non-isothermal nucleation and crystallization of 12HSA in vegetable oil. Cryst. Grow. Des. 2008, 8, 4596–4601.

Rogers, M.A.; Marangoni, A.G. Solvent modulated nucleation and crystallization kinetics of 12-hydroxystearic acid: a non-isothermal approach. Langmuir 2009, 25, 5886–5896.

Rogers, M.A.; Smith, A.S.; Wright, A.J.; Marangoni, A.G. A Novel Imaging Technique for Vegetable Based Organogels. J. Am. Oil Chem. Soc. 2007, 84, 899–906.

Rogers, M.A.; Wright, A.J.; Marangoni, A.G. Nanostructuring Fiber Morphology and Solvent Inclusions for 12-Hydroxystearic Acid/Canola Oil Organogels. Curr. Opin. Coll. Inter. Sci. 2009, 14, 33–42.

Strange, J.H.; Rahman, M.; Smith, E.G. Characterization of porous solids by NMR. Phys. Rev. Lett. 1993, 71, 3589–3591.

Suzuki, M.; Nakajima, Y.; Yumoto, M.; Kimura, M.; Shirai, H.; Hanabusa, K. Effects of Hydrogen Bonding and van der Waals Interactions on Organogelation Using Designed Low-

Molecular-Weight Gelators and Gel Formation at Room Temperature. Langmuir 2003, 19, 8622–8624.

Tachibana, T.; Kambara, H. The Sense of Twist in the Fibrous Aggregates from 12-Hydroxystearic Acid and its Alkali Metal Soaps. J. Colloid In. Sci. 1968, 28, 173–174.

Tamura, T.; Ichikawa, M. Effect of Lecithin on Organogel Formation of 12-Hydroxystearic Acid. J. Am. Oil Chem. Soc. 1997, 74, 491–495.

Terech, P.; Rodigues, V.; Barns, J.D.; McKenna, G.B. Organogels and Aerogels of Racemic and Chiral 12-Hydroxyoctadecanoic Acid. Langmuir 1994, 10, 3406–3418.

Weiss, R.G.; Terech, P. Introduction. Molecular Gels Materials with Self-Assembled Fibrillar Networks. Weiss, R.G., Terech, P., Eds.; Springer: Dordrecht, Netherlands, 2007; 1–13.

Uzu, Y.; Sugiura, T. Electron microscopic and Thermal Studies of Optically Active 12-Hydroxystearic Acids in Soap Formation. J. Colloid Int. Sci. 1975, 51, 346–349.

Vesely, D.; Molecular Sorption Mechanism of Solvent Diffusion in Polymers. Polymer. 2001, 42, 4417.

Wang, R.; Lui, X.Y.; Xiong, J.; Li, J. Real-Time Observation of Fiber Network Formation in Molecular Organogel: Supersaturation-Dependent Microstructure and Its Related Rheological Property J. Phys. Chem. B. 2006, 110, 7275.

6

Candelilla Wax as an Organogelator for Vegetable Oils—An Alternative to Develop *Trans*-free Products for the Food Industry

Jorge F. Toro-Vazquez, Míriam A. Charó-Alonso, J. David Pérez-Martínez, and Juan A. Morales-Rueda

Facultad de Ciencias Químicas, Centro de Investigación y Estudios de Posgrado, Universidad Autónoma de San Luis Potosí, México

Introduction

Vegetable oils, the most important edible lipids worldwide, lack the required functional properties to fit consumer demands for texture and oxidation stability in food products. To achieve these functional properties the food industry relied for long time mainly on animal fats, and in lower proportion, on vegetable oils blended with high melting temperature triacylglyceride (TAGS) fractions (i.e., stearins) or vegetables oils with native high concentration of saturated TAGS (i.e., coconut oil, palm oil). Unfortunately, clinical studies established that consumption of animal fats or vegetable oils high in saturated fatty acids (i.e., palmitic and lauric acid), is undesirable because of their cholesterolemic effect on humans with the subsequent increase in the risk for cardiovascular disease (Ascherio et al., 1999b).

After the discovery of liquid phase hydrogenation and further patent issued to Normann in 1902 (Allen, 1960), its use in the production of fully and partially hydrogenated vegetable oil had a dramatic economic impact on the use of vegetable oils by the food industry. When the consumption of saturated fat was determined to be primarily responsible for the increased risk in cardiovascular disease, the use of partially hydrogenated vegetable oils became the "healthy" alternative for the production of shortenings, margarines, and confectionary coatings.

However, the use of partially hydrogenated vegetable oils results in the consumption of *trans* fatty acids, which once ingested have detrimental effects on cell membrane integrity, the production of biologically active metabolites derived from essential fatty acids (Mensink, 2005; Stender & Dyerberg, 2004), the high density to low density lipoprotein ratio (Ascherio et al., 1999a, 1999b), and the high consumption of *trans* fatty acids is a rate limiting factor for the Δ6-desaturase that converts the α-linoleic fatty acid into γ-linoleic (Manku et al., 1982) in the metabolic pathway of the n-6 series. Of upmost importance is that diets high in *trans* fat have been linked to the increased incidence of the metabolic syndrome (Mittal, 2008). Additionally, the ratio of low density to high-density lipoproteins as a function of the percentage of energy provided by *trans* or saturated fatty acids results in higher coronary heart disease risk derived from *trans* fatty acids consumption than the one obtained from saturated fatty acids (Ascherio et al., 1999b). In consequence, health agencies, such as the World Health Organization, the American Heart Association, and the Dietary Guidelines 2005, issued recommendations to limit *trans* fat intake. Within this framework, in March 2003 the Danish government issued Order number 160, restricting the presence of *trans* fats in food products to 2% or less. In the same way the FDA set January 1, 2006 as the date by which all marketed foods must include *trans* fat concentration in the Nutrition Facts panel. These events and others occurring in several countries worldwide are influencing manufacturers to reduce or even eliminate *trans* fatty acids from their products (Kodali & List, 2005).

However, replacing partially hydrogenated vegetable oils in food products has functional and technological limitations. The use of partially hydrogenated fats and oils provides higher crystallization temperatures followed by rapid crystallization rates in comparison with the ones observed by *trans*-free alternatives. Additionally, the crystallization process of fats with unsaturated *trans* TAGS results in the development of a three-dimensional crystal network with the viscoelastic properties required by the lipid-based food systems such as confectionary fats, shortenings, and margarines. In contrast, the crystallization of *trans*-free formulations results in products with limited plasticity, texture profile, and oxidation stability.

The industry uses different approaches to reduce or eliminate the use of *trans* fats in food production. Some of them include the use of vegetable oils with modified TAGS composition through conventional plant breeding or genetic engineering, tailored to provide specific functional properties to food systems. Others alternatives include the use of oils after interesterification with high melting temperature TAGS fractions to modify the native TAGS fatty acid composition and distribution, improving the solid fat content (SFC) and plasticity (Petraskaite et al., 1998a; 1998b). Another common alternative to avoid the use of *trans* fats involves the use of fractions of TAGS with particular crystallization and melting profiles, obtained from vegetable oils by fractional crystallization (Reddy & Jeyarani, 2001; Berger & Idris, 2005). Furthermore, new hydrogenation methods have been investigated to either reduce *trans* fatty acids' production (Macher & Holmquivst, 2001; Mondal & Lalvani, 2003) or to promote the production of conjugated fatty acids with recognized health benefits (Yang et al., 2005). Unfortunately, none of these approaches easily results in oils and/or fats with the physical and chemical properties required to manufacture food products where TAGS are the main functional components. In other cases, the use of these alternatives results in higher costs.

More novel options were reviewed by Pernetti et al. (2007a), and Rogers (2009). Some of these alternatives take advantage of the capacity of low molecular weight compounds (< 3000 Daltons) to develop self-assembly molecular organizations providing structure to vegetable oils. Thus, successful application systems for the food and pharmaceutical industry have been obtained with liquid crystalline phases developed by monoglycerides as a function of temperature and the water to apolar phase (i.e., oil) ratio in the solvent (Marangoni et al., 2008; Sagalowicz et al., 2006). Organogelation, using low molecular mass organic gelators (LMOG) that result in gelation of vegetable oils with a concentration lower than 2–3%, is another promising alternative to modify the physical properties of vegetable oils without the use of oil/fats with *trans*-fatty acids.

Organogels are bi-continuous colloidal systems that coexist as a micro-heterogeneous solid (i.e., gelator) that immobilizes an organic liquid phase (Abdallah et al., 1999) that results in a system with viscoelastic and thermo reversible properties. LMOGs include lipophylic [i.e., cholesteryl anthraquinone derivatives (Ostuni et al., 1996), sterols (Bot & Agterof, 2006), *n*–alkanes (Abdallah et al., 1999; Abdallah & Weiss, 2000)] and ambiphatic [i.e., lecithin (Kumar & Katare, 2005), sorbitan monostearate (Murdan, et al., 2000), fatty acids (Wright & Marangoni, 2007)] molecules that, through spontaneous self-assembly, develop three-dimensional networks of randomly entangled fiber-like, tube-like, or plate-like structures. This three-dimensional network, stabilized mainly through non-specific short-range forces (i.e., London dispersion forces and van der Waals forces), holds micro-domains of the liquid in a non-flowing state mainly through surface tension (Abdallah & Weiss, 2000).

Candelilla wax (CW) is a wax obtained from the leaves of a small shrub native to northern Mexico and the southwestern United States, *Euphorbia cerifera* and *Euphorbia antisyphilitica,* from the family Euphorbiaceae. CW is a globally recognized food additive approved by the FDA (under regulations 21CFR, 175.105, 175.320, 176.180), used mainly as a glazing agent and binder for chewing gums. It is also used in the manufacture of lip balms and lotion bars, and in the paint industry to make varnishes. Additionally, CW can be used as a substitute for carnauba wax and beeswax in different food systems. Reports on the general CW composition indicates the presence of 49–50% *n*-alkanes with 29 to 33 carbons, 20–29% esters of acids and alcohols with even-numbered carbon chains (C28 to C34), 12–14% alcohols and sterols, and 7–9% free acids (Grant, 2005; Instituto de la Candelilla, 2004; LipidBank, 2007). A typical composition of a micronized high purity CW used as additive by the food industry shows hentriacontane as the main component (78.9%; Table 6.A; Toro-Vazquez et al., 2007).

Hentriacontane is a *n*-alkane with the molecular formula $CH_3(CH_2)_{29}CH_3$, with a molecular weight of 436.85 Da, and a reported melting temperature (99.5% purity) of 67.05°C with a heat of melting (ΔH_M) of 73.3 kJ/mol (INFOTHERM, 2006). The self-assembly capability of *n*-alkanes has been established in several apolar solvents, observing in all cases the development of microcrystalline structures with a plate-like shape (Abdallah et al., 1999; 2000; Abdallah & Weiss, 2000). These microplatelet units are the primary structures that form the three-dimensional crystal network of *n*-alkanes organogels (Abdallah et al., 2000). Since CW has as its major component hentriacontane and as minor components other *n*-alkanes (i.e., nonacosane and tritriacontane) with odd number of carbons, this creates the possibility of using CW to develop edible

Table 6.A. GC-mass Spectroscopy Composition of a Candelilla Wax Additive Used for the Food Industry. ***Souce:*** **Toro-Vazquez et al., 2007.**

Name	Condensed Formula	%[a]
Nonacosane	$C_{29}H_{60}$	4.16 ± 0.11
Hentriacontane	$C_{31}H_{64}$	78.86 ± 0.11
Tritriacontane	$C_{33}H_{68}$	8.01 ± 0.15
Alcohols of penta-cyclic triterpenoids [b]	$C_{30}H_{49}OH$	7.38 ± 0.08
Unidentified	—	1.60 ± 0.07

[a] Mean and standard deviation of two determinations (n = 2). Values reported as a percentage of the total area.

[b] Any or all of the following isomers: germanicol, lupeol, or moretenol.

organogels using vegetable oils as the liquid apolar phase. In this chapter, we discuss information generated mainly in our laboratory, associated with the development of CW organogels using vegetable oils as the liquid apolar phase. In the same way, we discuss the thermal, rheological, and structural characteristics of CW organogels and organogels formed by mixtures of tripalmitin (TP) and CW developed under static conditions or with the application of a controlled shear rate at particular stages during cooling. The interaction between TP and the hentriacontane present in CW form mixed self-assembled structures that result in organogels with thermo-mechanical properties different from the ones observed by the organogels developed with the separated systems (Toro-Vazquez et al., 2009). We also discuss results associated to the few investigations regarding the use of CW as coating barrier in food products, either as a crystalline dispersion in water or dispersed in a bio-polymer matrix.

Thermal Properties of Candelilla Wax and Candelilla Wax Solutions in Vegetable Oils

The thermograms of CW with the composition shown in Table 6.A are simple (Morales-Rueda et al., 2009a; Toro-Vazquez et al., 2007). Thus, the CW crystallization thermogram shows just one major exotherm, beginning at 76.58°C (± 0.68°C) (T_g), two temperature peaks at ≈59°C and 53°C, and a heat of crystallization (ΔH_g) of 147.35 J/g (± 1.91 J/g). Both, T_g and ΔH_g describe the crystallization process of the *n*-alkanes present in CW, mainly hentriacontane. In contrast, the melting temperature (T_M) and particularly ΔH_M, are parameters directly associated with the structural order of the gelator molecules within the assembly units (i.e., micro-platelet crystals) that form the three-dimensional crystal structure of the organogel. The corresponding heating thermogram shows just one endotherm, with a T_M at 64.42°C (± 0.23°C) and ΔH_M of 149.75 J/g (± 1.20 J/g) (Fig. 6.1). Although T_M of CW is close to the melting temperature reported for 99.5% pure hentriacontane (i.e., 67.05°C), the ΔH_M calculated per unit mass of hentriacontane present in the

CW (82.9 kJ/mol) is greater than the corresponding value reported for 99.5% hentriacontane (i.e., 73.3 kJ/mole) (INFOTHERM, 2006). The differences observed in T_M and ΔH_M with the values reported for pure hentriacontane were associated with the development of a mixed molecular packing between the hentriacontane and the other *n*-alkanes present in minor concentrations (i.e., nonacosane and tritriacontane; Table 6.A). The formation of these mixed molecular packaging among *n*-alkanes of close hydrocarbon chain length and their orthorhombic crystal structure was established by Chevallier et al., (1999a; 1999b) through X-ray diffraction and differential thermal analysis. Based on these results the phase transitions observed in CW additives are associated mainly with the phase behavior of hentriacontane in mixed structures with nonacosane and tritriacontane. (Morales-Rueda et al., 2009a; Toro-Vazquez et al., 2007).

Organogelation of LMOG in apolar solvents is, in most cases, easily achieved by heating the gelator solution in a non-polar solvent until achieving full solubility of the gelator and then cooling the system below the solubility temperature of the LMOG in the solvent. This solubility temperature is T_g, which is mostly determined by DSC (Fig. 6.1). Thus, we evaluated the capacity of CW to provide structure to vegetable oils by heating (20 to 30 min at 90°C) and then cooling solutions of CW (0.5% to 6%) in high oleic safflower oil (HOSFO) to a particular T_{set}, where T_{set} is below the corresponding T_g. The visual evaluation of these CW solutions, after 24 h of storage at the T_{set} of 5°C or 25°C, showed the development of well-structured gels at CW concentrations ≥ 2%. However, at CW concentrations lower than 2% CW a T_{set} of 5°C is required to achieve gelation (Toro-Vazquez et al., 2007). These results showed that the minimum CW concentration required to develop a three-dimensional crystal network that physically entraps the vegetable oil (i.e., HOSFO) is temperature dependent.

We performed a more specific analysis by DSC to describe the variation of T_g, T_M, ΔH_g, and ΔH_M as a function of CW concentration in HOSFO using two different cooling rates, 1°C/min and 10°C/min (Toro-Vazquez et al., 2007; Morales-Rueda et al., 2009a). The thermograms for CW solution in HOSFO showed two peaks, where the peak below 0°C in both the cooling and heating thermograms was associated with the crystallization and melting of TAGS from HOSFO. The other peak, always present above 0°C in both the cooling and heating thermograms, was associated to the crystallization (Fig. 6.2) and melting (see the insert of Fig. 6.2) of CW components, mainly hentriacontane.

In those studies T_g and T_M showed a steady logarithmic decrease as CW concentration decreased in the HOSFO. In contrast, ΔH_g, and ΔH_M showed a linear direct relationship with CW concentration, with some hysteresis as the CW concentration increased in the system and more mass of hentriacontane crystals was developed (Fig. 6.3.A) (Toro-Vazquez et al., 2007; Morales-Rueda et al., 2009a). This hysteresis phenomenon was associated with the inclusion of the gelator's heat of dissolution within the ΔH_M value (Abdallah et al., 1999; Toro-Vazquez et al., 2007). The thermal parameters of organogels developed by several *n*-alkanes in different organic solvents (Abdallah et al., 1999; Abdallah & Weiss, 2000) and dotriancontane

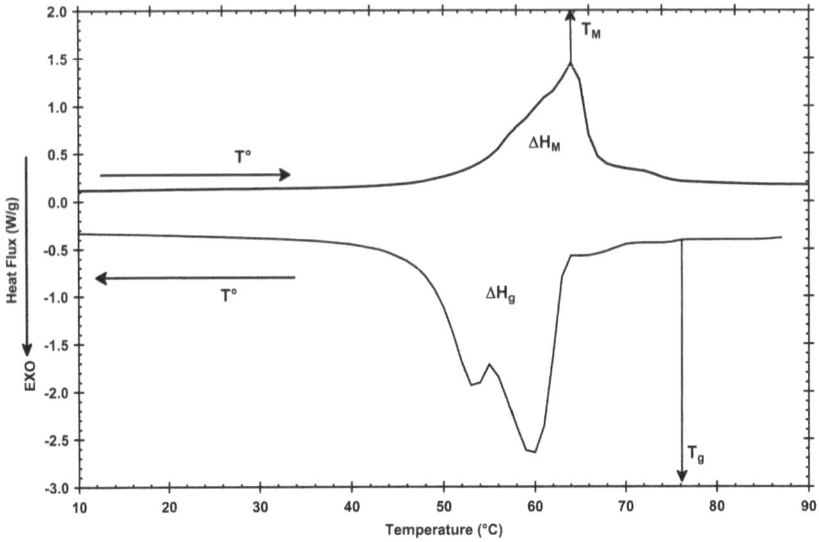

Fig. 6.1. Cooling and heating thermograms for a micronized high purity CW commercially used as additive for the food industry (composition described in Table 6.A) (modified from Toro-Vazquez et al., 2007). T_g indicates the beginning of crystallization and T_M the melting temperature of CW.

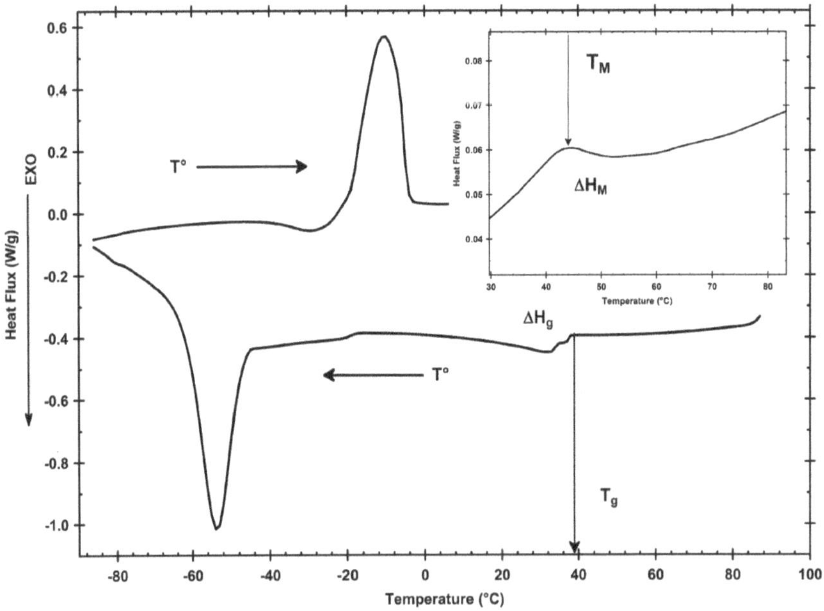

Fig. 6.2. Cooling and heating thermograms for 3% CW solution in HOSFO. The peak present below 0°C in both the cooling and heating thermograms is associated with the crystallization and melting of TAGS from HOSFO. The insert is the enlarged section between 30°C and 80°C for the heating thermogram, showing the endotherm for CW components. ΔH_g and ΔH_M is the heat of crystallization and melting, respectively (taken from Toro-Vazquez et al., 2007).

($C_{32}H_{66}$) in HOSFO (Figs. 6.3.B and 6.3.C; Morales-Rueda et al., 2009a) demonstrate similar behavior. Additionally, we established the thermoreversible behavior of CW organogels in HOSFO through heating-cooling cycles from 90°C to a particular T_{set} between –5°C and 30°C. Thus, for a given cooling rate (1°C/min or 10°C/min), the thermal parameters that describe organogel formation (i.e., T_g and ΔH_g) and melting (i.e., T_M, and ΔH_M) showed independence of both the number of thermal cycles applied and the T_{set} used (i.e., CW organogels have a thermoreversible behavior) (Toro-Vazquez et al., 2007).

We noted that the behavior of T_g as a function of gelator concentration was similar at 1°C/min or 10°C/min in dotriacontane and CW solutions in HOSFO (Morales-Rueda et al., 2009a). This occurred regardless of dotriacontane being in pure form and CW being a mixture of *n*-alkanes. This study showed that using a cooling rate of 1°C/min produces organogels with higher T_g than using 10°C/min (Fig.

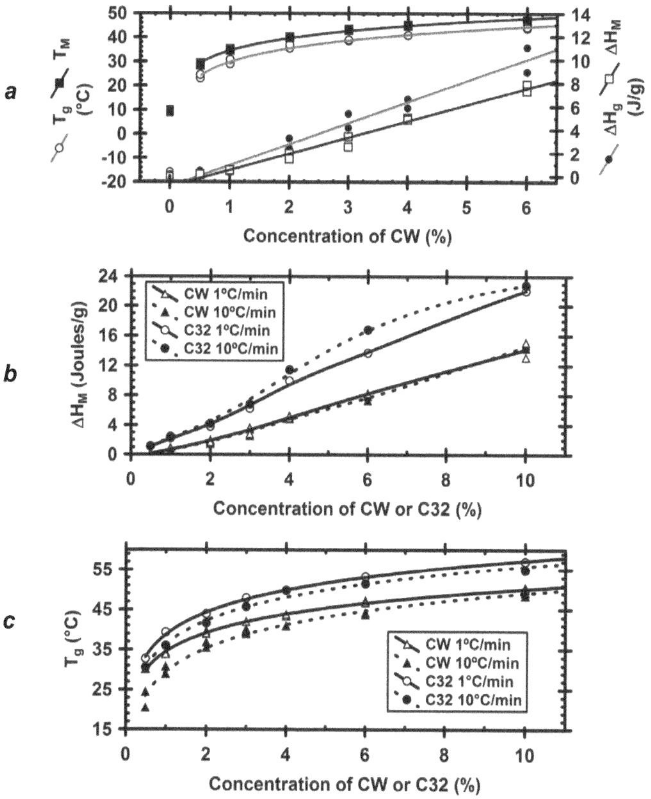

Fig. 6.3. Behavior of the thermal parameters that characterize the crystallization (T_g, ΔH_g) and melting (T_M and ΔH_M) of CW (a; taken form Toro-Vazquez et al., 2007) and dotriacontane (C32) (*b* and *c*; taken form Morales-Rueda et al., 2009a) components in HOSFO. In *a*, the T_g, T_M, ΔH_g and ΔH_M for the 0% CW concentration correspond to the thermal parameters that characterize the crystallization and melting of TAGS from the HOSFO.

6.3.C), a phenomenon associated with the time the gelator molecules have to achieve the required molecular packing for organogel formation during cooling (Toro-Vazquez et al., 2007; Morales-Rueda et al., 2009a). The cooling rate also has important implications in the micro-structural organization of the gel and, therefore, in the melting and rheology properties of the CW organogels (Morales-Rueda et al., 2009b).

Cooling Rate and Storing Temperature Effect

In CW-HOSFO solutions cooled at 1°C/min, the solubility temperature of the gelator (i.e., temperature of crystallization T_g) and the corresponding ΔH_g of the organogel are higher than at 10°C/min (Fig. 6.4). This phenomenon is explained by considering that during cooling, the thermodynamic driving force for gel formation changes faster as cooling rate increased. Then, at a high cooling rate (i.e., 10°C/min) gelator molecules, such as the hentriacontane present in CW, have less time to organize in a solid phase than at 1°C/min. As a result, gelator molecules require lower temperature (i.e., lower T_g) to achieve the molecular packing for crystal formation. Then, for organogels developed at a given T_{set}, where T_{set} is lower than T_g, those formed using a higher cooling rate (i.e., 10°C/min) have lower thermodynamic driving forces (i.e., $T_g - T_{set}$) for crystallization than organogels developed at a lower cooling rate (i.e., 1°C/min) (Toro-Vazquez et al., 2007; Morales-Rueda et al., 2009b). For CW organogels the cooling rate effect on T_g and its subsequent impact on supercooling is more evident as CW concentration decreased (Fig. 6.4.A). Thus, CW organogels developed at 10°C/min showed lower levels of molecular organization in their assembly units, and therefore lower ΔH_g than CW organogels developed at 1°C/min (Toro-Vazquez et al., 2007; Morales-Rueda et al., 2009a; 2009b).

From the above, the extent of supercooling (and supersaturation) and its change as a function of temperature are evidently cooling rate dependent. These two factors have a significant effect on the topological structure of the gel crystal network and, therefore, on the organogel's rheological properties (i.e., the rheology of the organogel is cooling rate dependent). Within this framework, Wang et al., (2006) showed that the proportion of transient (i.e., entanglement of fibers or plates) to permanent (i.e., branching of fibers or plates) junction zones in a three-dimensional crystal network

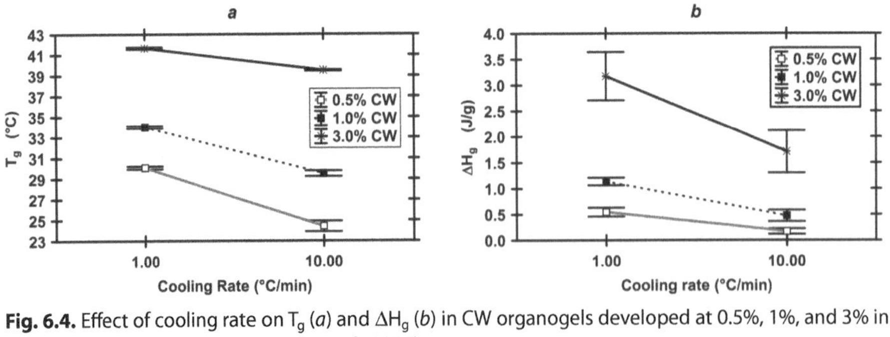

Fig. 6.4. Effect of cooling rate on T_g (*a*) and ΔH_g (*b*) in CW organogels developed at 0.5%, 1%, and 3% in HOSFO (modified from Toro-Vazquez et al., 2007).

is supersaturation/supercooling dependent. In the same way, Lam et al., (2010) showed that high cooling rates (i.e., higher than 5–7°C/min) produce crystal imperfections during 12 hydroxystearic acid crystallization that result in shorter fibers with branches (i.e., permanent junction zones) in contrast with the long non-branched fibers obtained at low cooling rates. These authors concluded that organogels with fibrillar network structures with a high extent of transient junction zones have higher elasticity than organogels with a high extent of permanent junction zones (i.e., spherulitic network structures). In the case of *n*-alkanes, organogels developed with dotriacontane (i.e., C32) at 1% and 3% (w/w) in HOSFO at the T_{set} of 5°C and 25°C showed that independent of the gelator concentration and T_{set}, higher elastic properties were obtained at 1°C/min than at 10°C/min (Morales-Rueda et al., 2009b). This was measured by the elastic modulus determined after 180 min at the corresponding T_{set} (G'_{180}) (Morales-Rueda et al., 2009b). The polarized light microphotographs of 3% C32 organogels showed that independent of T_{set}, at 1°C/min larger needle-like crystals with a higher extent of fiber bundling (Fig. 6.5.A) were obtained than at 10°C/min (Fig. 6.5.B). Fig. 6.6 shows the G'_{180} as a function of the corresponding solid fat content (SFC) for 1% and 3% C32-HOSFO (Fig. 6.6.A) and CW-HOSFO (Fig. 6.6.B) organogels developed at the T_{set} of 5°C and 25°C at a cooling rate of 1°C/min or 10°C/min. Except the 1% CW organogels at the T_{set} of 5°C and 25°C (i.e., SFC < 1.0%; Fig. 6.6.B), CW and C32 organogels developed at 1°C/min showed higher elasticity than the organogels developed at 10°C/min. This occurred despite

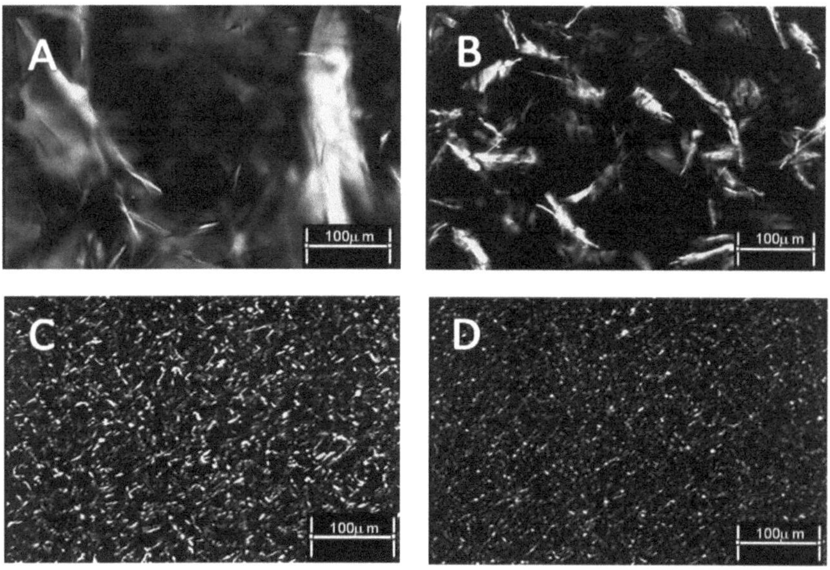

Fig. 6.5. Polarized light microphotographs of 3% C32 (*a* and b) and 3% CW (*c* and *d*) organogels in HOSFO developed at T_{set} of 25°C using a cooling rate of 1°C/min (*a* and *c*) and 10°C/min (*b* and *d*) (Adapted from Morales-Rueda et al., 2009b).

the SFC obtained at 1°C/min being similar or lower than the SFC present in the organogels developed at 10°C/min (Fig. 6.6). Unfortunately, with the microscope available, the size of the crystals was too small, particularly with the CW organogels (Figs. 6.5.C and 6.5.D), to determine if branching occurred in the crystals developed at the higher cooling rate (Morales-Rueda et al., 2009b). Although hentriacontane is the major component in CW, the presence of other *n*-alkanes (i.e., nonacosane and tritriacontane) resulted in the development of mixed self-assembled structures. This resulted in smaller crystals than the ones achieved by pure dotriacontane. Schaink et al., (2007) and Gandolfo et al., (2004) showed the strong influence that gelator composition has on crystal size, shape, and rheology of organogels developed in sunflower oil by different stearic acid to stearyl alcohol ratios. These authors developed organogels with smaller crystals and higher rheological profiles at a 3:7 stearic acid to stearyl alcohol ratio than with the pure compounds (Gandolfo et al., 2004; Schaink et al., 2007). These results show that the micro-topological structure of gel networks,

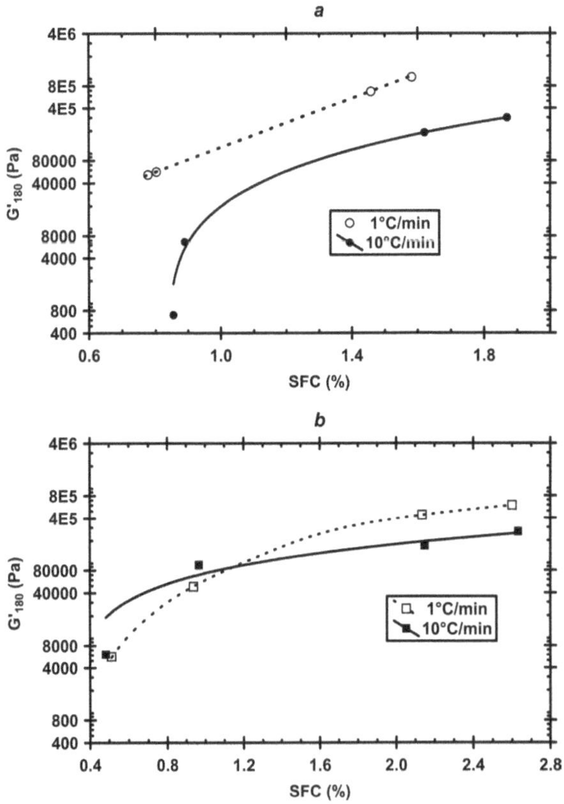

Fig. 6.6. G′ values after 180 minutes (G'_{180}) for 1% and 3% C32 (*a*) and CW (*b*) organogels in HOSFO plotted as a function of the solid fat content (SFC %). The organogels were developed at the T_{set} of 5°C and 25°C and achieved using a cooling rate of 1°C/min or 10°C/min (Adapted from Morales-Rueda et al., 2009b).

and therefore the rheological properties of gel materials, might be engineered through controlling time-temperature conditions and gelator composition.

On the other hand, CW-HOSFO organogels kept under isothermal conditions showed a steady increase in G′ and ΔH_M as a function of time, a process that indicated the development of higher structural order in the crystal network during storage (Figs. 6.7 and 6.8) (Toro-Vazquez et al., 2007; Morales-Rueda et al., 2009b).

After molecules attain a three-dimensional organization, an appropriate combination of time and temperature results in additional ordering of the three-dimensional molecular structure (i.e., annealing or perfection of the crystalline regions)

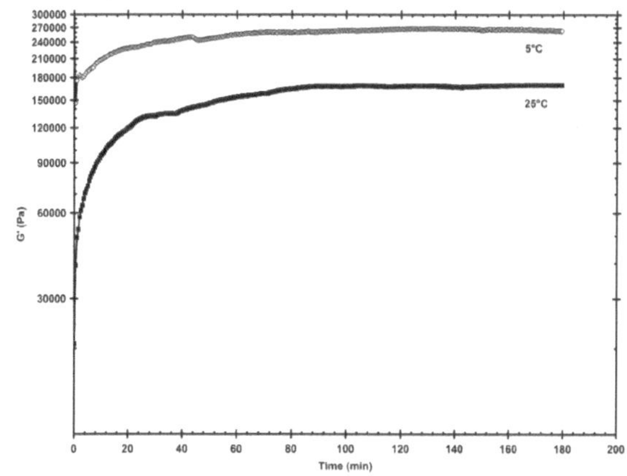

Fig. 6.7. Rheological profiles (G′ profiles) for 3% CW organogels. The G′ profiles were obtained under isothermal conditions at the T_{set} of 5°C and 25°C, achieved using a cooling rated of 10°C/min. The solid fat content at T_{set} of 5°C was 2.64 (± 0.14) and at T_{set} of 25°C was 2.15 (± 0.12) (modified from Morales-Rueda et al., 2009b).

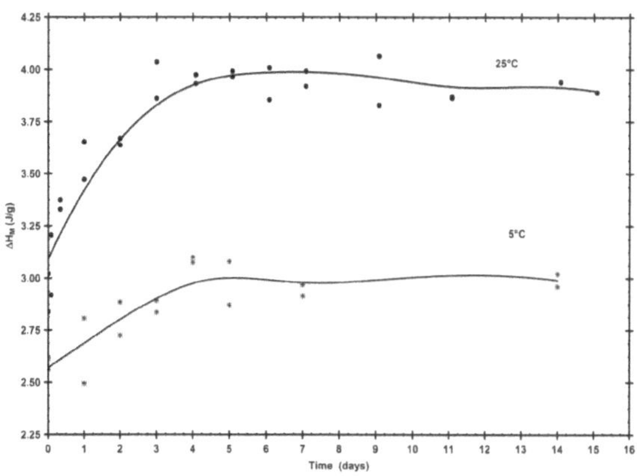

Fig. 6.8. Heat of melting (ΔH_M) as a function of storage time for 3% CW organogels developed and stored at the T_{set} of 5°C and 25°C. The solid fat content at T_{set} of 5°C was 2.63 (± 0.14) and at T_{set} of 25°C was 2.15 (± 0.12) (modified from Toro-Vazquez et al., 2007).

and/or ordering of amorphous regions. In polymers with crystalline structures such as starch, annealing is more effective at storing temperatures just below the melting temperature (i.e., as storing temperature approaches T_M), resulting in structures with higher crystallinity than non-annealed systems (Hoover & Vasanthan, 1994; Köhler et al., 2004). It seems that a molecular reorganization within the assembly units of the CW organogel structure (i.e., annealing) takes place during storage, particularly when the storage temperature is close to the melting temperature of the system. We proposed (Toro-Vazquez et al., 2007; Morales-Rueda et al., 2009b) that during storage at high temperature, the *n*-alkanes crystals evolve from a rotator phase into a higher state order as a function of time through a solid → solid transition. Rotator phases are commonly observed in *n*-alkanes and are characterized by a crystalline lattice of the molecular centers while molecules still rotate about their chain axes (i.e., structural disorder) (Espeau & White, 1997; Tozaki et al., 2003; Zgardzinska et al., 2006). The G′ and ΔH_M increments as a function of time observed in CW organogels could not be associated to changes in the solid content, mainly because the solid phase developed by the organogels right after attaining isothermal conditions remained constant during the whole period of rheological and calorimetric measurements (Toro-Vazquez et al., 2007; Morales-Rueda et al., 2009b). This annealing effect was more evident in 3% CW organogels than in 1% CW organogels, suggesting that a concentration effect is involved in the development of a better-structured three-dimensional crystal network (Morales-Rueda et al., 2009b). Rogers et al., (2008) have observed a similar annealing process in 12-hydroxystearic acid organogels. The annealing process in CW organogels also resulted in an increase in hardness, occurring after 1 to 2 days of storage at 5°C and 25°C (Toro-Vazquez et al., 2007).

From the above results we conclude that for a given CW concentration, the structural organization of CW organogels depends on the thermodynamic drive force for gelation at a given T_{set}. This is affected by the cooling rate and the annealing process occurring during organogels storage, particularly as storage temperature approaches the gelator's melting temperature.

Mixed Self-Assembled Structures Developed by Candelilla Wax and Tripalmitin

Several investigations had shown that particular combinations of LMOGs might result in the development of organogels formed by crystals with mixed self-assembled structures that outperform the physical properties of the organogels developed by the separated LMOGs. Examples of such gelator mixtures used to gel vegetable oils are β-sitosterol + γ-oryzanol (Bot & Agterof, 2006), stearic acid + stearyl alcohol (Schaink et al., 2007; Gandolfo et al., 2004), lecithin + sorbitan tristearate (Pernetti et al., 2007b), mixtures of ceramides derived from milk and eggs (Rogers, 2009), and mixtures of CW + tripalmitin (Toro-Vazquez et al., 2009; Chopin-Doroteo et al., 2011).

Vegetable oils contain different families of TAGS, some of them able to crystallize under similar temperature conditions as LMOGs, such as the *n*-alkanes. Thus, the interaction between TAGS and gelator molecules might result in the development of mixed self-assembled crystals that structure vegetable oils and provide particular physical and functional properties. Within this framework, Martini et al., (2008) showed that the addition of sunflower oil waxes modifies the crystallization of TAGS present in anhydrous milk fat. However, these authors did not measure any physical (i.e., rheological) or functional (i.e., textural) property in the mixed systems. In the case of CW, Dassanayake et al., (2009) and Chopin-Doroteo et al., (2011) established the orthorhombic (O⊥) subcell packing of CW with wide-angle diffraction peaks at 4.07–4.14 Å and 3.65–3.74 Å (Fig. 6.9.A). When a mixture of CW and tripalmitin (TP) in a proportion 3:1 was melted (90°C for 30 min), thoroughly mixed, and cooled (1°C/min) until achieving the T_{set} of 15°C, we developed co-crystals (i.e., crystals that contain two or more different molecules) with *d* values of 4.13 Å and 3.72 Å (Fig. 6.9.C), similar to those observed by the β′ polymorph of TP (Figs. 6.9b) (Chopin-Doroteo et al., 2011). CW-TP organogels developed in HOSFO at the T_{set} of 15°C, using the same 3:1 ratio of CW (3%) to TP (1%) showed the same O⊥ subcell packing (Fig. 6.9.D), with slightly smaller *d* values (i.e., 4.04–4.05 Å and 3.64–3.65 Å) than the *d* values observed in the CW:TP mixture in the bulk state (Fig. 6.9.C). The decrease in the *d* values might be associated with the presence of a large proportion

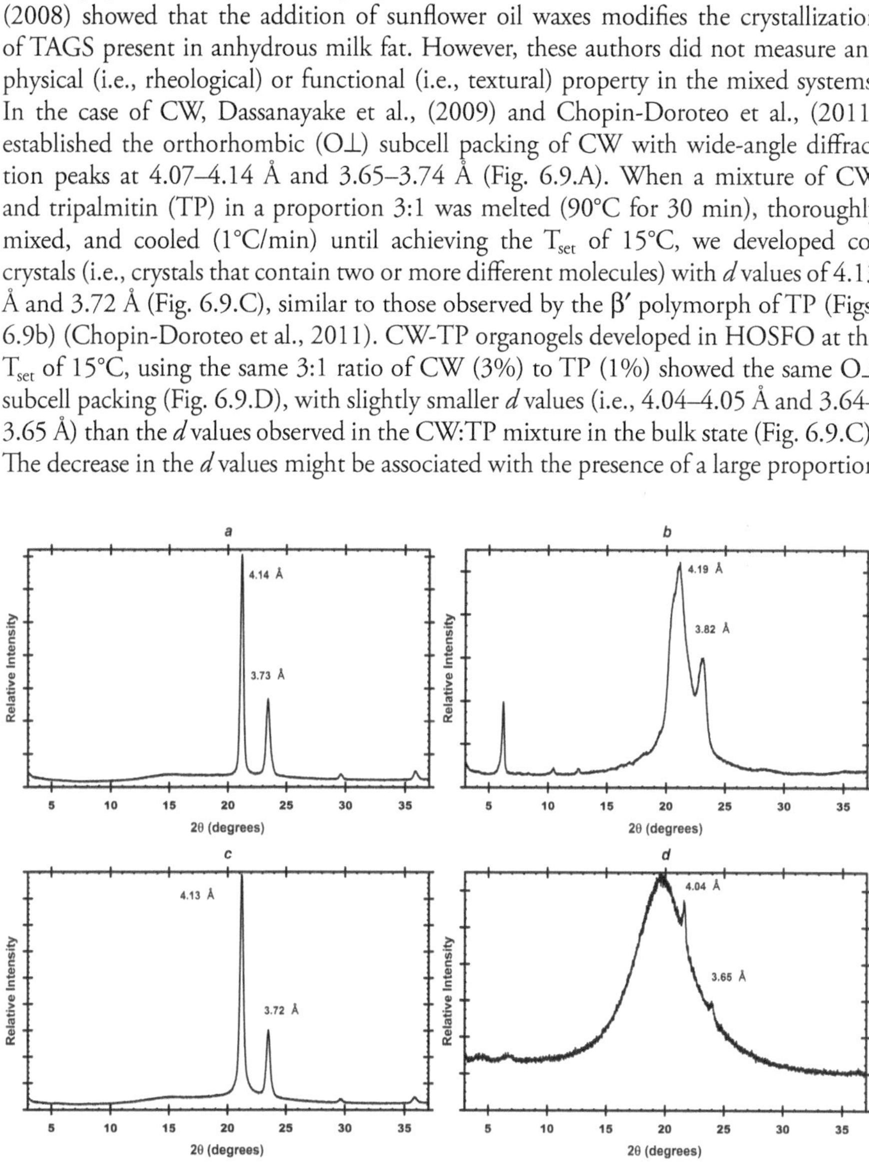

Fig. 6.9. X-ray diffractograms for candelilla wax (*a*), tripalmitin (*b*), a 3:1 mixture of CW and TP in the bulk state, and a 3% CW-1% TP organogel (*d*) (modified from Chopin-Doroteo et al., 2011).

of TAGS in the liquid state (i.e., amorphous phase) and its effect on the diffraction pattern of CW-TP crystals. Larsson (1972) pointed out that TAGS in the liquid state (i.e., from HOSFO) show wide angle X-ray scattering that peaks centered at 4.55 Å (≈ 19.5 2θ). Considering that only 3.69% of solids in the organogel were associated to CW and TP, TAGS in the liquid state represented the larger fraction (i.e., 96% – 97%). From here the presence of the amorphous diffraction zone observed in the diffractogram for the CW-TP organogel (Fig. 6.9.D) (Chopin-Doroteo et al., 2011).

Given the structural similarities between the palmitic acid and the aliphatic chains of the hentriacontane, main *n*-alkane present in CW, we proposed a tentative organization of hentriacontante and TP in the co-crystal (Fig. 6.10) (Chopin-Doroteo et al., 2011). This organization would result in minor alterations in the O⊥ subcell packing of hentriacontane in CW (Fig. 6.9.A) and the β′ polymorph of TP (Fig. 6.9.B) as was observed in the X-ray analysis of crystallized 3:1 mixture of CW and TP (Fig. 6.9.C) and the 3% CW-1% TP organogels (Fig. 6.9.D) (Chopin-Doroteo et al., 2011). Through short angle X-ray measurements we corroborated that in the 3% CW-1% TP organogel, CW and TP co-crystallized in a O⊥ subcell packing with TP in a 2L stacking organization (Chopin-Doroteo et al., 2011).

Additionally, our group established that the organogels resulting from co-crystallization of TP and CW in HOSFO show distinctive thermo-mechanical properties when compared with the organogels developed by the separated components, i.e., CW or TP (Toro-Vazquez et al., 2009; Chopin-Doroteo et al., 2011). Thus, when 0.5% and 1% TP solutions in HOSFO were cooled (10°C/min) from 90°C to a particular T_{set} between –10°C and 25°C, TP did not develop a three-dimensional crystal network that provided a significant elasticity (i.e., solid-like behavior) or structure vegetable oil (Toro-Vazquez et al., 2009). On the other hand, 0.5% to 3% CW solutions in HOSFO (i.e., no TP present) processed under the

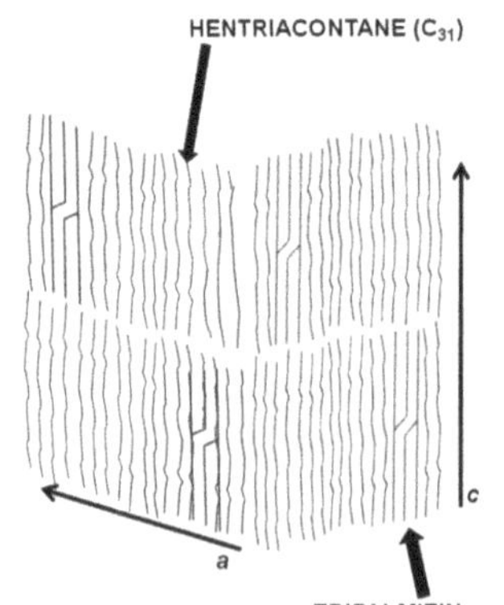

Fig. 6.10. Tentative organization of hentriacontane (C_{31}) and tripalmitin (1, 2, 3-Trihexadecanoylglycerol) in a co-crystal. The *a* and *c* axis of the subcell are shown for orientation purposes. Tripalmitin is included in the hentriacontane organization as a double chain length lamellar structure (2*L*) (from Chopin-Doroteo et al., 2011).

same time-temperature conditions showed, independent of T_{set} used, the typical T_M behavior previously described (Fig. 6.4.A), i.e., a steady logarithmic decrease from ≈42.5°C to ≈30.5°C as CW concentration decreased (Fig. 6.11.A). However, comparing the T_M of CW organogels developed with 0.5% and 1% TP shows that 3% CW-0.5% TP organogels had lower T_M, particularly at CW concentrations lower that 1.0%. In the presence of 1% TP, T_M remained constant independent of the CW concentration but T_M increased as the T_{set} to develop the organogel was raised (i.e., 35.81°C at T_{set} = –5°C, 36.2°C at T_{set} = 5°C, and 37.77°C at T_{set} = 15°C; Fig. 6.11.B). The T_M of the 0.5% to 3% CW organogels developed in the presence of 1% TP at all T_{set}'s investigated was lower than the T_M of 3% CW-0.5% TP organogels, but higher than the T_M of the 0.5% CW-0.5% TP, 0.75% CW-0.5% TP, and the 1% CW-0.5% TP organogels ($P < 0.05$; Fig. 6.11.B) (Toro-Vazquez et al., 2009). These results show the presence of an eutectic point in the 3% CW-1% TP mixture. The corresponding organogels might be useful in the development of

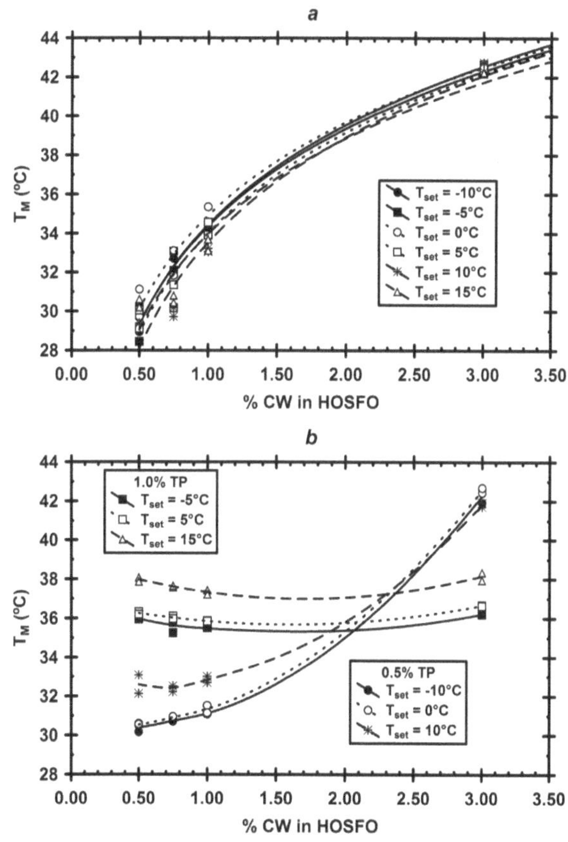

Fig. 6.11. Melting temperature (T_M) as a function of CW concentration for organogels developed at the T_{set}'s indicated with 0% TP (a) and with 0.5% and 1.0% TP (b) (taken from Toro-Vazquez et al., 2009).

products for the pharmaceutical and food industry where the need for a melting profile close to human body temperature is required.

Rheological Properties of CW and CW-TP Organogels

Organogels Developed Under Static Conditions

Fig. 6.12 shows the melting thermograms of the 3% CW-0.5% TP and 3% CW-1% TP organogels developed using a cooling rate of 10°C/min and T_{set}'s of 10°C (Fig. 6.12.A) and 15°C (Fig. 6.12.B) (Toro-Vazquez et al., 2009). The DSC analysis indicated that TP and CW crystallized concomitantly at these T_{set} values during organogel formation, developing co-crystals (Toro-Vazquez et al., 2009). In particular, the 3% CW-1% TP organogel developed at the T_{set} of 15°C using a cooling rate of 10°C/min (Fig. 6.12.B) showed a distinctive melting behavior (i.e., T_M and ΔH_M), which

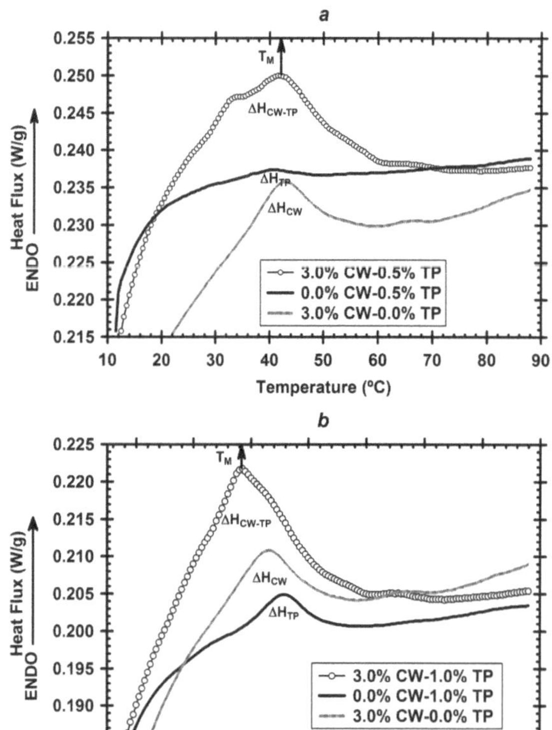

Fig. 6.12. Melting thermograms of the organogels developed at a given T_{set} with the mixed CW-TP system in HOSFO. *a*: 3% CW-0.5% TP organogel developed at T_{set} of 10°C. *b*: 3% CW-1% TP organogel developed at T_{set} of 15°C. The thermal behavior of control systems (i.e., 0% CW with 0.5% or 1% TP and 3% CW with 0% TP) are also included (modified from Toro-Vazquez et al., 2009).

could not be associated with the thermal behavior observed by the separated systems (i.e., 0% CW-1% TP and 3% CW-0% TP; Fig. 6.12.B) (Toro-Vazquez et al., 2009). Similar results have been obtained with 3% CW-1% TP organogels at the same T_{set} achieved using a cooling rate of 1°C/min (Chopin-Doroteo et al., 2011).

Regarding the mechanical properties of CW and CW-TP organogels using HOSFO as the liquid phase, Fig. 6.13 shows the rheograms for 3% CW-0% TP and 3% CW-1.0% TP organogels developed at the T_{set} of –5°C, 15°C, and 25°C using a cooling rate of 10°C/min to achieve the corresponding T_{set} (Toro-Vazquez et al., 2009). The rheograms were determined with a mechanical spectrometer (Paar Physica MCR 301, Stuttgart, Germany) using a cone-plate geometry (CP50-2/TG) equipped with a true-gap system. The use of a true-gap system is necessary when the rheology measurement program involves changing temperature conditions (i.e., high cooling rates) (Morales-Rueda et al., 2009b). This is because the true-gap system

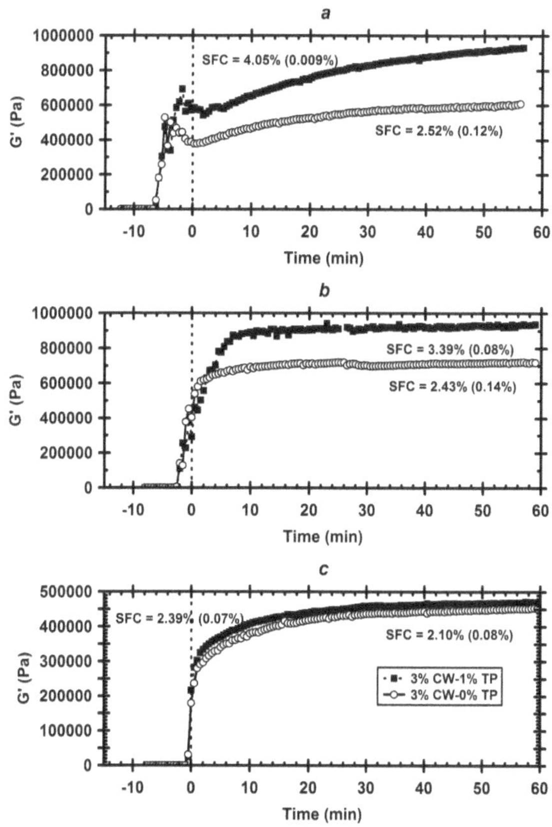

Fig. 6.13. G′ profiles as a function of time for organogels developed with 3% CW-1.0% TP and 3% CW-0% TP in HOSFO developed at the T_{set} of –5°C (*a*), 15°C (*b*), or 25°C (*c*). The dotted line indicates the beginning of the isothermal conditions. SFC is the solid fat content present in the organogel (modified from Toro-Vazquez et al., 2009).

automatically makes the corrections in gap size associated with the expansion or shrinkage of the sample and/or the rheometer geometry during measurements. The use of a fixed-gap system to measure rheological properties during changing temperature conditions results in measurements not necessary associated with structural properties of the sample (Morales-Rueda et al., 2009a; 2009b).

At all temperature conditions investigated, the 3% CW-1.0% TP organogels achieved higher G′ than the 3% CW-0% TP organogels, and this difference increased as T_{set} decreased. At the T_{set} investigated, the SFC was higher in the organogels developed in the presence of 1% TP than in the 3% CW-0% TP organogels, and this resulted in higher G′ in the 3% CW-1.0% TP organogels. The logarithmic increase in G′ observed before T_{set} was achieved (indicated as the dotted line in Fig. 6.13) shows the development of the three-dimensional structure by CW and CW-TP crystals, that resulted in organogelation of HOSFO (Toro-Vazquez et al., 2009). In this region, the G′ of the 3% CW-0% TP and 3% CW-1% TP organogels developed at the T_{set} of 15°C showed a higher increase rate, and achieved a plateau at higher values (Fig. 6.13.B) than the organogels developed at the T_{set} of –5°C (Fig. 6.13.A). This occurred regardless of the lower SFC obtained at T_{set} = 15°C than at T_{set} = –5°C ($P < 0.05$), particularly in the 3% CW-1.0% TP organogel (Table 6.B). As already discussed, the 3% CW-1.0% TP system develops a co-crystal structures during organogelation at T_{set} of 15°C (Toro-Vazquez et al., 2009; Chopin-Doroteo et al., 2011). The particular behavior of the organogels developed at T_{set} of 15°C was also evident through yield stress (σ*) measurements. Thus, the σ* in the 3% CW-0% TP and 3% CW-1.0% TP organogels developed at the T_{set} of 15°C and –5°C was

Table 6.B. Percentage of Solid Fat Content (% SFC) and Yield Stress (σ*) for Organogels Developed with 3% CW Dispersions in HOSFO with 0% or 1% of Tripalmitin (TP) at the Temperatures Indicated (T_{set}). Modified from Toro-Vazquez et al., 2009.

	% SFC[D]		σ* (Pa)[E]	
T_{set} (°C)	3% CW-0% TP	3% CW-1% TP	3% CW-0% TP	3% CW-1% TP
–5	$2.52^{A,A}$ (0.12)	$4.05^{A,B}$ (0.009)	$378.15^{A,A}$ (2.25)	$682.04^{A,B}$ (292.54)
15	$2.43^{A,A}$ (0.14)	$3.39^{B,B}$ (0.08)	$402.97^{A,A}$ (66.38)	$932.34^{A,B}$ (462.70)
25	$2.10^{B,A}$ (0.08)	$2.39^{C,B}$ (0.07)	$228.06^{A,A}$ (24.94)	$230.98^{B,A}$ (54.72)

[A, B, C] For the same column, the same first letter indicates that the values are statistically equal. A different letter indicates that the values for the same column are significantly different ($P < 0.05$) (i.e., the T_{set} effect was significant).

[A, B, C] For the same row and parameter, the same second letter indicates that the values are statistically equal. A different letter indicates that the values for the same row and parameter are significantly different ($P < 0.05$) (i.e., the TP effect was significant).

[D] The % SFC values after 60 min and 120 min were statistically the same in all cases but for the 3% CW-0% TP at T_{set} = –5°C. When the % SFC values were statistically the same, the values were averaged and the mean and standard deviation reported.

[E] Determined after 120 min at the corresponding T_{set} by applying a strain sweep between 0% and 100%. σ* was calculated from the log-log plot of shear stress vs. % strain at the corresponding upper limit of stress.

statistically the same (Table 6.B). In consequence, the 3% CW-1.0% TP organogels developed at the T_{set} of 15°C have higher elasticity (i.e., higher G′) but similar resistance to permanent (i.e., plastic) deformation than organogels developed at the T_{set} of −5°C. This occurred regardless of the lower SFC present in the organogels developed at 15°C than in the organogels developed at −5°C (Table 6.B).

Organogels Developed Under Shear Rate Conditions

Exploratory experiments done in our laboratory showed that the application of a continuous shear (i.e., 600 s^{-1}) during organogelation of CW in HOSFO resulted in the formation of sol-type system with weaker gel structure that the organogels developed under quiescent conditions. Similar results have been observed with 5% monoglyceride organogels developed applying shear rates between 50 to 2,000 s^{-1} (Da Pieve et al., 2010). It seems that the application of shear at temperatures below the nucleation temperature (i.e., where the three-dimensional crystal network is being developed) does not permit the establishment of permanent junction among the crystals. The development of gel structure (i.e., G′ > G″) requires an appropriate balance between permanent and transient junctions zones throughout the three-dimensional crystal network. Within this framework, we considered that the application of shear to LMOG solutions under metastable conditions and then allowing the development of the organogel under static conditions might enhance nucleation and crystal growth and still enable the formation of permanent junction zones during the development of the three-dimensional crystal network (Chopin-Doroteo et al., 2011). The modification of the crystal network organization, particularly in the proportion between permanent and transient junction zones, might result in organogels with particular thermo-mechanical properties different from the ones shown by organogels developed under static conditions or with the application of continuous shear.

To evaluate this hypothesis we developed 3% CW-0% TP and 3% CW-1% TP organogels in HOSFO following the time-temperature program show in Fig. 6.14. This was done using a mechanical spectrometer (Paar Physica MCR 301, Stuttgart, Germany) equipped with a Peltier system for temperature control and a cone-plate geometry equipped with a true-gap system. This set-up allowed the application of a controlled shear rate at particular stages during the time-temperature program used to develop the organogels. The application of stirring was evaluated under two temperature conditions established based on the T_g for CW (i.e., T_g for the 3% CW solution in HOSFO was 42.2°C ± 0.3°C). Thus, the gelator solutions were stirred at 100 rpm (i.e., shear rate of 600 s^{-1}) during cooling from 90°C to either 5°C (i.e., 47°C) or 10°C (i.e., 52°C) above the T_g for 3% CW in HOSFO. After this, stirring was stopped and cooling (1°C/min) continued under static conditions until achieving 5°C. After 30 min at this temperature, the system was heated (5°C/min) until it achieved the T_{set} of 15°C. After a set time period at 15°C, the melting (i.e., T_M, and ΔH_M) and mechanical (i.e., G′, G″, and σ*) properties of the organogels were measured. The same thermo-mechanical properties were evaluated in 3% CW-0% TP and 3% CW-1% TP organogels developed under static conditions using the same time-temperature program (Chopin-Doroteo et al., 2011).

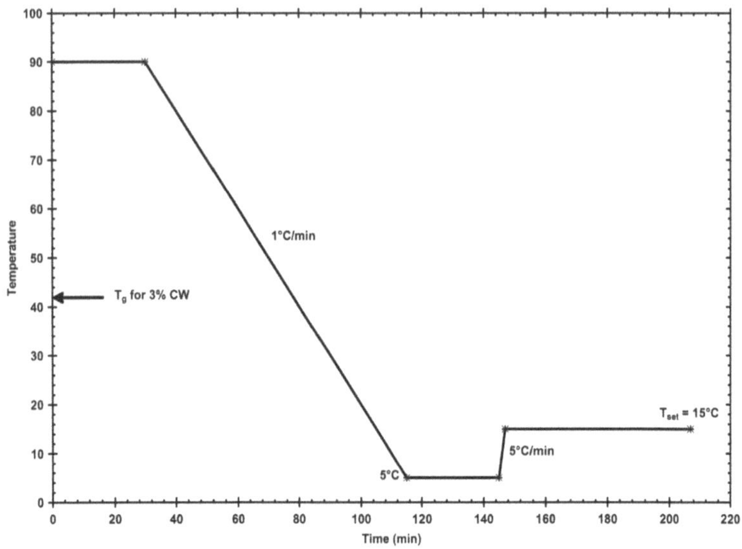

Fig. 6.14. Time-temperature relationship applied to develop CW and CW-TP organogels. The arrow indicates the T_g under static conditions for a solution of 3% CW in SFO.

These experiments considered that since under static conditions T_g for 3% CW in HOSFO is 42.2°C, shearing was applied to the 3% CW-0% TP and 3% CW-1% TP solutions until they reached near nucleation conditions (i.e., stirring was applied until achieving 47°C) or until metastable conditions were attained before CW nucleation (i.e., stirring was applied until achieving 52°C). The results showed that under static or stirring conditions the presence of 1% TP produced a decrease in T_M and an increase in ΔH_M in comparison with the 3% CW-0% TP organogels (Table 6.C). These results confirmed that the formation of CW-TP co-crystals modifies the melting properties of CW organogels. Additionally, the T_M and ΔH_M (Table 6.C) and the X-ray diffractograms of the 3% CW-0% TP and 3% CW-1% TP organogels developed statically or with the application of shearing did not show significant differences. Thus, the application of stirring at temperatures above T_g does not modify the O⊥ subcell packing of the crystals developed by CW and the CW-TP mixture (Chopin-Doroteo et al., 2011).

From the rheological point of view, the application of stirring during cooling just until reaching a temperature 10°C above T_g (i.e., 52°C) and then continuing cooling under static conditions, resulted in organogels (i.e., CW 52°C and CW-TP 52°C) with higher elastic properties that the organogels developed under static conditions (i.e., CW and CW-TP) or with the application of shear until reaching 47°C (i.e., 5°C above T_g; CW 47°C and CW-TP 47°C) (Table 6.D). Thus, the application of stirring in both the CW and the CW-TP systems produced a similar effect on the organogels elastic properties (Table 6.D). However, because of the higher SFC present in the 3% CW-1% TP organogels (Table 6.C) under the same stirring conditions, G' was larger in these organogels than in the 3% CW-0% TP organogels (Table 6.D). Nevertheless, independent of the stirring conditions the σ^* (i.e., the resistance to

Table 6.C. Melting Parameters for 3% CW-0% TP and the 3% CW-1% TP Organogels Developed in HOSFO Under Static and Stirring Conditions. Modified from Chopin-Doroteo et al., 2011.

Sample	T_M (°C)	ΔH_M (J/g)	SFC (%)
3% CW–0% TP	44.66[a] (0.06)	1.51[a] (0.46)	2.25[a] (0.16)
3% CW–0% TP 47°C	44.94[a] (0.19)	1.90[a] (0.01)	2.40[b] (0.11)
3% CW–0% TP 52°C	44.76[a] (0.04)	1.76[a] (0.22)	2.42[b] (0.90)
3% CW–1% TP	40.60[b] (0.06)	3.60[b] (0.45)	3.69[c] (0.12)
3% CW–1% TP 47°C	40.64[b] (0.11)	2.91[b] (0.65)	3.66[c] (0.09)
3% CW–1% TP 52°C	40.36[b] (0.21)	3.28[b] (0.26)	3.67[c] (0.14)

[a, b, c] For the thermal parameter values with the same letter, there was no significant difference. Values with a different letter indicate a significant difference ($P < 0.10$).

Table 6.D. Yield Stress (σ^*) and G′ for the 3% CW-0% TP and the 3% CW-1% TP Organogels Developed in HOSFO. Modified from Chopin-Doroteo et al., 2011.

Sample	σ^* (Pa)	G′(Pa)[e] × 10^4
3% CW–0% TP	55.64[a] (10.36)	4.05[a] (0.78)
3% CW–0% TP 47°C	57.27[a] (13.0)	4.21[a] (0.09)
3% CW–0% TP 52°C	118.13[b] (11.19)	8.26[b] (0.50)
3% CW–1% TP	62.37[a] (0.52)	7.54[c] (0.04)
3% CW–1% TP 47°C	76.71[a] (9.45)	7.43[c] (1.03)
3% CW–1% TP 52°C	108.37[b] (1.70)	14.10[d] (1.69)

[a, b, c, d] $\sigma^*$and G′ values with the same letter indicate no significant difference. Values with a different letter indicate a significant difference ($P < 0.10$).

[e] Mean calculated from G′ values of the last 20 to 30 min of the G′ profile.

permanent deformation) was the same in the organogels with or without TP (Chopin-Doroteo et al., 2011).

These results were explained by the effect of shearing in the proportion between permanent and transient junction zones present in the organogels' crystal network (Chopin-Doroteo et al., 2011). Thus, the development of CW organogels requires the presence of a minimum amount of permanent junctions among the microplatelet units (i.e., *n*-alkane crystals) to achieve a gel structure. Applying a constant shear even at temperatures below T_g where nucleation is occurring does not allow the establishment of permanent junction zones among the microplatelet units, and therefore weak gel structures are developed. In contrast, the application of shear just until reaching nucleation conditions (i.e., until reaching 47°C) favors the entanglement of plates (i.e., development of transient junction zones), developing a crystal network with higher elasticity and resistance to deformation than the one achieved under static conditions (Table 6.D). The application of shearing in the metastable zone (i.e., until reaching 52°C) enhances even more the entanglement of plates. Additionally, since under this temperature condition CW nucleation has yet to occur, the development of the three-dimensional crystal network would include the formation of permanent junction zones. Polarized light microphotographs of the organogels showed that when shearing is applied until reaching 47°C (Figs. 6.15.B and 6.16.B) or 52°C (Figs. 6.15.C and 6.16.C), the organogels showed the presence of larger crystals (i.e., fiber-like or plate-like) than those observed under static conditions (Figs. 6.15.A and 6.16.A). This was even more evident when compared with the crystal network developed under static conditions (Figs. 6.15.A and 6.16.A). Unfortunately, with the magnification utilized we could not assess the effect of shearing on the proportion between the permanent and the transient junction zones throughout the crystal network. We concluded that at the cooling rate of 1°C/min, the application of shear until reaching the metastable zone of CW (i.e., at 52°C) results in the development of a three-dimensional crystal network formed by larger crystals (Figs. 6.15.C and 6.16.C) and with higher extent of transient and permanent junction zones than the organogels develop under static conditions. As a consequence, the CW 52°C and the CW-TP 52°C organogels have higher elasticity (i.e., high G′) and resistance to permanent deformation (i.e., high σ^*) in comparison with the organogels developed under static conditions (i.e., CW and CW-TP organogels) or with the application of shearing under near nucleation conditions for CW (i.e., CW 47°C and CW-TP 47°C organogels).

Candelilla Wax Dispersions as Coating Barrier of Food Products

The use of CW as a coating barrier against oxygen and water vapor diffusion in food products is based on its non-polar nature and capacity to develop a compact network of crystals. For this purpose, CW is used as a disperse aqueous phase to control the respiration rate of fruits and vegetables and reduce moisture loss from foods (Bourlieu et al., 2008; Callegarin et al., 1997). The adequate dispersion of CW in aqueous systems is

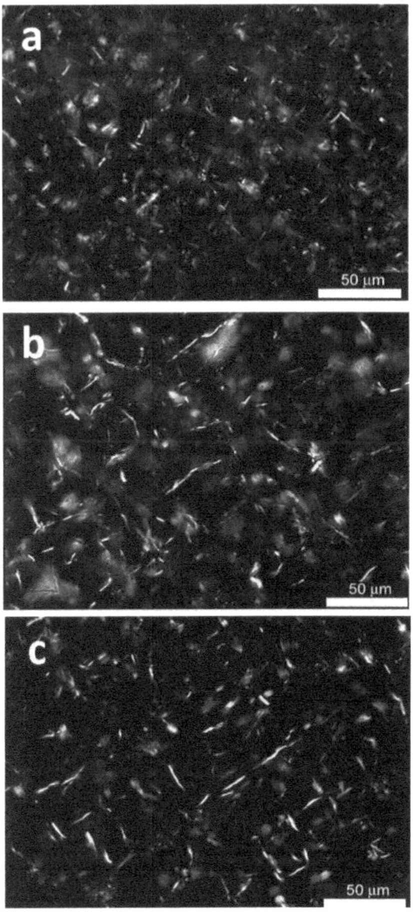

Fig. 6.15. Microphotographs for the 3% CW organogels in HOSFO under static (*a*) and stirring conditons (100 rpm) applied during the cooling stage until arriving at 47°C (*b*) or 52°C (*c*) (from Chopin-Doroteo et al., 2011).

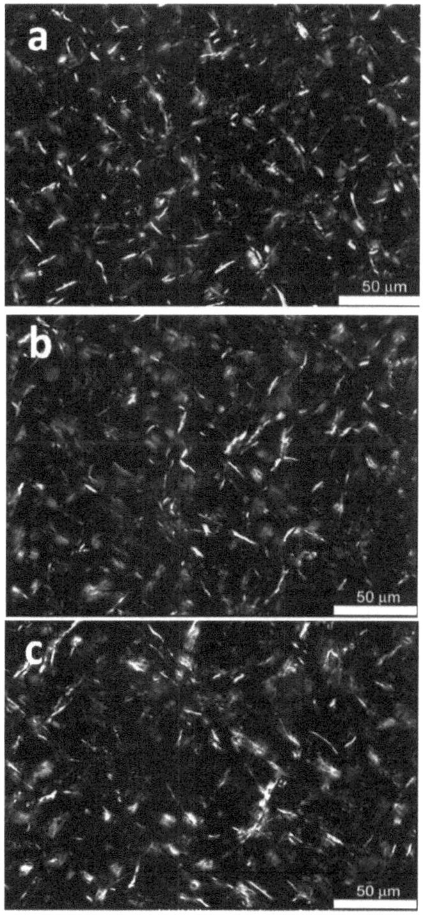

Fig. 6.16. Microphotographs for the 3% CW-1% TP organogels in HOSFO under static (*a*) and stirring conditions (100 rpm) applied during the cooling stage until arriving at 47°C (*b*) or 52°C (*c*) (from Chopin-Doroteo et al., 2011).

achieved by emulsification of the melted CW with surfactant agents such as fatty acids, glycerol, fatty acids' derivatives, or surface-active hydrocolloids (Bourlieu et al., 2008).

Since water vapor permeability depends on hydrocarbon chain length and concentration of ambiphatic compounds like fatty acids, fatty alcohols, and esters (Martini et al., 2006; Donhowe & Fennema, 1993), waxes have lower water and oxygen permeability than other lipids, such as hydrogenated and non-hydrogenated vegetable oils and fatty acids (Callegarin et al., 1997). Thus, Donhowe and Fennema (1993) showed that both CW and carnauba wax have lower oxygen permeability than bee wax, microcrystalline wax (composed entirely of hydrocarbons, C41 to C50), and several films made with non-edible polymeric compounds. In the same way, CW showed superior resistance to water vapor permeability than carnauba, bee, and microcrystalline wax. These authors associated the higher barrier properties of CW with its higher melting profile. However, CW in its pure state is rarely used as a barrier for foods, mainly because the high temperature required (>85°C) to disperse it appropriately on the product. As a consequence, the use of CW dispersions is preferred because they can be applied at room temperature without damaging the food product.

Two main types of dispersions have been developed: dispersions where the continuous phase is an aqueous solution and dispersions where the continuous phase is a bio-polymeric matrix. The aqueous dispersions are used as coating agents while the bio-polymeric dispersions are used to cast films (Bourlieu et al., 2008; Hagenmaier, 2004).

Crystalline Dispersions of Candelilla Wax in Water

Although frequently referred to as emulsions or micro emulsions, coatings based on CW or other waxes are mostly crystalline emulsions in water prepared from the melted wax, stabilized with ionic surfactants, and cooled below the wax's melting temperature (Hagenmaier, 2004). In general, water dispersions of CW with 20–25% of total solids are emulsified with 8–20 g of fatty acids/100 g of CW using morpholine or ammonia solutions to ionize the fatty acids (pH between 8.6 and 10.1). Pure oleic acid or in mixture with palmitic and/or myristic acid are used as emulsifying agents. Many CW dispersions produced under such conditions are translucent or clear with a mean globule diameter between 0.011 and 0.120 μm without showing phase separation for up to 18 months at 22°C (Hagenmaier & Baker, 1996; Hagenmaier, 2004). This high stability against sedimentation and flocculation is associated to the sub-micrometric CW crystal size and the electrostatic repulsion provided by the fatty acids adsorbed on the crystal's surface (McClements, 2005). The application of CW dispersions to coat citrus and bananas produced a reduction in fruit weight loss (i.e., moisture barrier), and reduced the respiration rates and the development of off-flavors in comparison with the use of resin coatings (Hagenmaier & Baker, 1996). The functionality of these highly stable CW dispersions might be appealing to other areas of the food industry. Nevertheless, the high pH required to produce stable CW dispersions limit their application beyond coatings. If this disadvantage were surpassed, aqueous dispersion of CW crystals might provide stabilization and modify the texture of water in oil emulsions without the use of *trans* fatty acids.

Bio-polymeric Dispersions of Candelilla Wax

Edible films are formed by a matrix of high molecular weight bio-polymers, designed to have high cohesive strength and elasticity. Proteins and polysaccharides develop good films, but because of their hydrophilic character, the resulting films have poor barrier properties against moisture. This drawback is partially overcome using hydrophobic compounds, such as CW, dispersed throughout the matrix (Callegarin et al., 1997; Chick & Hernandez, 2002).

Unlike CW water dispersions, edible films containing CW are less well-studied with only a few papers having been published in recent years (Saucedo-Pompa et al., 2009; Saucedo-Pompa et al., 2007), and just a few address the micro-structure of the system (Ruíz-Ramos et al., 2006; Bósquez-Molina et al., 2003; Chick & Hernández, 2002). Composite films with CW are produced through wax emulsification in a biopolymer solution. Overall, stable emulsions of CW in casein solutions are achieved in alkaline conditions (Chick & Hernández, 2002), while in polysaccharides solutions (i.e., mesquite gum, chitosan) stability is achieved under acid conditions (Ruíz-Ramos et al., 2006; Bósquez-Molina et al., 2003). According to Chick and Hernández (2002), casein films containing CW and sorbitol show higher phase separation from the matrix as wax concentration increased in comparison with the films developed without CW. However, regardless of the phase separation, the behavior of the tensile strength and elongation of films showed that CW remains dispersed in the casein matrix (Chick & Hernández, 2002). These results indicated that sorbitol was the main component that went out of the matrix phase.

In polysaccharide-based films, a study involving emulsions of mesquite gum solutions with pure CW or with blends of CW-lipid (i.e., CW-mineral oil, CW-oleic acid, or CW-bee wax) showed a mean volume-surface globule size ($D_{3,\,2}$) between 1.67 and 2.55 μm (Bósquez-Molina et al., 2003), a value 10-fold higher than the globule size observed in the CW aqueous emulsions studied by Hagenmaier (2004). The main reason for such difference might be associated with the lower concentration of surface active compound (i.e., ionized fatty acids) present in the emulsions with mesquite gum, in comparison with the CW aqueous emulsions studied by Hagenmaier (2004). According to Bósquez-Molina et al., (2003), the moisture barrier capacity of the coatings is not associated to the droplet diameter (i.e., $D_{3,\,2}$) of the emulsion but rather to the surface morphology. Thus, coatings developed by emulsions of mesquite gum with pure CW ($D_{3,\,2}$ = 2.55 μm) or by emulsions of mesquite gum with CW-mineral oil ($D_{3,\,2}$ = 1.67 μm) provided the best barrier properties and showed the most uniform surface (i.e., absence of cracks) of all films investigated (Bósquez-Molina et al., 2003). Finally, it is important to point out that the use of CW in water-in-oil emulsions resulted in the development of a gel-like structure. This behavior was associated to oil droplet aggregation produced by CW crystallization, resulting in a gel-like structure throughout the emulsion (Ruíz-Ramos et al., 2006). However, emulsion with gel-like structures not always result in an improvement in the film barrier properties (Ruíz-Ramos et al., 2006).

Conclusions

Candelilla wax has as its main components *n*-alkanes with hentriacontane as its major constituent. The *n*-alkanes present in CW show self-assembly properties in organic solvents, such as vegetable oils, developing organogels with a thermoreversible behavior independent of the cooling rate (i.e., 1°C/min and 10°C/min) and gel setting temperature (T_{set}). The three-dimensional structural organization of the *n*-alkane crystals, and therefore, the mechanical properties of the organogel, depend on the cooling rate, the thermodynamic drive force for gelation, and the annealing process occurring at high T_{set} (i.e., 25°C). TP and CW might interact, developing mixed crystal systems (i.e., co-crystals) that result in organogels with thermal and rheological properties different from the ones obtained by the individual components. The proportion between CW and TP in the organogel primary crystal structure depends on T_{set}, and this modifies the organogel's thermal behavior. Therefore, the interaction between TP, and overall symmetrical TAGS containing palmitic or stearic acid (i.e., POP and POSt), and CW components must be investigated as a possible tool to tailor organogels with particular thermal, rheological, and textural properties. Additional experiments showed that the application of shearing during cooling of CW and CW-TP solutions just until attaining the metastable zone favors the development of fiber-like crystals of larger size and tentatively with higher extent of transient and permanent junction zones than the organogels develop under static conditions. Therefore, the application of stirring to CW and CW-TP solutions up to temperatures well above T_g modifies the three-dimensional organization of the *n*-alkanes from microplatelets into fiber-like crystalline structures. This opens the possibility of tailoring particular functional properties to develop *trans*-free vegetable-oils-based spreads and coatings, and novel food products with new textural perceptions for the consumers. In addition, the 3% CW organogels show remarkable stability given that no phase separation is observed even after 12 months of storage at either 5°C or 25°C.

References

Abdallah, D.J.; Lu, L.; Weiss, R.G. Thermoreversible Organogels from Alkane Gelators with One Heteroatom. Chem. Mater. 1999, 11, 2907–2911.

Abdallah, D.J.; Weiss, R.G. *n*-alkanes Gel *n*-alkanes (an many other organic liquids). Langmuir 2000, 16, 352–355.

Abdallah, D.J.; Sirchio, S.A.; Weiss, R.G. Hexatriacontane Organogels. The First Determination of the Conformation and Molecular Packing of a Low-Molecular-Mass Organogelator in Its Gelled State. Langmuir 2000, 16, 7558–7561.

Allen R.R. Hydrogenation. J. Am. Oil Chem. Soc. 1960, 37, 521–523.

Ascherio A.; Stampfer, M.J.; Willett, W.C. Trans Fatty Acids and Coronary Heart Disease. http://www.hsph.harvard.edu/reviews/transfats.pdf. 1999a.

Ascherio, A.; Katan, M.; Zock, P.L.; Stampfer, M.J.; Willett, W.C. *Trans* Fatty Acids and Coronary Heart Diseases, N. Engl. J. Med. 1999b, 340, 1994–1998.

Berger, K.G.; Idris, N.A. Formulation of Zero-trans Acid Shortenings and Margarine and Other Food Fats with Products of the Oil Palm. J. Am. Oil Chem. Soc. 2005, 82, 775–782.

Bósquez-Molina, E.; Guerrero-Legarreta, I.; Vernon-Carter, J. Moisture Barrier Properties and Morphology of Mezquite Gum-Candelilla Wax Based Edible Emulsion. Food. Res. Int. 2003, 36, 885–893.

Bot, A.; Agterof, W.G.M. Structuring of Edible Oils by Mixtures of γ-oryzanol with β-sitosterol or Related Phytosterols. J. Am. Oil Chem. Soc. 2006, 83, 513–521.

Bourlieu C.; Guillard, V.; Vallès-Pamiès, B.; Gontard, N. Foods Materials Science, 9; Aguilera J.M. Lillford, P.J., Eds.; Food Engineering Series. Springer: NY, 2008; 547–575.

Callegarin, F.; Quezada-Gallo, J.A.; Debeaufort, J.; Voilley, A. Lipids and Biopackaging. J. Am. Oil Chem. Soc. 1997, 74, 1183–1192.

Chick J.; Hernández, R.J. Physical, Thermal, and Barrier Characterization of Casein-Wax-Based Edible Films. J. Food Sci. 2002, 67, 1073–1079.

Chevallier, V.; Provost E.; Bourdet, J.B.; Bouroukba, M.; Petitjean, D.; Dirand, M. Mixtures of Numerous Different *n*-alkanes: 1. Structural Studies by X-ray Diffraction at Room Temperature—Correlation Between the Crystallographic Long *c* Parameter and the Average Composition of Multi-alkane Phases. Polymer 1999a, 40, 2121–2128.

Chevallier, V.; Petitjean, D.; Bouroukba, M.; Dirand, M. Mixtures of Numerous Different *n*-alkanes: 2. Studies by X-ray Diffraction and Differential Thermal Analysis with Increasing Temperature. Polymer 1999b, 40, 2129–2137.

Chopin-Doroteo, M.; Morales-Rueda, J. A.; Dibildox-Alvarado, E.; Charó-Alonso, M. A.; Toro-Vazquez, J. F. The Effect of Shearing in the Thermo-Mechanical Properties of Candelilla Wax and Candelilla Wax-Tripalmitin Organogels. Food Biophysics 2011, In press.

Da Pieve, S.; Calligaris, S.; Nicoli, M.C.; Marangoni, A.G. Shear Nanostructuring of Monoglyceride Organogels. Food Biophysics 2010, 5, 211–217.

Dassanayake, L.S.K.; Kodali, D.R.; Ueno, S.; Sato, K. Physical Properties of Rice Bran Wax in Bulk and Organogels. J. Am. Oil Chem. Soc. 2009, 86, 1163–1173.

Espeau, P.; White, J.W. Thermodynamic Properties of n-alkanes in Porous Graphite. J. Chem. Soc. Faraday Trans. 1997, 93, 3197–3200.

Donhowe, I.G.; Fennema, O. Water Vapor Permeability and Oxygen Permeability of Wax Films. J. Am. Oil Chem. Soc. 1993, 70, 867–873.

Gandolfo, F.G.; Bot, A.; Flöter, E. Structuring of Edible Oils by Long-Chain FA, Fatty Alcohols, and their Mixtures. J. Am. Oil Chem. Soc. 2004, 81, 1–6.

Grant, D.L. Candelilla Wax, First draft. WHO Food Additive Series 30, http://www.inchem.org/documents/jecfa/jecmono/v30je12.htm. 2005.

Hagenmaier, R.D. Fruit Coatings Containing Ammonia Instead or Morpholine. Proc. Fla. State Hort. Soc. 2004, 117, 396–402.

Hagenmaier, R.D.; Baker, R. Edible Coatings from Candelilla Wax Microemulsions. J. Food Sci. 1996, 61, 562–565.

Hoover, R.; Vasanthan, T. The effect of annealing on the physicochemical properties of wheat, oat, potato and lentil starches. J. Food Biochem. 1994, 17, 303–325.

INFOTHERM. http://www.fiz-chemie.de/infotherm/servlet/infothermSerch. 2006.

Instituto de la Candelilla, http://www.candelilla.org/es/propiedades.htm. 2004.

Kodali R. D.; List. G.R. *Trans Fats Alternatives.* AOCS Press: Champaign, IL, 2005.

Köhler, K.; Förster, G.; Hauser, A. Dobner, B.; Heiser, U.F.; Ziethe, F.; Richter, W.; Steiniger, F.; Drechsler, M.; Stettin, H.; Blume, A.; Temperature-Dependent Behavior of a Symmetric Long-Chain Bolaamphiphile with Phosphocholine Headgroups in Water: From Hydrogel to Nanoparticles. J. Am. Chem. Soc. 2004, 126, 16804–16813.

Kumar, R.; Katare, O.P. Lecithin Organogels as a Potential Phospholipid-Structured System for Topical Drug Delivery: A Review. AAPS PharmSciTech 2005, 6, E298–E310.

Lam, R.; Quaroni, L.; Pederson, T.; Rogers, M.A. A molecular insight into the nature of crystallographic mismatches in self-assembled fibrillar networks under non-isothermal crystallization conditions. Soft Matter 2010, 6, 404–408.

Larsson, K. Molecular Arrangement in Glycerides. Fette Seifen Anstrichm. 1972, 74, 136–142.

LipidBank. http://lipidbank.jp/cgi-bin/detail.cgi?id=WWA1101. 2007.

Macher, M.; Holmquivst. A. Hydrogenation of Palm Oil in a Near-critical and Supercritical Propane. Europ. J. Lipid Sci. Technol. 2001, 101, 81–84.

Manku, M.S.; Horrobin, D.F.; Morse, N.; Kyte, V.; Jenkins, K.; Wright, S.; Burton. J.L. Reduced level of Prostaglandin Precursors in the Blood of Atopic Patients: Defective Δ-6-desaturase Function as a Biochemical Base for Atopy. Prostaglandins Leukot. Med. 1982, 9, 615–628.

Martini, S.; Kim, D.A; Ollivon, M.; Marangoni, A.G. The Vapor Permeability of Polycrystalline Fat Barrier Films. J. Agric. Food Chem. 2006, 54, 1880–1886.

Marangoni, A. G.; Idziak, S.H.J.; Rush. J.W. Controlled release of Food Lipids Using Monoglyceride Gel Phases Regulates Lipid and Insulin Metabolism in Humans, Food Biophysics. 2008, 3, 241–245.

Martini, S.; Carelli, A.A.; Lee, J. Effect of the Addition of Waxes on the Crystallization Behavior of Anhydrous Milk Fat, J. Am. Oil Chem. Soc. 2008, 85, 1097–1104.

McClements, D.J. Food Emulsions Principles, Practices, and Techniques. CRC Press: Boca Raton, FL, 2005.

Mensink, R.P. Metabolic and Health Effects of Isomeric Fatty Acids. Curr. Opin. Lipidol. 2005, 16, 27–30.

Mittal, S. The Metabolic Syndrome in Clinical Practice. Springer: London, UK, 2008.

Mondal, K.; Lalvani, S.B. Electrochemical Hydrogenation of Canola Oil Using Hydrogen Transfer Agent. J. Am. Oil Chem. Soc. 2003, 80, 1135–1141.

Morales-Rueda, J.A.; Dibildox-Alvarado, E.; Charó-Alonso, M.A.; Weiss, R.G.; Toro-Vazquez, J.F. Thermo-mechanical Properties of Candelilla Wax and Dotriacontane Organogels in Safflower Oil. European J. Lipid Sci. Technol. 2009a, 111, 207–215.

Morales-Rueda, J.A.; Dibildox-Alvarado, E.; Charó-Alonso, M.A.; Toro-Vazquez, J.F. Rheological Properties of Candelilla Wax and Dotriacontane Organogels Measured with a True-Gap System. J. Am. Oil Chem. Soc. 2009b, 86, 765–772.

Murdan, S.; Gregoriadis, G.; Florence, A.T. Novel Sorbitan Monostearate Organogels. J. Phar. Sci. 2000, 88, 608–614.

Ostuni, E.; Kamaras, P.; Weiss, R.G. Novel X-ray Method for in situ Determination of Gelator Strand Structure: Polymorphism of Cholesteryl Anthraquinone-2-carboxylate. Angew. Chem. Int. Ed. Engl. 1996, 35, 1324–1326.

Pernetti, M.; van Malssen, K.F.; Flöter, E.; Bot. A. Structuring of Edible Oils by Alternatives to Crystalline Fat. Current Opn. Colloid Interface Sci. 2007a, 12, 221–231.

Pernetti, M.; van Malssen, K.F.; Kalnin, D.; Flöter, E. Structuring Edible Oil with Lecithin and Sorbitan Tri-sterarate. Food Hydrocolloids 2007b, 21, 855–861.

Petraskaite, V.; de Greyt, W.F.; Kellens, M.J.; Huyghebaert, A.D. Physical and Chemical Properties of trans-Free Products by Chemical Interestrification of Vegetable Oil Blends. J. Am. Oil Chem. Soc. 1998a, 75, 489–493.

Petraskaite, V.; de Greyt, W.F.; Kellens, M.J.; Huyghebaert, A.D. Chemical Interesterification of Vegetable Oil Blends: Optimization of Process Parameters. Oleagineuz Corp Gras Lipides 1998b, 5, 65–69.

Reddy, S.Y.; Jeyarani, T. *Trans*-free Bakery Shortenings from Mango Jernel and Mahua Fats by Fractionation and Blending. J. Am. Oil Chem. Soc. 2001, 78, 635–640.

Rogers, M.; Wright, A.J.; Marangoni, A.G. Post-crystallization Increases in the Mechanical Strength of Self/Assembled Fibrillar Networks is Due to an Increase in Network Supramolecular Ordering. J. Phys. D 2008, 41, 215501.

Rogers, M.A. Novel Structuring Strategies for Unsaturated Fats. Meeting the Zero-Trans, Zero-Saturated Fat Challenge: A Review. Food Research Inter. 2009, 42, 747–753.

Ruíz-Ramos, J.O.; Pérez-Orozco, J.P.; Báez-González, J.G.; Bósquez-Molina, E.; Pérez-Alonso, C.; Vernon-Carter, J. Interrelationship Between the Viscoelastic Properties and Effective Moisture Diffusivity of Emulsions with the Vapor Permeability of Edible Films Stabilized by Mesquite Gum-Chitosan Complexes. Carb. Polym. 2006, 64, 355–363.

Sagalowicz, L.; Leser, M.E.; Watzke, H.J.; Michel, M. Monoglyceride Self-Assembly Structures as Delivery Vehicles. Trends Food Sci. Technol. 2006, 17, 204–214.

Saucedo-Pompa, S.; Rojas-Molina, R.; Aguilera-Carbó, A.F.; Saenz-Galindo A.; de la Garza, H.; Jasso-Cantú, D.; Aguilar, C.N. Edible Film Based on Candelilla Wax to Improve the Shelf Life and Quality of Avocado. Food Res. Int. 2009, 42, 511–515.

Saucedo-Pompa, S.; Jasso-Cantú, D.; Ventura-Sobrevilla, J.; Saenz-Galindo A.; Rodríguez-Herrera, R.; Aguilar, C. N. Effect of Candelilla Wax with Natural Antioxidants on the Shelf Life Quality of Fresh-Cut Fruits. J.Food Qual. 2007, 30, 823–836.

Schaink, H.M.; van Malssen, K.F.; Morgado-Alves, S.; Kalnin, D.; Van der Linden, E. Crystal Network for Edible Oil Organogels: Possibilities and Limitations of the Fatty Acid and Fatty Alcohol Systems. Food Res. Int. 2007, 40, 1185–1193.

Stender, S.; Dyerberg, J. Influence of *trans* Fatty Acids on Health. Ann. Nutr. Metab. 2004, 48, 61–66.

Toro-Vazquez, J.F.; Morales-Rueda, J.A.; Dibildox-Alvarado, E.; Charó-Alonso, M.A.; Alonzo-Macías, M.; González-Chávez, M.M. Thermal and Textural Properties of Organogels Developed by Candelilla Wax in Safflower Oil. J. Am. Oil Chem. Soc. 2007. 84, 989–1000.

Toro-Vazquez, J.F.; Alonzo-Macías, M.A.; Dibildox-Alvarado, E.; Charó-Alonso, M.A. The Effect of Tripalmitin Crystallization on the Thermo-Mechanical Properties of Candelilla Wax Organogels. Food Biophysics 2009, 4, 199–212.

Tozaki K.; Inaba, H.; Hayashi, H.; Quan, C.; Nemoto, N., Kimura, T. Phase Transitions of n-$C_{32}H_{66}$ measured by means of high resolution and super-sensitive DSC. Therm. Acta. 2003, 397, 155–161.

Wang, R.; Liu, X.; Xiong, J.; Li, J. Real-Time Observation of Fiber Network Formation in Molecular Organogel: Supersaturation Dependent Microstructure and Its Related Rheological Property. J. Phys. Chem. B 2006, 110, 7275–7280.

Wright, A.J.; Marangoni, A.G. Time, Temperature, and Concentration Dependence of Ricinelaidic acid-canola Oil Organogelation. J. Am. Oil Chem. Soc. 2007, 84, 3–9.

Yang, E.S.; Jung, M.Y.; Min, D.B. Hydrogenation for Low Trans and High Conjugated Fatty Acids. J. Food Sci. 2005, 1, 22–30.

Zgardzinska B.; Pietrow, M.; Goworek, T.; Wawryszczuk, J. *Ortho*-Positronium in Some *n*–Alkanes; Influence of Temperature and Pressure. Acta Physica Pol. A 2006, 110, 747–753.

7

Physical Properties of Organogels Made of Rice Bran Wax and Vegetable Oils

Lakmali Samuditha K. Dassanayake[1], Dharma R. Kodali[2,3], Satoru Ueno[1], and Kiyotaka Sato[1]

[1]Laboratory of Food Biophysics, Graduate School of Biosphere Science, Hiroshima University, Kagamiyama 1-4-4, 739-8528, Japan; [2]Global Agritech, Inc., 710 Olive Ln., Minneapolis, Minnesota; and [3]Department of Bioproducts and Biosystems Engineering, University of Minnesota, 1390 Eckles Avenue, St. Paul, Minnesota

Introduction

The term "oleogel" or "organogel" has been significant in food research field recently as a proper alternative for triglyceride hardstocks in oil structuring purposes. Many sensory attributes (e.g., spreadability, mouth-feel, snap, and texture) of foods depend on the fat crystal network, which is comprised mainly of high-melting fats (Narine & Marangoni, 2005). Thus far, high-melting fats containing *trans* fatty acid and saturated fatty acid moieties have been employed. However, it has recently been claimed that the intake of high levels of *trans* and saturated fats contribute to global epidemics related to metabolic syndrome and cardiovascular disease (Schrimpf-Moss & Wilkening, 2005). These negative health implications can be reversed by altering intake and replacing unhealthy fats with healthier alternatives. Novel technologies for structuring edible oils must be implemented to achieve these requirements (Rogers et al., 2008; Rogers et al., 2009; Rogers, 2009; Hughes et al., 2009; Pernetti et al., 2007). Organogels were introduced to address this challenge as alternative oil-structuring materials that have physical, chemical, and organoleptic properties similar to those of conventional high-melting fats; they are also cost-effective and food-grade and avoid negative health effects (Rogers et al., 2008; Rogers et al., 2009; Rogers, 2009; Hughes et al., 2009; Pernetti et al., 2007; Abdallah & Weiss, 2000; Bot & Agterof, 2006; Bot et al., 2009; Wright & Marangoni 2006; Wright & Marangoni 2007; Pernetti et al., 2007; Toro-Vazquez et al., 2007; Morales-Rueda et al., 2009; Dassanayake et al., 2009).

Organogels have attracted considerable attention as promising oil-structuring systems for *trans*-fat alternative hardstocks (Terech & Weiss, 1997; Abdallah & Weiss, 2000; Abdallah et al., 2000; Pernetti et al., 2007; Toro-Vazquez et al., 2007; Huang

& Weiss, 2007; Pernetti et al., 2007; Bot et al., 2009). These are viscoelastic materials comprised of organic gelators and liquid oils (Abdallah et al., 2000) and are semi-solid systems in which the liquid oil phase is immobilized by a three-dimensional network composed of self-assembled, intertwined gelator fibers (Vintiloiu & Leroux, 2008). Recently, different types of organogels have been employed in cosmetics, deodorant, and hair-care materials (Pernetti et al., 2007). Such organogels are also significant for food materials because they may reveal softness, easy handling, sharp melting, spreadability, easy film formation, water-barrier properties, etc., as observed in the organogels using phytosterols and sterol esters (Bot & Agterof, 2006; Bot et al., 2008), fatty acids and fatty alcohols (Gandolfo et al., 2004; Schaink et al., 2007), 12-hydroxy-stearic acid (Tamura & Ichikawa, 1997; Rogers & Marangoni, 2008; Rogers et al., 2008; Rogers et al., 2009), ricinoelaidic acid (Wright & Marangoni, 2006; Wright & Marangoni, 2007), sorbitan monostearate (Murdan et al., 1999; Daniel & Rajasekaran, 2003) and ceramides (Rogers, 2009).

In this chapter, we present physical properties of organogel made of rice bran wax and vegetable oils, which were examined with thermal, structural, and optical methods.

Rice Bran Wax as an Oil Structuring Material

Wax is defined as a fatty substance that contains long hydrocarbon chains with or without a functional group (Small, 1986; Parish et al., 2002; Martini & Añón, 2003; Vali et al., 2005). The functional groups that are often present in waxes include alcohol, ester, ketone, and aldehyde. The wax esters derived from plants, insects, marine animals, and petroleum are used in various industrial applications such as cosmetics, lubricants, polishes, surface coatings, inks, and food products (Parish et al., 2002; Ghosh & Bandyopadhyay, 2005). Current consumers' demands are moving toward the use of bio-based waxes rather than mineral-based waxes for the utilization in the food, cosmetics, and pharmaceutical industry. One group of the bio-based waxes available on a massive scale is the plant wax group, which contains the esters of long-chain fatty acids and long-chain aliphatic alcohols. Candellila and carnauba waxes are two common plant waxes used as food ingredients as well as for cosmetics (Vali et al., 2005). Research of organogel, however, using plant waxes is rather limited. Toro-Vazques et al. and Morales-Rueda et al. recently reported physical properties of organogels using candellila wax (CLX) and dotriacontane in safflower oil (Toro-Vazquez et al., 2007; Morales-Rueda et al., 2009).

Rice (*Oryza sativa*) bran wax (RBX) is a natural plant wax derived from rice bran, which is a byproduct of rice milling. It is a major wax resource in East Asia where rice is the main food (MD Idea Exporting Division, 2008). The content of wax in rice bran oil varies with extraction conditions. Wax in crude rice bran oil is removed via a dewaxing step in refining processes. The refining of rice bran oil usually includes dewaxing, degumming, deacidification, bleaching, deodorization, and winterization (Vali et al., 2005; Martini & Añón, 2003; Ghosh & Bandyopadhyay, 2005). RBX has potential applications in cosmetics, pharmaceuticals, food, polymer, and leather industries as costly and efficiently as other plant waxes such as carnauba wax (CRX) and candellila wax (Vali et al., 2005). Because RBX have relatively high melting points, around 78–81°C, they easily crystallize at ambient temperatures, resulting in finely dispersed mixtures together with liquid oil (Global Agritech Inc., 2009). The present study shows that RBX crystals easily form organogels with different types of liquid oil.

Thermal and Structural Properties

The principles behind organogel formation, crystal network formation, and their functionality are the most important areas to be studied, as they are unique to the organogelators used as well as playing crucial roles in the structure of the product. The identification and quantification of the structural organization of food systems provide insight into the relationships among composition, processing, structure, and mechanical properties of the networks formed. However, comprehensive knowledge of structural network characteristics is still lacking (Narine & Marangoni, 2005), and organogels have been underutilized thus far, due to the inability to modify or predict the microstructure of the fibrous network structure (Wang et al., 2006).

From the perspective of structuring liquid oil using RBX, continued in-depth studies of the physico-chemical properties of organogels at the microscopic level are of utmost importance. The best approach to obtain detailed knowledge of physical properties is to start from the molecular arrangement adopted by the components of the system (Larsson et al., 2006; Flöter 2009). Fundamental research on the molecular characteristics of RBX organogels would help guide the modification of the macroscopic properties of the gels.

We have compared the properties of RBX with the other plant waxes that are commonly used in foods and cosmetics. RBX, CRX, and CLX were supplied by Global Agritech, Inc., Minneapolis, MN USA. Table 7.A.1 lists the specifications of RBX and Table 7.A.2 presents the chemical compositions of three waxes. The melting point of RBX is 78 ± 1°C. The acid value is 6.3mg KOH/g, the iodine value is 11.4, and the saponification value is 86.3mg KOH/g (Table 7.A.1).

The approximate chemical composition of CRX (wt%) is aliphatic and aromatic esters, 84–85% (containing aliphatic esters 40%, ω-hydroxy esters 13%, cinnamic aliphatic diesters 8%), free fatty acids 3–3.5%, alcohols 2–3%, lactides 2–3%, hydrocarbons 1–3%, resins 4–6%, and moisture and inorganic residues 0.5–1% (Letcher, 1983). The composition of CLX is 49–50% *n*-alkanes, 20–29% esters of acids and alcohols with even-numbered carbon chains (C_{28-34}), 12–14% alcohols and sterols, and 7–9% free acids (Morales-Rueda et al., 2009).

The organogels, were prepared by blending RBX, CLX, and CRX with salad oil (canola:soybean oil = 50:50) (C16:0 8.3%, C18:1 38.1%, C18:2 40.3%, C18:3

Table 7.A. Chemical Properties of Rice Bran Wax (RBX).

(1) Specifications of RBX		
Property	Typical Analysis	Range
Melting point (°C)	78	77–79
Acid value (mg KOH/g)	6.3	3–8
Iodine value	11.4	8–15
Saponification value (mg KOH/g)	86.3	80–90
(2) Chemical composition		
Carbon Number	Fatty Acid wt. %	Fatty Alcohol wt. %
16	3.6	—
18	2.3	—
20	5.3	—
22	26.1	—
24	40.5	3.5
26	11.5	8.4
28	6.6	14.0
30	3.1	26.4
32	1.0	19.6
34	—	16.5
36	—	9.2
38	—	2.4

7.9%) (Nisshin OilliO, Japan), olive oil (J-Oil Mills, Japan), and liquid paraffin (Wako Chemicals, Japan) at 0.2 to 4.0 wt% levels. Samples were prepared by dissolving the weighed solid waxes in a weighed amount of liquid oil at 80°C, which was sufficient to obtain full melting of all components. The heated solutions were subsequently cooled to room temperature (20°C) and stored at this temperature. All concentrations discussed here are expressed as a weight percentage (wt%). As a reference, we employed tripalmitoyl-glycerol (PPP), which was purchased from Sigma Chemicals, Japan (purity, 99%).

Thermal Behavior

Most of the wax esters in their natural form contain small amounts of sterols, sterol esters, fatty alcohols, fatty acids, and resinous matters. The plant wax esters after refining mainly contain esters of fatty acids and fatty alcohols of different chain lengths. The chemical composition of fatty acid and fatty alcohol moieties of RBX is shown in

Table 7.A.2. The RBX measured in this study contains esters of fatty acids and fatty alcohols of carbon numbers varying from 16 to 32, and from 24 to 38, respectively. Taking into account the high concentrations of carbon numbers of 22 and 24 for the fatty acid moiety, and 30 and 32 for the fatty alcohol moiety, we may assume that the most abundant wax esters contain the total carbon numbers of 52, 54 and 56 (Fig. 7.1). The high-temperature gas chromatography analysis reported by Vali et al. (Vali et al., 2005) corroborates this observation with the molecular species having carbon numbers 52, 54, and 56 as the major peaks in the chromatogram. This property may be a unique feature of RBX having long-chain fatty acids and fatty alcohol esters. CLX contains hydrocarbons (*n*-alkanes) to be 50–55% with C_{31}, the principal component—hentriacontane (Morales-Rueda, 2009; Tulloch, 1973). In addition, it contains esters of fatty acids with carbon numbers varying from 16 to 34, and fatty alcohols with carbon numbers varying from 22 to 34 with free fatty acids and free fatty alcohols (Table 7.A.2). The large range of chain lengths gives rise to the large range in the esters (Tulloch, 1973). However, Tulloch reported that CRX contains 36% monoesters, 11% free alcohols, and 47% total volatiles (Tulloch,1973). Tada et al. reported the presence of fatty acids, carbon numbers varying from 18 to 28, and fatty alcohols, carbon numbers varying from 28 to 34 with an abundance of C_{32} fatty alcohols (Tada et al., 2007). As noted by Letcher and others, CRX contains fatty esters (80–85%), free alcohols (10–15%), acids (3–6%), and hydrocarbons (1–3%) (Letcher, 1983; Cyber lipid center, 2009). As a peculiarity, carnauba wax contains esterified fatty dialcohols (diols, about 20%), hydroxylated fatty acids (about 6%), and cinnamic acid (about 10%). Plant waxes are derived from natural sources; therefore, these waxes have different compositions and exhibit various physical behaviors, which are unique to each material.

The melting and crystallization temperatures of waxes and wax-liquid oil mixtures were determined by DSC using a Rigaku DSC-8240 calorimeter (Rigaku,

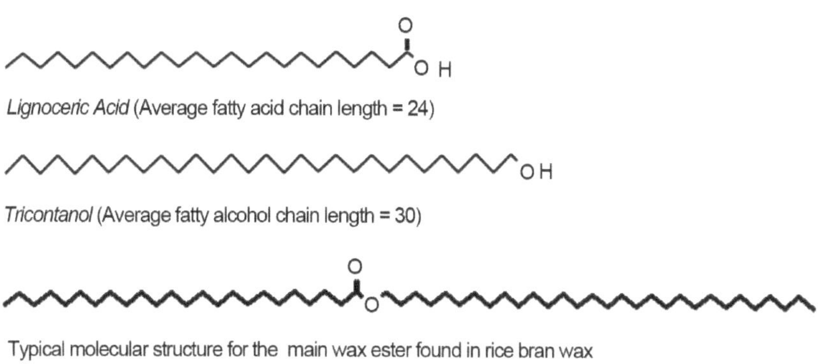

Fig. 7.1. Schematic illustration of RBX chemical structure.

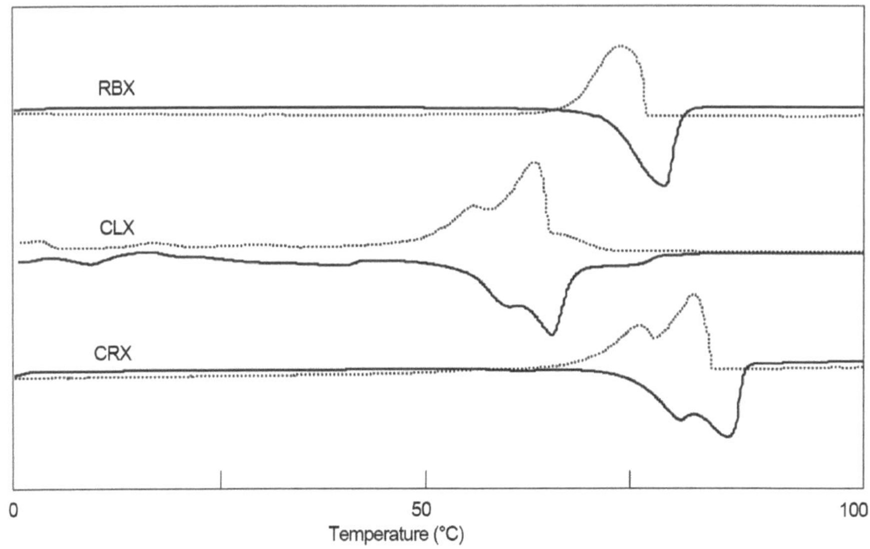

Fig. 7.2. DSC patterns taken during cooling (dotted line) and heating (solid line) processes of rice bran wax (RBX), carnauba wax (CRX), and candellila wax (CLX) in bulk states.

Tokyo, Japan). Fig. 7.2 shows DSC cooling and heating thermopeaks of RBX, CLX, and CRX in pure bulk states examined during cooling to 0°C after the melting at 100°C, and re-heating (0°C to 100°C) from the first-crystallized materials. Irrespective of the gelator concentration and the cooling and heating rates employed in this study, the three waxes showed broader DSC melting peaks with indefinite start and blunt maximum, indicating the presence of a series of different solid substances. However, RBX shows only single melting and crystallization profiles, while multiple melting and crystallization profiles are observed for CLX and CRX. The molecular compositional differences explain the thermal behavior of these waxes. RBX, containing a single chemical class of wax esters of aliphatic fatty acids esterified to fatty alcohols, gives more uniform single melting and crystallization peaks. CRX and CLX contain multiple chemical classes and hence exhibit complex melting and crystallization behavior.

To characterize the melting and crystallization behavior of CLX and CRX that revealed the multiple DSC patterns, we took the values of melting and crystallization temperatures at the peak tops of the highest-temperature endothermic and exothermic patterns taken during the heating and cooling processes, respectively. The values of enthalpy and entropy of melting were obtained for a single endothermic peak for RBX, and for the summation of major endothermic peaks. Table 7.B summarizes the melting temperatures, T_m, and the enthalpy and entropy of melting (ΔH_m and ΔS_m) of RBX, CLX, and CRX. The thermal stability expressed as T_m increases is an order

Table 7.B. Thermal Properties of Plant Waxes (RBX, CLX, and CRX) and Tripalmitoyl-glycerol (PPP).

	T_m (°C)	ΔH_m (J/g)	ΔS_m (mJ/g/K)
RBX	76	190.5	559.0
CLX	64	129.0	376.1
CRX	84	137.6	384.6
PPP	66.4	209.6	609.3

of CRX > RBX > CLX. However the values of ΔH_m and ΔS_m do not follow the same trend: the value of ΔH_m of RBX is larger than that of CRX by 40% and larger than that of CLX by 50%. A similar trend was observed for ΔS_m values. Quite interestingly, ΔH_m and ΔS_m of RBX are close to those of PPP, rather than the CRX and CLX. The ΔH_m and ΔS_m values indicate that cohesive energy for the packing of molecules in the crystal structure of RBX is large. The similarity of ΔH_m and ΔS_m between RBX and PPP can be explained by similarities in the crystal packing between the two lipids.

Toro-Vazquez et al. described that the presence of different *n*-alkanes (nonacosane, tritriacontane, etc.) in CLX developed a mixed molecular packing with hentriacontane during cooling, which caused multiple melting and crystallization profiles (Toro-Vazquez et al., 2007). Therefore we can assume that the long chain fatty acid and fatty alcohol esters present in RBX do not tend to form such mixed self-assembled structures with a lower extent of three-dimensional molecular structure (i.e., lower crystallinity and ΔH_m). In addition to that, RBX exhibited comparatively high self-assembly capability and this is confirmed by the longer crystal dimension with birefringence of the structures (see Fig. 7.3).

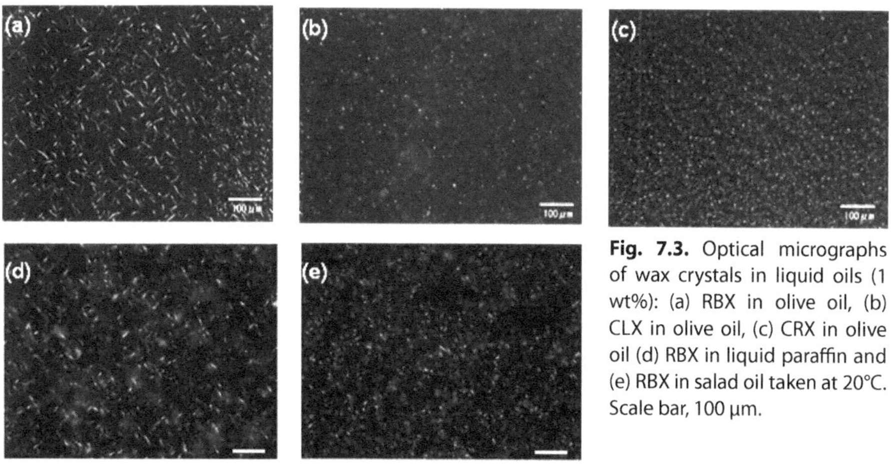

Fig. 7.3. Optical micrographs of wax crystals in liquid oils (1 wt%): (a) RBX in olive oil, (b) CLX in olive oil, (c) CRX in olive oil (d) RBX in liquid paraffin and (e) RBX in salad oil taken at 20°C. Scale bar, 100 μm.

RBX was blended with salad oil (canola: soybean oil = 50:50) (C16:0 8.3%, C18:1 38.1%, C18:2 40.3%, C18:3 7.9%) (Nisshin OilliO, Japan), olive oil (J-Oil Mills, Japan), and camellia oil (Yokozeki Oil and Fat Industries Co., Japan) at 1, 3, 6, and 10 wt% levels.

The melting and crystallization temperatures (T_m and T_c) of RBX-vegetable oil organogels were determined by differential scanning calorimetry (DSC) using a Rigaku DSC-8240 calorimeter (Rigaku, Tokyo, Japan). Peak T_m, T_c, melting enthalpy (ΔH_m), and crystallization (ΔH_c) were determined using software provided with the instrument. DSC measurements were carried out in triplicate for RBX in liquid oils. The average endothermic peak value was taken as the T_m.

The cooling and heating thermograms of RBX in salad oil, olive oil, and camellia oil mixtures at a concentration of 1 wt% are presented in Fig. 7.4. The thermal behavior of the RBX-vegetable oil mixtures was examined during cooling to 0°C after melting at 100°C, and re-heating of the first crystallized material from 0°C to 100°C. The cooling and heating rates were 5°C/min. Regardless of the type of vegetable oil used, RBX exhibited broad DSC melting peaks with indefinite start and a blunt maximum, as we observed in earlier investigations for bulk RBX (Fig. 7.2) and RBX-salad oil mixtures in different RBX concentrations at a rate of 2°C/min for cooling and heating (see Fig. 7.5).

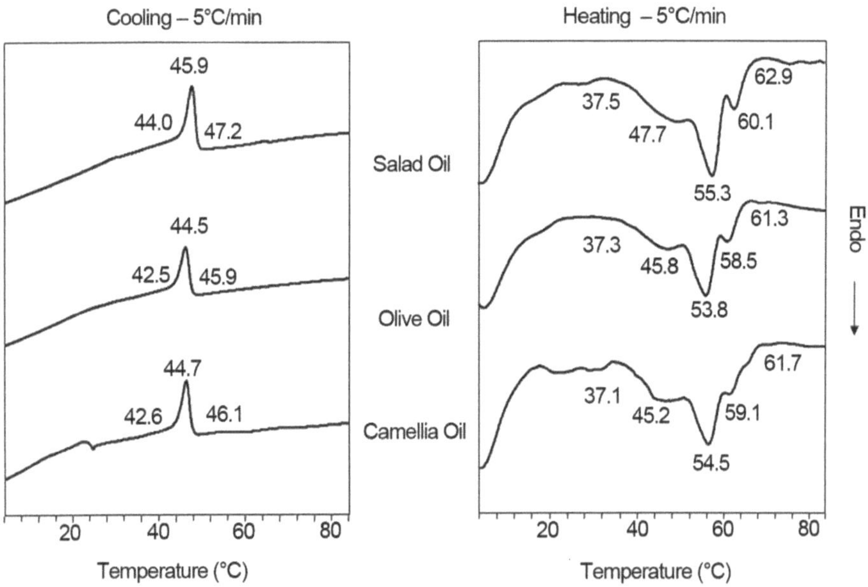

Fig. 7.4. DSC measurements of mixtures of RBX in different oils (1 wt%).

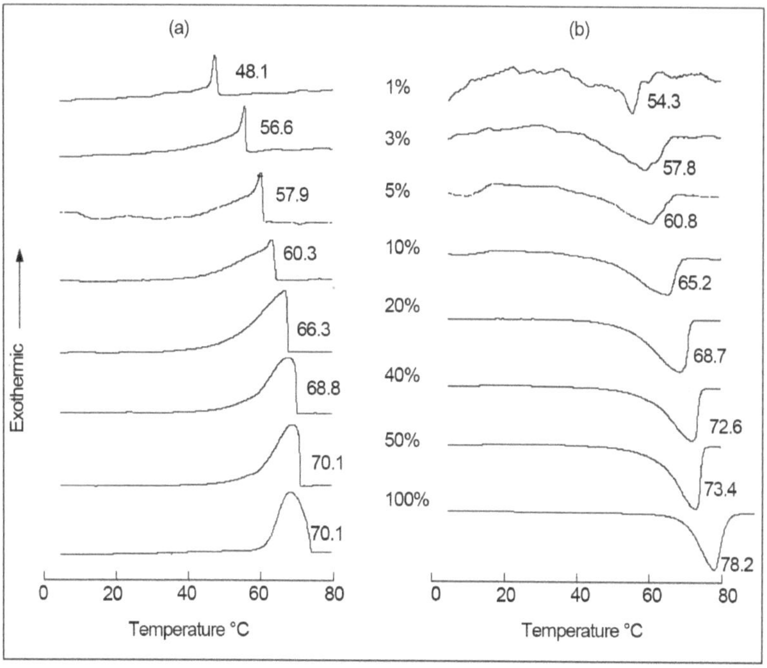

Fig.7.5. DSC (a) cooling and (b) heating thermopeaks of rice bran wax (RBX) and salad oil mixtures at different concentrations of RBX.

RBX is a natural product with a complex composition created by esterified long-chain fatty acids and long-chain fatty alcohols of different chain lengths, along with various other minor components, which explains the presence of broader exothermal and endothermal peaks (Dassanayake et al., 2009). RBX exhibited clear single exotherms in all tested oil mixtures, indicating that all of the components in RBX form crystals in a narrow temperature range (3.3±0.2°C). The onset of crystallization, peak top, and the end set of crystallization temperatures are indicated in the figure. RBX-olive oil and RBX-camellia oil mixtures have more similar melting and crystallization temperatures than RBX-salad oil mixtures.

However, during heating, RBX-vegetable oil mixtures exhibited much broader (24.6±0.8°C) endotherms, with one prominent peak and two shoulder peaks on either sides of the endothermic peak. This behavior is quite different from what we observed previously with bulk RBX (Fig. 7.2) and higher concentrations (>1 wt%) of RBX in salad oil at a heating rate of 2°C/min (see Fig. 7.5). Fig. 7.4 presents the onset of melting, three peak top temperatures, and the end set of melting temperatures. These multiple peaks in the melting profile of RBX tend not to be visible at higher RBX concentrations (>1 wt%). Thus, melting of the less dominant fatty acid and fatty alcohol esters becomes less prominent, due to the melting of more abundant fatty acid and fatty alcohol esters in the higher levels of RBX-vegetable oil mixtures. Almost all of the combinations of RBX-vegetable oil mixtures in Fig. 7.4 exhibit similar thermal behavior, although RBX-salad oil mixtures deviate from the rest. This result indicates that the liquid component does not affect the thermal properties of the organogel, and it is solely governed by the solid component or the organogelator that causes the network to entrap the liquid oil to produce the organogel. Burkhardt et al. studied the correlation between the polarity of the solvent and the amount of gelator required for gelation in organogels prepared using 12-hydroxystearic acid (HSA) (Burkhardt et al., 2009). The required amount of HSA for gelation increased with more polar solvents. The appearance of formed organogels differs with HSA concentration and solvent type (Burkhardt et al., 2009). However, the appearance of RBX-vegetable oil organogels differs only with RBX concentration and not with solvent type.

Crystal Morphology

The wax crystals formed in olive oil, liquid paraffin, and salad oil were observed with polarized optical microscopy. The crystal morphology of RBX, CRX, and CLX were compared at the same solute concentration of 1 wt%. The mixtures of the three waxes and liquid oils were prepared by adding an appropriate amount of wax to liquid oils and heating to 80°C to form a homogenous solution. The solutions were cooled to room temperature without stirring until wax crystals formed.

Fig. 7.3 presents optical micrographs of the crystals of RBX in olive oil, liquid paraffin, and salad oil, and CRX and CLX (1 wt%) in olive oil taken at room temperature. In the first micrograph, we compare the crystal morphology of RBX in three liquid oils prepared by moderate cooling without stirring. Although details of the organogel-forming properties will be described later, it must be mentioned here that at the 1 wt% level of wax in olive oil, only RBX was in a gel state, whereas CLX and CRX were in a viscous sol state. Accordingly, the crystal morphologies of the three waxes in olive oil were quite different. The morphology of RBX crystals was long (20 to 50μm) needles, which is a desirable feature for gel formation (Terech et al., 1997; Kodali, 2009). The crystal morphology of CRX and CLX differs from RBX, with spherulitic structures having a diameter of less than 10μm, which are not conducive to organogel formation. The long needle-like structures of RBX form good crystal matrices that mesh well at the intercrystalline interfaces to form organogels. This enables RBX to entrap large volumes of liquid oils in crystalline scaffolding. Similar observations were reported by Morales-Rueda et al., for CLX and pure *n*-alkanes, i.e., dotriacontane (C_{32}) (Morales-Rueda et al., 2009). Regarding the effects of liquid oils on the crystal morphology of RBX, no difference was observed in crystal morphology between olive oil and salad oil; however, the RBX crystals in liquid paraffin were round and therefore not conducive to organogel formation. Several researchers reported the strong influences of the gelator composition on crystal size, shape, and rheology of organogels developed in edible oils (Bot et al., 2008; Daniel & Rajasekaran, 2003; Gandolfo et al., 2004; Morales-Rueda et al., 2009). The presence of different molecular components in the wax seemed to have a profound effect on the crystal habit of the plant wax-vegetable oil organogels. Also the conditions of organogel preparation (e.g., stirring or rate of cooling) and storage conditions can affect the crystal morphology of the gelators (Toro-Vazquez et al., 2007; Bot & Agterof, 2006). Therefore the influence of rapid cooling with mechanical disturbance on the crystal morphology and gel formation of plant wax-vegetable oil organogels needs to be evaluated.

The RBX crystals formed in olive oil were observed with polarized optical microscopy. The crystal morphology was compared at different solute concentrations of RBX at 1, 6, and 10 wt%. The mixtures of RBX and olive oil were prepared by adding an appropriate amount of wax to liquid oils and heating to 80°C to form a homogenous solution. The solutions were cooled to room temperature until wax crystals formed.

Fig. 7.6 presents optical micrographs of the crystals of RBX in olive oil taken at room temperature. We compared the crystal morphology of different RBX concentrations in olive oil. The RBX crystals had a long, needle shape, which is a desirable feature for gel formation (Terech & Weiss, 1997; Kodali, 2009). The size of the crystal increased proportionately with increased RBX concentration. RBX crystals at 10 wt% in olive oil mixture exhibited thick crystals, 100μm or longer, whereas the crystals at 1 wt% in olive oil mixture were 20 to 50μm long. The long needle-like structures of RBX form a good crystal matrix that meshes well at the crystal-crystal interfaces to form organogels. Therefore, we can expect that the interaction between RBX crystals and gel strength may improve in proportion to the solute concentration. Regarding the effects of liquid oils on the crystal morphology of RBX, no difference was observed in crystal morphology between olive oil and salad oil (data not shown).

Crystal Structure

X-ray diffraction (XRD) studies of bulk RBX, CLX, and CRX and wax-olive oil mixtures were made by using a RINT-2000 X-Ray Diffractometer (Rigaku, Tokyo, Japan).

Fig. 7.7 displays the XRD patterns of RBX crystals in bulk and wax-oil mixtures (olive oil, RBX concentration of 1 wt%). The wide-angle diffraction peaks of 0.415nm and 0.373nm indicating orthorhombic perpendicular ($O_\perp$) subcell packing are observed in the bulk state and wax-oil mixtures. The crystals of triacylglycerols

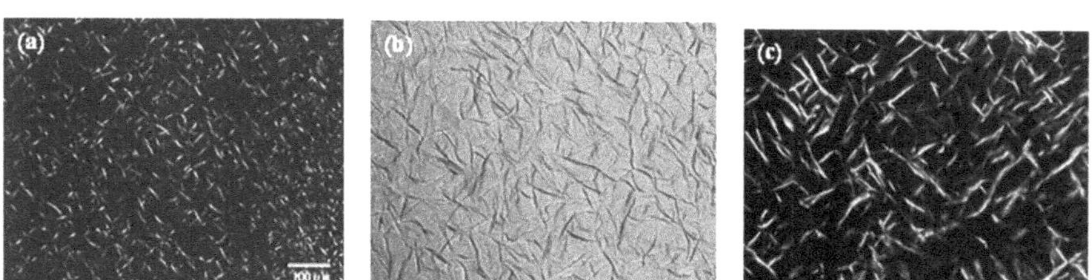

Fig. 7.6. Optical micrographs of RBX crystals (a) 1, (b) 6, (c) and 10 wt% in olive oil. Scale bar, 100 μm.

are typically packed in three different polymorphs (α, β', and β) with hexagonal, $O_{\perp}$, and triclinic parallel ($T_{//}$) subcell packings (Small, 1986; Parish et al., 2002). Among the three polymorphs, the fat crystals of the beta-prime polymorph exhibit more functional properties, such as smooth texture and better mouth-feel in margarine and shortening, because of its crystal morphology, dispersability, and optimal melting (Parish et al., 2002; Sato et al., 2001). The wide-angle XRD patterns of the three wax crystals have peaks of 0.41nm and 0.37nm, similar to those of the β' form of triacylglycerols. However, only RBX revealed clear and prominent small-angle diffraction peaks compared to CRX and CLX, both in bulk state and wax-oil mixtures. The long spacings were calculated from the (003) reflections of the small-angle diffraction patterns as 6.9nm. The XRD patterns of RBX-olive oil mixture exhibited the same small-angle and wide-angle peaks as those in the bulk state, although a broad peak due to liquid oil appeared in the wide-angle region.

The intense XRD diffraction peaks shown in Fig. 7.7 may be related to the morphology of the RBX crystals. The intensity of the long-spacing patterns was remarkably weaker than that of the short-spacing patterns. In general, such a contrast in the XRD diffraction peak intensity between the long and short-spacing patterns is related to strong anisotropy in crystal growth rates between the directions perpendicular to the lamellar plane and parallel to the lamellar plane. If the growth rate parallel to the lamellar plane is slower that that perpendicular to the lamellar plane, thin crystals having needle or platelet morphology are formed, showing weak long-spacing patterns and stronger short-spacing patterns. In the case of RBX crystals, the

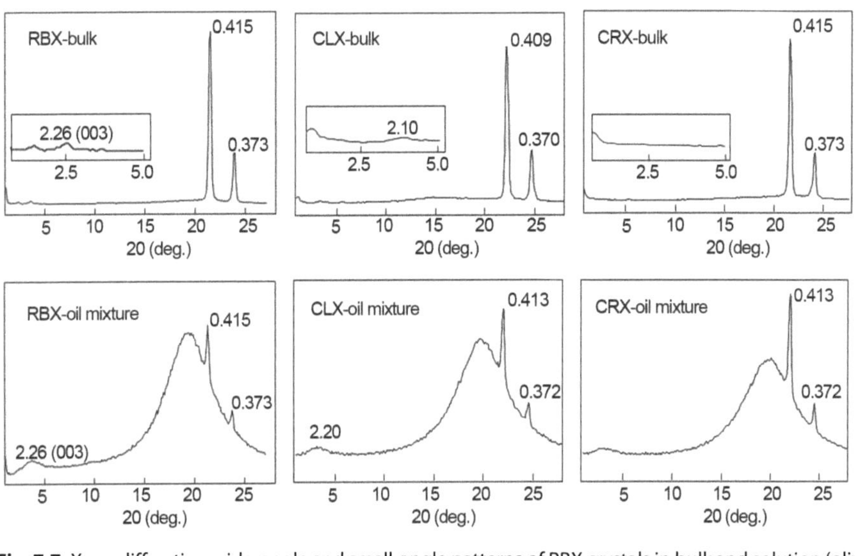

Fig. 7.7. X-ray diffraction wide-angle and small-angle patterns of RBX crystals in bulk and solution (olive oil) states taken at 20°C. The small-angle pattern of the bulk data is enlarged in a box.

long linear molecular structure, containing an average of 54 carbons together with polar interactions at the ester-bond region, may provide strong attractive interactions within the lamellar plane and weak interactions between lamellar planes through methyl end-groups. This property may explain the difference in the XRD peaks of the long and short-spacing patterns of the RBX crystals.

Synchrotron radiation X-ray diffraction (SR-XRD, λ = 0.15nm) studies were conducted at the Photon Factory (PF) at the National Laboratory for High-Energy Physics, Tsukuba, Japan. The temperature of the sample was controlled by two water baths and recorded in combination with XRD data collection. The 2 wt% RBX-salad oil organogel was heated to melt, and the XRD spectra were recorded during cooling from 80°C to 20°C.

Fig. 7.8 depicts small and wide-angle SR-XRD patterns that were simultaneously measured with DSC, taken during the cooling process of 2 wt% RBX-salad oil gels. Fig. 7.8.B depicts a DSC cooling thermogram that was initially heated to 100°C and then cooled to 0°C at a rate of 5°C/min. The onset of the endothermic peak was 52°C, and the peak top temperature was 51°C.

The powerful X-ray beam obtained from the synchrotron source enabled a complete and detailed X-ray diffraction spectra compared to the conventional powder XRD studies, especially in the small-angle region. The short-spacing spectra agree with the conventional powder XRD studies of bulk RBX and 8 wt% RBX-olive oil gels

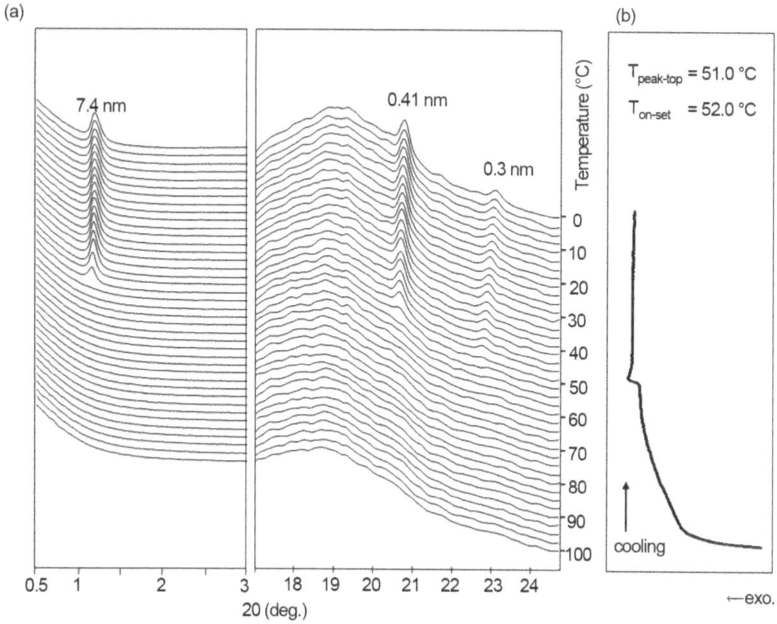

Fig. 7.8. Synchrotron radiation X-ray diffraction spectra (a) and DSC cooling thermogram (b) of rice bran wax (2 wt%) in salad oil.

observed earlier (Fig. 7.7). However, the long-spacing spectra obtained from SR-XRD do not agree with the conventional powder XRD observations (Fig. 7.7). Considering the reliability and the potential of SR-XRD over the conventional XRD in producing diffraction spectra in small-angle region, the d-spacings obtained by SR-XRD could be accepted.

Fig. 7.8.A depicts small and wide-angle SR-XRD spectra taken at 30-second intervals with a 10-second exposure time. Soon after 52°C was reached, a small peak appeared in the small-angle region, having long spacing of 7.4nm. Upon cooling, the peak constantly remained up to 0°C. Correspondingly, short-spacing spectra of 0.41nm and 0.38nm appeared at 52°C in the wide-angle region. This result indicates orthorhombic perpendicular ($O_{\perp}$) subcell packing that was characterized by RBX in Fig. 7.8 and described by Kodali (Kodali, 2009). Similar structures are found for most natural plant waxes and beeswax (Kameda, 2005; Ritter et al., 2001; Ensikat et al., 2006; Mellema, 2009). Remarkably weaker intensity of long-spacing patterns than that of short-spacing patterns demonstrates strong anisotropy in crystal growth rates between directions perpendicular to the lamellar plane and parallel to the lamellar plane. Kodali stated that the molecular interactions within the lamellar planes are far greater than those between methyl end-groups due to van der Waal's interactions between the long hydrocarbon chains and polar ester functional groups (Kodali, 2009). This property leads to faster crystal growth rate perpendicular to the lamellar plane and an extremely slow crystal growth rate parallel to the methyl planes. This unique crystal growth of RBX leads to either needle or platelet crystal morphology, as observed in Fig. 7.3 and 7.6. Recent cryo-TEM (transmission electron microscopic) analysis of fat crystals in liquid oil revealed such crystal morphology as thin plate and long needle plate on the nanometer-scale (Acevedo & Marangoni, 2010).

Phase Diagrams

The melting points of RBX crystals in liquid paraffin, salad oil, and olive oil were measured in mixtures with different RBX concentrations of 1, 2, 3, 4, 5, 10, 15, 20, 25, 30, 40, 50, and 100 wt% to construct the phase diagram of RBX in liquid oil. The heated solutions were subsequently cooled to room temperature (20°C) and stored at this temperature.

The melting temperature and crystallization temperature of three waxes in pure bulk states and in RBX-salad oil organogel states were determined by differential scanning calorimetry using the Rigaku DSC-8240 calorimeter (Rigaku, Tokyo, Japan). The rate of temperature variation was 2(C/min for heating and cooling.

Fig. 7.5 shows graphs of the crystallization (T_c) and melting (T_m) temperaturesof RBX in salad oil-RBX mixtures at different RBX concentrations. The peak temperatures of DSC cooling and heating thermopeaks were chosen as the T_c and T_m of RBX throughout the entire range of RBX concentrations in the RBX-salad oil mixtures. As the melting point of salad oil (canola:soybean oil = 50:50) is –22°C, all the exothermic and endothermic peaks depicted in Fig. 7.5 were due to the crystallization and melting of RBX. T_m decreased from 78.2 to 73.4°C from pure bulk to 50% solution, and a further decrease in T_m was observed with lower concentrations of RBX. However, T_c did not differ from that of the pure bulk sample at RBX concentrations down to 40% and gradually decreased with lower RBX concentrations. Although the DSC patterns exhibited a high noise level because of the increase of salad oil, a sharp exothermic peak of crystallization was detected at 48.1°C and an endothermic peak of melting was detected at 54.3°C, even at an RBX concentration of 1%. The molecular asymmetry between linear wax molecules compared to the oil molecules (triacylglycerols), which take a non-linear chair configuration in the solid state, is different from each other. This molecular difference is responsible for rapid phase separation and crystallization of wax solute from oil solvent.

Fig. 7.9 presents the average values of T_m and T_c of RBX in salad oil, liquid paraffin, and olive oil. It is clear that T_c did not decrease significantly from that of pure RBX with lower RBX concentrations down to 40%, whereas T_c started to decrease

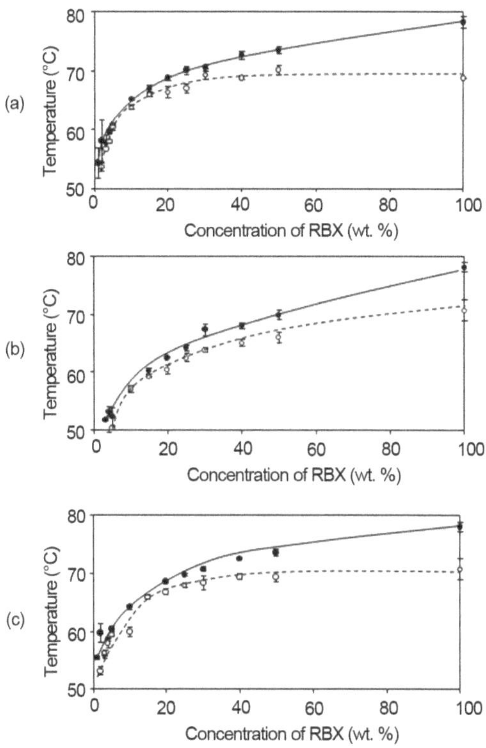

Fig. 7.9. Crystallization temperatures (open circle) and melting temperatures (closed circle) of rice bran wax (RBX) in (a) salad oil, (b) liquid paraffin, and (c) olive oil.

below 30% in the three RBX-oil mixtures. In contrast, T_m gradually decreased with lower RBX concentrations; the values of supercooling ΔT (T_m–T_c) decreased with lower RBX concentrations from 10 (pure) to 4°C (20%) and reached as low as less than 2°C at RBX concentrations below 20% for RBX-salad oil mixtures. The same tendency was observed in the mixtures of RBX-liquid paraffin and RBX-olive oil. This result indicates that the rate of crystallization of RBX at low concentration in salad oil is relatively high. Below the plotted T_c line, the RBX-liquid oil mixture is in a complete gel state, while above the T_m line it is in a complete sol state. Therefore, we conclude that the RBX-liquid oil mixtures tend to form a gel structure as soon as the temperature of the mixture decreases to the melting point below the RBX concentration of 20%. In particular, the gelation ability at room temperature (20°C) was high even at very low concentrations of 1 wt% in salad oil and at the concentration of 0.5 wt% in olive oil.

Gel Strength

Organogel hardness is a critical factor in industrial applications. Neither harder nor softer states are considered a favorable quality for organogels. We measured the hardness of the different levels of the three plant waxes and olive oil mixtures at 20°C (Fig. 7.10). An EX–120–E Penetrometer (Elex Scientific, Tokyo, Japan) was used to measure the hardness of the organogels made of RBX, CRX, and CLX in olive oil. The penetration depth was measured three times for each sample at 20°C, and average values were calculated. The depth (mm) at which the penetration probe of the

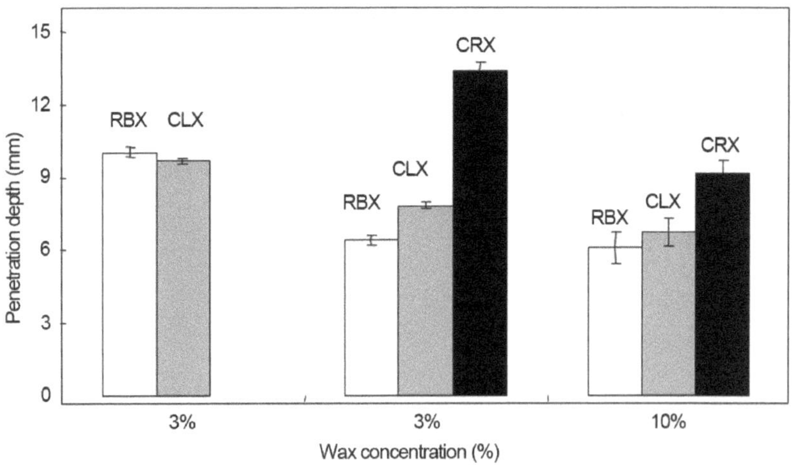

Fig. 7.10. Penetration depth of the organogels of three waxes in olive oil at different concentrations measured at 20°C.

instrument penetrated the gel was used to evaluate hardness, and large values of penetration depth indicate softer gel structure.

At a concentration level of 3%, the penetration depths of RBX-olive oil and CLX-olive oil mixtures were almost the same. However, RBX exhibited greater hardness than CRX and CLX, with increasing wax concentrations of 6% and 10%. In particular, the penetration depth of the RBX-olive oil mixture was half that of the CRX-olive oil mixture and 20% less than that of the CLX-olive oil mixture.

The effect of wax type on hardness and viscosity can be explained by crystal behavior, shape size, and thermal kinetics. Fiber-like, smaller, needle crystals lead to stronger gel network formation, followed by high melting and crystallization points, which cause high viscosity and hard gels. The comparatively longer, fibrous crystals of RBX form a better gelator network that entraps the liquid oil, as evidenced in microscopy (Fig. 7.6) and gelation behavior (Tables 7.C and 7.D).

Table 7.C. Induction Time for Crystallization (τ_c) and Gelation (τ_g) of Three Plant Waxes at Different Concentrations in Olive Oil at 20°C.

Concentration of wax (%)		τ_c (min:sec)			τ_g (min:sec)	
		RBX	CLX	CRX	RBX	CRX
0.2	6:1	*	4:5	•	•	•
0.5	2:4	*	2:1	10:5	•	•
1.0	1:4	7:4	1:3	7:2	•	•
2.0	1:0	4:4	1:0	6:2	13:5	•
4.0	0:3	3:1	0:4	4:4	9:2	13:5

* No crystals observed in 2 days

• No gelation observed in 2 days

Table 7.D. Induction Time for Crystallization (τ_c) and Gelation (τ_g) of Rice Bran Wax at Different Concentrations in Liquid Paraffin and Salad Oil at 20°C.

Concentration of wax (%)		liquid paraffin	salad oil	
	τ_c (min:sec)	τ_g (min:sec)	τ_c (min:sec)	τ_g (min:sec)
0.5	7:2	*	3:0	*
1.0	4:0	14:1	2:5	10:5
2.0	3:5	13:5	0:5	8:1
4.0	1:1	6:6	0:2	4:2
8.0	0:2	5:6	0:2	3:2

* No gelation

Kinetics of Gel Formation

We measured the induction time of crystallization (τ_c) and gelation (τ_g) of RBX, CLX, and CRX in olive oil to determine the kinetic properties of organogel formation (Table 7.C). The induction time of crystallization (τ_c) was defined as the duration between the time when the temperature of the wax-olive oil mixture reached 20°C and the time when crystallization was confirmed by optical microscopic observation. Similarly, the τ_g was defined as the time necessary for the oil to stop flowing after the sample was tilted to 45° at 20°C.

The plant waxes (RBX, CRX, and CLX) and olive oil mixtures were heated to 80°C and then cooled to 20°C; they were visual observed until the appearance of crystals was confirmed under quiescent conditions.

The effects of liquid oils on the crystallization of RBX were observed by using salad oil, olive oil, and liquid paraffin. In addition, the formation of organogels was observed by using the three liquid oils with RBX concentrations of 0.2, 0.5, 1.0, 2.0, and 4.0 wt% in the solutions. The mixtures were heated to 80°C, cooled, and kept at 20°C to observe gelation. The gelation time was measured as the duration until the oil stopped flowing.

We found that τ_c and τ_g were highly dependent on the wax concentration. However, τ_c of RBX and CRX was much shorter than that of CLX at any wax concentration. Although CRX formed crystals relatively faster than other waxes, it exhibited a weaker ability to form gel, and the minimum wax concentration that could form gel was 4.0 wt%. The lowest concentration of CLX that exhibited crystallization was 1.0 wt%, and the lowest concentration exhibiting gelation was 2.0 wt%. RBX had the strong ability to form a gel at a concentration as low as 0.5 wt%, which is the lowest among the three waxes, since the minimum concentration of CLX that formed gels was 2.0 wt% and the minimum concentration of CRX that formed gels was 4.0 wt%. Below these levels, no gelation was observed even after two days at 20°C, even though crystallization was detectable in the wax-solvent mixtures. Although RBX formed crystals more slowly than CRX, it was much faster and more capable in gel formation. The gelation time of RBX in different oils decreased in the order of olive oil < salad oil < paraffin oil.

It has been reported that the nature of the organic solvent can also influence the aggregation of wax molecules (Terech & Weiss, 1997). Therefore, we measured τ_c and τ_g values in RBX-liquid paraffin and RBX-salad oil mixtures at 20°C (Table 7.D). Crystallization and gelation occurred more rapidly in the RBX-salad oil mixture than in the RBX-liquid paraffin mixture at the same RBX concentration. No gelation was observed at a concentration of 0.5 wt% in paraffin and salad oil, although gelation occurred in olive oil at this level of RBX (Table 7.C). This result indicates that the properties of the solvent play an important role in plant wax-vegetable oil organogel formation.

As another kinetic property of gel formation, it is interesting to relate the occurrence of crystallization and gel formation in the two mixtures of RBX-olive oil and

CLX-olive oil, in which the wax concentration was 1 wt%. The viscosity of the plant wax-vegetable oil organogels was measured during cooling using a Vibro Viscometer SV–10 (A & D, Japan) with a Thermo-mate BF 400 (Yamato, Japan). This viscometer uses a frequency of 30Hz at a constant frequency and amplitude. The cooling and heating rates were not constant and depended on the viscometer itself. The effect of the type of plant wax on the viscosity of the organogels was measured for 1 wt% of RBX and CLX in olive oil.

The prepared gel samples were heated to 80°C for complete melting. The melted sample was placed in the sample cell of the viscometer at 80°C, and two gold-plated paddle sensors were immersed in the sample. The viscosity change of the sample was then measured with respect to the temperature during cooling from 80°C to 20°C. The measured viscosity values for each sample were plotted against the temperature, and these viscosity profiles were analyzed by comparison with the DSC thermographs.

CRX was not examined, since its gel formation ability was too low. Crystallization was observed by DSC, and gel formation was examined by viscosity measurement (Fig. 7.11). Upon cooling, the two wax gels revealed significant differences in viscosity-crystallization features. For the CLX-olive oil mixture, an exothermic DSC peak appeared at 33°C, yet no increase in viscosity was observed until the temperature

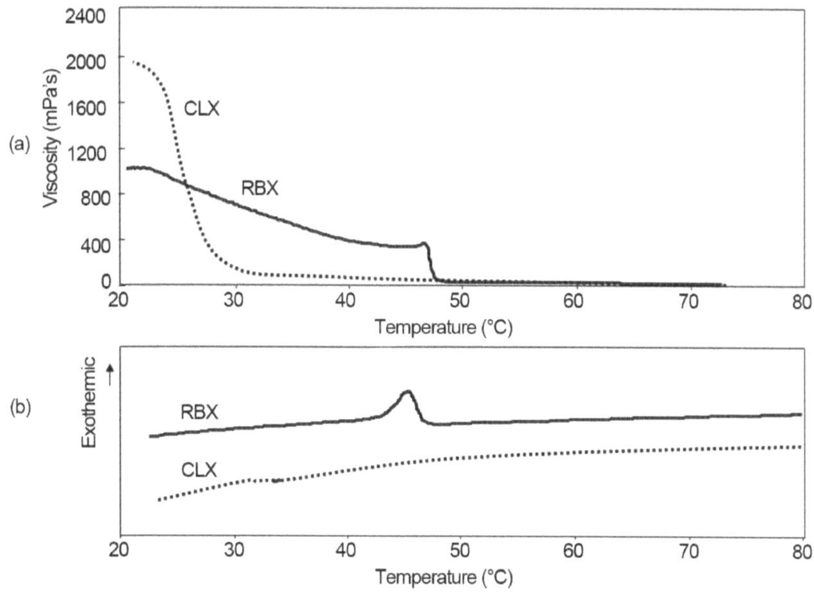

Fig. 7.11. Changes in viscosity during cooling from 80°C to 20°C along with DSC exothermic curves of different wax-olive oil mixtures.

declined below 30°C, whereas viscosity rose above 1600 mPa.s below 25°C. In contrast, the viscosity of the RBX-olive oil mixture started to increase and reached 400mPa.s soon after the DSC exothermic appeared at 46°C, and its value gradually increased with decreasing temperature, reaching 1100mPa.s at 20°C. Therefore, it is evident that viscosity changes in the gels occurred parallel to the onset of crystallization for the RBX-olive oil mixture. This result is highly consistent with the results of visually examined kinetic properties of gel formation (Tables 7.C and 7.D). For the CRX-olive oil 1 wt% mixture, no gelation was observed, although an increase in viscosity was observed at 20°C (not shown). From these results, we conclude that the network formation of RBX crystals to reveal organogel properties occurred more rapidly, coinciding with temperatures of crystallization higher than that of the mixtures of CLX and CRX in olive oil.

The viscosity of the RBX in different vegetable oil and different concentrations of RBX in salad oil organogels was measured during cooling using the same method described earlier. Fig. 7.12 presents the changes in viscosity with temperature of organogels prepared with 1 wt% RBX in the three vegetable oils. The viscosity of the organogels was measured during cooling from 80°C to 20°C. RBX-olive oil had the highest viscosity, and RBX-salad oil organogels had the lowest viscosity during the cooling process. RBX-olive oil organogel had nearly three times higher viscosity than the maximum viscosities of RBX-salad oil organogel. Furthermore, no significant difference was observed between the maximum viscosities presented by the olive oil and camellia oil with RBX at the end of cooling. In Fig. 7.11, we demonstrated that gel viscosity changes occurred at the same temperature as the onset of crystallization. Similar viscosity changes were observed during heating. All wax-oil mixtures reached minimum viscosities at temperatures equal to the end temperatures of the DSC endotherms.

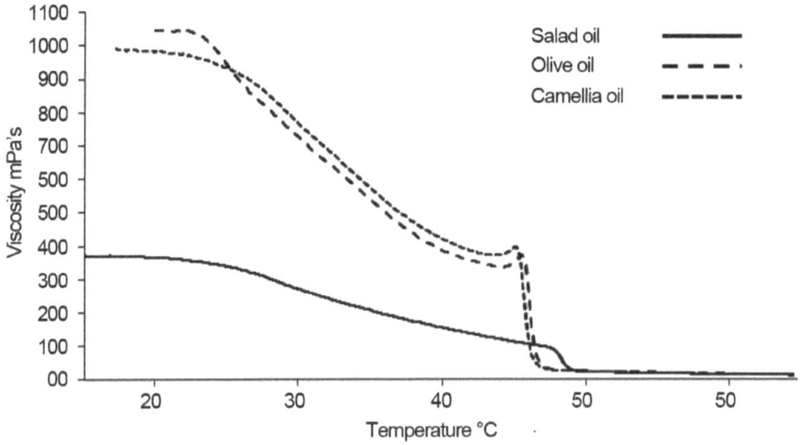

Fig. 7.12. Changes in viscosity during the cooling process of RBX (1 wt%) in three liquid oils.

As for the effects of liquid oil type, RBX-salad oil mixture displayed rather different characteristics in gelation behavior (Fig. 7.4), gel strength (Fig. 7.10), and viscous properties (Fig. 7.12), compared to RBX-olive oil and RBX-camellia oil mixtures. The major fatty acid composition of olive oil is Oleic (18:1) – 78.0%, Linoleic (18:2) – 8.3%, Linolenic (18:3) – 0.8%, Palmitic (16:0) – 8.4%, and Stearic (18:0) – 2.5% (Firestone, 2005). Camellia oil possesses similar fatty acid composition: Oleic (18:1) – 82.0%, Linoleic (18:2) – 6.8%, Linolenic (18:3) – 0.4%, Palmitic (16:0) – 7.9%, Stearic (18:0) – 1.8%, and Eicosaenoic (20:1) – 0.6% (Haiyan et al., 2007). The salad oil used in this study was a blend of canola oil containing Oleic (18:1) – 61.6%, Linoleic (18:2) – 21.7%, Linolenic (18:3) – 9.6%, Palmitic (16:0) – 3.6%, Stearic (18:0) – 1.5%, and Eicosaenoic (20:1) – 1.4% (Firestone, 2005), and soybean oil containing Oleic (18:1) – 23.8%, Linoleic (18:2) – 53.3%, Linolenic (18:3) – 7.6%, Palmitic (16:0) – 10.8%, and Stearic (18:0) – 4.0% (Firestone, 2005). The viscosity of RBX-salad oil mixture is lower than that of the other two oils (Fig. 7.12). The liquid oil type should therefore have some effect, since organogels using salad oil are softer than those made using olive oil and camellia oil. This result may be due to the increased concentration of oleic acid and the decreased concentration of linoleic acid for olive oil and camellia oil, compared with salad oil; therefore, RBX crystals are more soluble in salad oil than in the other two oils, and the strength of the crystal network of RBX in salad oil organogel may be weakened at the crystal junctions.

Conclusion

The physical properties of organogels formed by RBX and liquid oils in comparison with other typical plant waxes of CLX and CRX can be summarized as follows.

(a) The rate of formation of organogels was highest for RBX gels, as determined by visual observation and viscosity-temperature relationships.
(b) The hardness assessed by the penetration depth measurement was highest for RBX gels.
(c) Thermal stability expressed as the T_m of wax crystals in organogels was highest for RBX gels.

These properties may be ascribed to the high crystallinity of RBX crystals having the highest ΔH_m and ΔS_m, whose values are similar to those of the PPP β form. As PPP crystals are tightly packed in the $T_{//}$ subcell, it is reasonable to assume that the RBX crystals are also tightly packed in the $O_{\perp}$ subcell structures. The differences in the subcell structure may be responsible for the molecular differences between RBX and PPP structures. Furthermore, the crystal morphology of the needle-shaped RBX crystals in organogels may indicate a greater ability to entrap the liquid oils to form organogels.

Crystallization kinetics and physical properties of organogels prepared by RBX and different liquid oils were studied as a possible candidate for structured oils, namely organogels, without TAGs. High thermal stability, crystallinity, and strength of the RBX organogels were emphasized by the high melting temperatures and crystal morphology of the thin needle-shaped crystals of RBX, which caused a sharp increase in viscosity during the cooling process. The physical and functional properties of organogels were somewhat affected by the type of liquid oil used: salad oil had softer organogels than olive oil and camellia oil. However, RBX concentration exhibited the most significant effect on organogel characteristics, as evidenced by microscopy, viscosity, and hardness evaluations. SR-XRD results supported the evidence of tight packing of RBX crystals in the orthorhombic perpendicular ($O_{\perp}$) subcell structure.

References

Acevedo, N.C.; Marangoni, A.G. Characterization of the Nanoscale Triacylglycerol Crystal Networks. Cryst Growth Des. 2010, 10, 3327–3333.

Abdallah, D.J.; Weiss, R.G. Organogels and Low Molecular Mass Organic Gelators. Adv Mater. 2000, 12, 1237–1247.

Abdallah, D.J.; Sirchio, S.A.; Weiss, R.G. Hexatriacontane Organogels. The First Determination of the Conformation and Molecular Packing of a Low-Molecular-Mass Organogelator in Its Gelled State. Langmuir 2000, 16, 7558–7561.

Bot, A.; Agterof, W.G.M. Structuring of Edible Oils by Mixtures of γ-Oryzanol with β-Sitosterol or Related Phytosterols. J. Am. Oil Chem. Soc. 2006, 83, 513–521.

Bot, A.; Adel, R.; Roigers, E.C. Fibrils of γ-Oryzanol + β-Sitosterol in Edible Oil Organogels. J. Am. Oil Chem. Soc. 2008, 85, 1127–1134.

Bot, A.; Veldhuizen, Y.S.J.; den Adel, R.; Roijers, E.C. Non-TAG Structuring of edible Oils and Emulsions. Food Hydrocoll. 2009, 23, 1184–1189.

Burkhardt, M.; Kinzel, S.; Gradzielski, M. Macroscopic Properties and Microstructure of HSA Based Organogels: Sensitivity to Polar Additives. J. Colloid Interface Sci. 2009, 331, 514–521.

Cyber Lipid Center. Waxes and Other Esters. http://www.cyberlipid.org/wax/wax0001.htm#2 (accessed July 2009).

Daniel, J.; Rajasekaran, R. Organogelation of Plant Oils and Hydrocarbons by Long-Chain Saturated FA, Fatty Alcohols, Wax Esters, and Dicarboxylic Acids. J. Am. Oil Chem. Soc. 2003, 80, 417–421.

Dassanayake, L.S.K.; Kodali, D.R.; Ueno, S.; Sato K. Physical Properties of Rice Bran Wax in Bulk and Organogels. J. Am. Oil Chem. Soc. 2009, 86, 1163–1173.

Ensikat, H.J.; Boese, M.; Mader, W.; Barthlott, W.; Koch, K. Crystallinity of Plant Epicuticular Waxes: Electron and X-ray Diffraction Studies. Chem. and Phys. Lipids. 2006, 144, 45–59.

Firestone, D. Olive Oil. Bailey's Industrial Oil & Fat Products, 6th Ed. Shahidi, F., Ed.; John Wiley & Sons: New Jersey, USA, 2005; Vol. 2, 317.

Flöter, E. The Role of Physical Properties Data in Product Development. Eur. J. Lipid Sci. Tech. 2009, 111, 219–226.

Gandolfo, F.G.; Bot, A.; Flöter, E. Structuring of Edible Oils by Long-Chain FA, Fatty Alcohols and Their Mixtures. J. Am. Oil Chem. Soc. 2004, 81, 1–6.

Ghosh, M.; Bandyopadhyay, S. Studies on the Crystal Growth of Rice Bran Wax in a Hexane Medium. J. Am. Oil Chem. Soc. 2005, 82, 229–231.

Global Agritech Inc. Stabilization of Long Chain Polyunsaturated Oils. U.S. Patent (PTC/US2008/071178), 2009.

Haiyan, Z.; Bedgood Jr., D.R.; Bishop, A.G.; Prenzler, P.D.; Robards, K. Endogenous Biophenol, Fatty Acid and Volatile Profiles of Selected Oils. Food Chem. 2007, 100, 1544–1551.

Huang, X.; Weiss, R.G. Molecular Organogels of the Sodium Salt of (R)-12-hydroxystearic Acid and their Templated Syntheses of Inorganic Oxides. Tetrahedron 2007, 63, 7375–7385.

Hughes, N.E.; Marangoni, A.G.; Wright, A.J.; Rogers, M.A.; Rush, J.W.E. Potential Food Applications of Edible Oil Organogels. Trends Food Sci. Tech. 2009, 20, 470–480.

Kameda, T. C-13 Solid-state NMR Analysis of Heterogeneous Structure of Beeswax in Native State. J. Phys. D. Appl. Phys. 2005, 38, 4313–4320.

Kodali, D.R. The Utilization of Rice Bran Wax to Stabilize Long Chain ω-3 Polyunsaturated Fatty Acid Esters. Lipid Tech. 2009, 21, 254–256.

Larsson, K.; Quinn, P.; Sato, K.; Tiberg, F. Basic Concepts, Lipids: Physical Properties and Functionality. The Oily Press: England, 2006; 5–6.

Letcher, C.S. Waxes. Kirk-Othmer Encycl. Chem. Technol., 3rd Ed. Wiley: NewYork, 1983; 24, 466–81.

Martini, S.; Añón, M.C. Crystallization of Sunflower Oil Waxes. J. Am. Oil Chem. Soc. 2003, 80, 525–532.

MD Idea Exporting Division. Uses and Application Rice Bran Wax, from Octacosanol to Phytosterol. http://www.mdidea.com/products/proper/proper059.html. (accessed Nov. 2008).

Mellema, M. Co-crystals of Beeswax and Various Vegetable Waxes with Sterols Studied by X-ray Diffraction and Differential Scanning Calorimetry. J. Am. Oil Chem. Soc. 2009, 86, 499–505.

Morales-Rueda, J.A.; Dibilox-Alvarado; Charó-Alonso, M.; Weiss, R.G.; Toro-Vazquez, J.F. Thermo-mechanical Properties of Candelilla Wax and Dotriacontane Organogels in Safflower Oil. Eur. J. Lipid Sci. Technol. 2009, 111, 207–215.

Murdan, S.; Gregoriadis, G.; Florence, A.T. Novel Sorbitan Monostearate Organogels. J. Pharm. Sci. 1999, 88, 608–614.

Narine, S.S.; Marangoni, A.G. Microstructure. Fat Crystal Networks. Marangoni, A.G., Ed.; Marcel Dekker: New York, 2005; 179–181.

Parish, E.J.; Boos, T.L.; Li, S. The Chemistry of Waxes and Sterols, Food Lipids, Chemistry, Nutrition & Biotechnology, 2nd ed.; Marcel Dekker, Inc.: New York and Basel, 2002; 103–106.

Pernetti, M.; Kees van Malssen, F.; Flöter, E.; Bot, A. Structuring of Edible Oils by Alternatives to Crystalline Fat. Curr. Op. Coll. Inter. Sci. 2007a, 12, 221–231.

Pernetti, M.; Kees van Malssen, F.; Kalnin, D.; Flöter, E. Structuring Edible Oil with Lecithin and Sorbitan Tri-stearate. Food Hydrocoll. 2007b, 21, 855–861.

Ritter, B.; Schulte, J.; Schulte, E. Detection of Coating Waxes on Apples by Differential Scanning Calorimetry. Eur. Food Res. Technol. 2001, 212, 603–607.

Rogers, M.A.; Marangoni A.G. Non-Isothermal Nucleation and Crystallization of 12-Hydroxystearic Acid in Vegetable Oils. Cryst. Growth Des. 2008, 8, 4596–4601.

Rogers, M.A.; Wright, A.J.; Marangoni A.G. Crystalline Stability of Self-assembled Fibrillar Networks of 12-hydroxystearic Acid in Edible Oils. Food Res. Int. 2008, 41, 1026–1034.

Rogers, M.A.; Wright, A.J.; Marangoni A.G. Nanostructuring Fiber Morphology and Solvent Inclusions in 12-hydroxystearic Acid / Canola Oil Organogels. Curr. Opin. Coll. Inter Sci. 2009, 14, 33–42.

Rogers, M.A. Novel Structuring Strategies for Unsaturated Fats—Meeting the Zero-trans, Zero-saturated Fat Challenge: A Review. Food Res. Int. 2009, 42, 747–753.

Sato, K.; Goto, M.; Yano, J.; Honda, K.; Kodali, D.R.; Small, D.M. Atomic Resolution Structure Analysis of β′ Polymorph Crystals of Triacylglycerol: 1,2-dipalmitoyl-3-myristoyl-*sn*-glycerol. J. Lipid Res. 2001, 42, 338–345.

Schaink, H.M.; Kees van Malssen, F.; Morgado-Alves, S.; Kalnin, D.; van der Linden, E. Crystal Network for Edible Oil Organogels: Possibilities and Limitations of Fatty Acid and Fatty Alcohol Systems. Food Res. Intern. 2007, 40, 1185–1193.

Schrimpf-Moss, J.; Wilkening, V. Trans Fat—New FDA Regulations. Trans Fats Alternatives. Kodali, D.R.; List, G.R., Eds.; AOCS Press: Champaign, IL, 2005; 26–28.

Small, D.M. Introduction. Handbook of Lipid Research 4. The Physical Chemistry of Lipids From Alkanes to Phospholipids; Plenium Press: New York and London, 1986; 3.

Tada, A.; Masuda, A.; Sugimoto, N.; Yamagata, K.; Yamazaki, T.; Tanamoto, K. Analysis of Constituents of Ester Type Gum Bases Used as Natural Food Additives. Food Hygiene 2007, 48, 179–185.

Tamura, T.; Ichikawa, M. Effect of Lecithin on Organogel Formation of 12-Hydroxystearic Acid. J. Am. Oil Chem. Soc. 1997, 74, 491–495.

Terech, P.; Weiss, R.G. Low Molecular Mass Gelators of Organic liquids and the Properties of Their Gels. Chem. Rev. 1997, 97, 3133–3159.

Toro-Vazquez, J.F.; Morales-Rueda, J.A.; Dibildox-Alvarado, E.; Charó-Alonso, M.; Alonzo-Macias, M.; González-Chávez, M.M. Thermal and Textural Properties of Organogels Developed by Candelilla Wax in Safflower Oil. J. Am. Oil Chem. Soc. 2007, 84, 989–1000.

Tulloch, A.P. Comparison of Some Commercial Waxes by Gas Liquid Chromatography. J. Am. Oil Chem. Soc. 1973, 50, 367–371.

Vali, S.R.; Yi-Hsu, J.; Kaimal, T.N.B.; Yaw-Terang, C. A Process for the Preparation of Food-Grade Rice Bran Wax and the Determination of its Composition. J. Am. Oil Chem. Soc. 2005, 82, 57–64.

Vintiloiu, A.; Leroux, J.C. Organogels and their Use in Drug Delivery—A Review. J. Control Release 2008, 125, 179–192.

Wang, R.; Xiang-Yang Lui.; Xiong, J.; Li, J. Real-time Observation of Fiber Network Formation in Molecular Organogel: Supersaturation Dependent Microstructure and its Related Rheological Property. J. Phys. Chem. B 2006, 110, 7275–7280.

Wright, A.J.; Marangoni, A.G. Formation, Structure, and Rheological Properties of Ricinoelaidic Acid-Vegetable Oil Organogels. J. Am. Oil Chem. Soc. 2006, 83,497–503.

Wright, A.J.; Marangoni, A.G. Time, Temperature, and Concentration Dependence of Ricinoelaidic Acid-Canola Oil Organogelation. J. Am. Oil Chem. Soc. 2007, 84, 3–9.

8

Monoglycerides in Oils

Chia-Hung Chen and Eugene M. Terentjev
Cavendish Laboratory, University of Cambridge, Cambridge, United Kingdom

Introduction

Natural and synthetic polar lipids are used as emulsifiers in foods, cosmetics, plastics, and pharmaceuticals. Understanding the physical properties of polar lipids is the key to selecting the optimal emulsifier, or combinations thereof, for given applications. The molecular organization of polar lipids is governed by their amphiphilic structure (Garti & Sato, 2001). The polar head groups are oriented in sheets, allowing involving hydrogen bonding or ionic interactions. The layers of polar head groups are separated or are forced to be packed orderly by the hydrocarbon chains, forming bimolecular layers similar to the structure of biomembranes. Food-grade emulsifiers are usually esters of polyvalent alcohols (glycerol, propylene glycerol, sorbitol/sorbitan, or sucrose) and other edible lipids. Monoglycerides (MG) and their organic acid derivatives are the most commonly used polar lipids in food products and their phase behavior is associated with their functional effects in many products (Larsson, 1966; Heertje et al., 1998). The first descriptions of the mesomorphic behavior of MG were given by Lutton (1965), Larsson, (1966), Lutton and Jackson (1966), Krog and Larsson (1968). Comprehensive reviews of liquid crystalline structures of polar lipids have been given in Small (1986) and Larsson (1994). Apart from several studies in MG and MG-aqueous systems, there have been surprisingly few studies in MG-oil, which is an important substance in the food industry, especially for forming cream. Our current study is primarily devoted to the phase behavior of MG in oils and discusses the results of mixing additional component colloids, such as water, with these composite systems. Some fundamental discussions of general lipid morphology are also presented.

In this chapter we briefly review the fundamentals of pure MG and MG dispersed in water, and then concentrate on the mixtures of MG and oils. We will also examine relevant theories addressing the phase behavior. In addition to the classical questions raised in surfactant solutions, such as phase diagrams, rheological properties and the microstructure of phases, we also focus on aging and metastability of phases, quantified by changes in melting enthalpy and in infrared spectra that reflect the changes of hydrogen bonding patterns (Chen & Terentjev, 2009).

Lipids are commonly present as building blocks in biological tissues, forming the major component of biomembranes and fat tissues. Although the term lipids is sometimes used as a synonym for fats, fats are a particular subgroup of lipids called triglycerides. In general, lipids can be broadly defined as synthetic or naturally-occurring molecules with amphiphilic structure, such as fats, waxes, fat soluble vitamins, mono-, di- and triglycerides, phospholipids, and others (see Fig. 8.1). Due to their pronounced amphiphilic nature, lipids form aggregated structures such as vesicles, liposomes, or membranes in solvents. The aggregated forms of lipids in various conditions are highly important from a biological point of view. For instance, lateral packing and the conformation of lipids influence the fluidity of biomembranes, which closely relates to the functional activity of biomembranes (Silver, 1985). On the other hand, the lyotropic liquid crystalline state of lipids in aqueous phases is widely used in the food industry (Cassin et al., 1998; Chen et al., 2009; Chen & Terentjev, 2009).

The physical properties of lipids, such as demixing point, melting point, heat capacity, rheological properties, etc. are considerably influenced by the structure of aggregates (Small, 1986). Therefore the technology to make a desirable crystalline or liquid crystalline state is of great importance. Although the microstructures of lipids is fundamental for various fields, the structure of these compounds is often difficult to identify. Furthermore, the metastable features of crystalline or liquid crystalline phases in the lipid systems also increase the difficulty to analyze these structures. Only a limited number of structures have been determined (Garti & Sato, 2001). Therefore the packing, conformation, and dynamical properties of many lipid systems still remain a mystery.

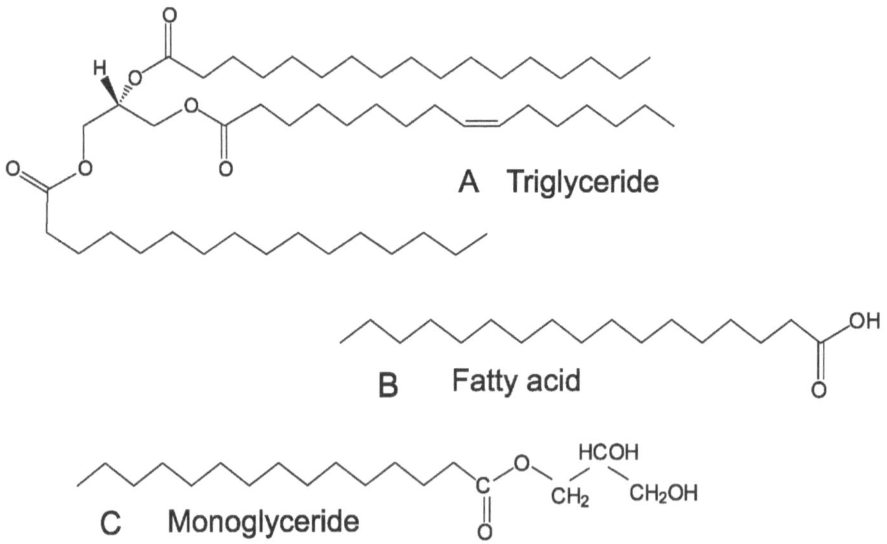

Fig. 8.1. Composition of typical lipids. Schematics of the chemical structure of a typical triglyceride (A), a free fatty acid (B), and a monoglyceride (C), which dominates the composition of the food emulsifiers.

Despite some differences, different lipid systems show similar phase behavior, common to many other surfactant solutions (Marangoni & Narine, 2001). In most situations, on increasing a lipid concentration in solution, the phase transforms an initially homogeneous disordered solution to spherical micelles and then usually to a lamellar phase. It does not mean, however, that the lipid systems could not have the other liquid crystalline phases. Some of them also form hexagonally packed cylindrical micelles, bicontinuous cubic phases, or other phases. Comprehensive reviews of microstructures and rich phase behavior found in lipid systems can be found (Jones, 2001; Larson, 1999; Laughlin, 1994). Some insight into the factors that determine which type of structure is formed by a given amphiphile is obtained from a simple geometrical argument comparing the size of the head-group area of a molecule, a_0, the aliphatic tail chain length, l_c, and the total hydrocarbon volume of molecule, v. A brief summary of lyotropic liquid crystalline phases with their packing parameter vl_c/a_0 is provided in Table 8.A (Larsson, 1999).

Here we focus on a particular lipid, monoglyceride (MG). MG consists of a single fatty acid esterified with a 1-hydroxy glycerol group (Krog & Larsson, 1968; Morley & Tiddy, 1993) (see Fig. 8.2). They are distinguished by the length of their carbon chain. Unlike typical non-ionic surfactants, MG is an optically active

Table 8.A. Packing Shapes and Structures of Typical Surfactants.

Lipid Type	Packing Parameter	Packing Shape	Structure
Single chain lipid with a large polar head area	< 1/3	Cone	Spherical micelle
Single chain lipid with a medium polar head area	1/3 – 1/2	Truncated cone	Cylindrical micelle
Single chain lipid with a small polar head area	1/2 – 1	Truncated cone	Curved bilayers, cubic phase
Double chain lipid with a medium polar head area	~1	Cylinder	Planar bilayer
Lipids in hydrophobic solution	>1	Inverted truncated cone	Inverted structures

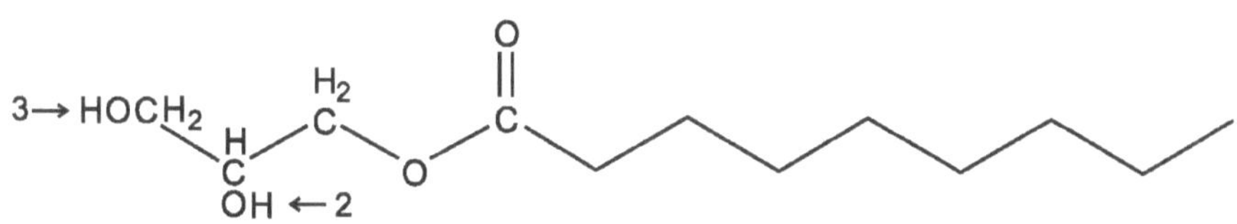

Fig. 8.2. 1-MG is a lipid molecule consisting of a fatty acid esterified with a 1-hydroxy glycerol group, which contains two hydro oxide groups (OH). One OH-group is at the end of the molecule head (labeled 3-OH); the other is attached to the middle carbon (2-OH). These two groups give the complex hydrogen bonding interaction in MG systems (Atovmyan et al., 1988).

molecule that exists in two chiral isomers, denoted by *D* and *L* (Cassin et al., 1998; Chen & Terentjev, 2009). Industrially produced MG usually have a certain distribution of mono-, di-, and triglycerides and also vary with respect to the length of their fatty acid tails. Such products are normally referred to as distilled MG (Krog et al., 1997); a typical composition would contain ~97 wt% MG (divided into 98 wt% 1-MG and around 2 wt% of 2-MG isomers) and less than 3 wt% of di- and triglycerides. Here the notation 1-MG and 2-MG refers to the isomers of MG in which the aliphatic tail is attached to the end or the middle carbon of the glycerol group, respectively. The fatty acid length of MG depends on the source of raw materials. The most common manufacturing process is interesterification of edible animal or vegetable fats, which may be fully or partly hydrogenated to achieve a specific fatty acid distribution. An alternative method is esterification of fatty acid with glycerol. By selecting fatty acids with a specific chain length, MG with a narrow fatty acid profile can be produced by the second method. The most commonly used fatty acids are palmitic (C16) and stearic (C18), which are present in products made from hydrogenated fats. The mixed fatty acid lengths of industrial MG affect their melting and crystallization properties, as well as their polymorphic behavior, in comparison with the MG with monodisperse fatty acid tails. As a general rule, the melting and crystallization temperatures of mixed fatty acid MGs are lower and the polymorphic behavior is more stable than that of pure components (Larsson, 1996; Krog & Larsson, 1968).

The polymorphic behavior of MG depends on the temperature, concentration of surfactant, co-surfactants used, as well as the background solvents. At high temperature MG would be melted to yield the expected isotropic fluid. Below a critical temperature, MG forms liquid crystalline phases with diverse structures (Heertje et al., 1998). The densely ordered, inverse lamellar packing of MG is formed below the demixing point to produce solid dispersions in the solvent (Garti & Sato, 2001; Krog et al., 1997) These dispersions can form a percolating network to sustain shear stress and retain the solvents inside, resulting in the creation of gel-like materials (see Fig. 8.3).

The microstructure of MG dispersions has a significant impact on the macroscopic mechanical properties (Chen et al., 2009). This microstructure is dependent on the molecular composition as well as its aggregation type. The fundamental polymorphic modifications of MG are well-described (Krog & Larsson, 1986; Pitzalis et al., 2000) and are schematically depicted in Fig. 8.4. There are three typical polymorphs: α, sub-α, and β, named in accordance with subcell structures, which correspond to cross-sectional packing modes of the lipid molecules (Garti & Sato, 2001). The α–gel contains the hexagonal packing of surfactant molecules in the lamellar planes. The sub-α crystalline state corresponds to the orthorhombic chain packing. In the β–crystalline state one finds the triclinic packing of aliphatic chains in the lamellae. In the mixing phase behavior the chain length plays critical roles (Krog & Larsson, 1968; Krog et al., 1997). The basic three polymorphic forms are modified, depending on the compositions of MG and other compounds present, so that multiple sub-α or β forms, or even new forms, occur. Such diversity in polymorphism has

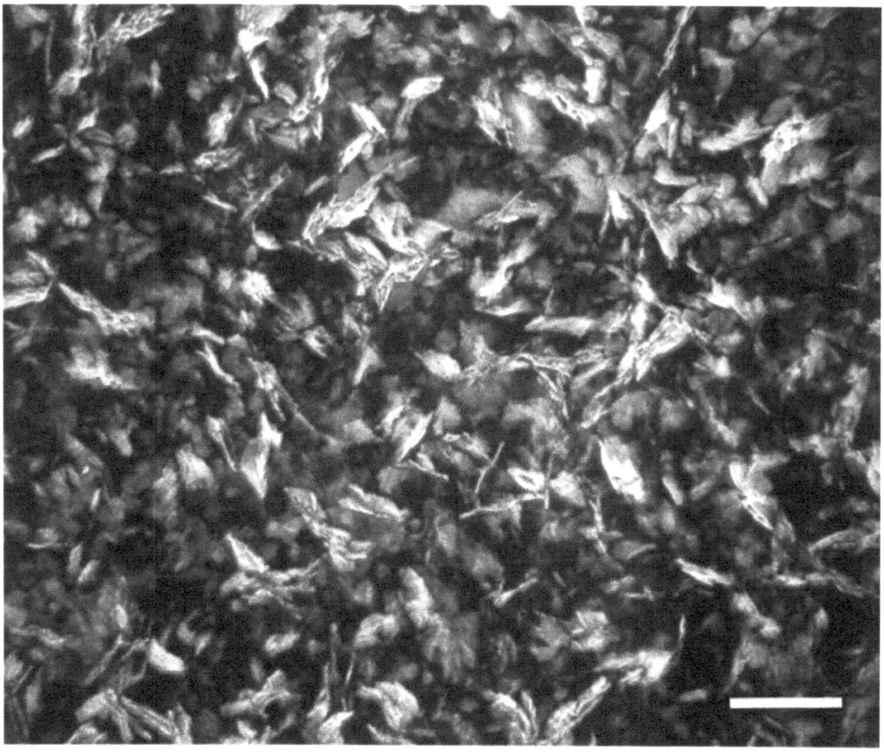

Fig. 8.3. Microstructure of MG-oil gel phase viewed between crossed polars of an optical microscope (scale bar is 50 μm). The image shows the network of highly birefringent domains. According to the morphology of the network, different texture and mechanical responses of the MG-oil gel is obtained.

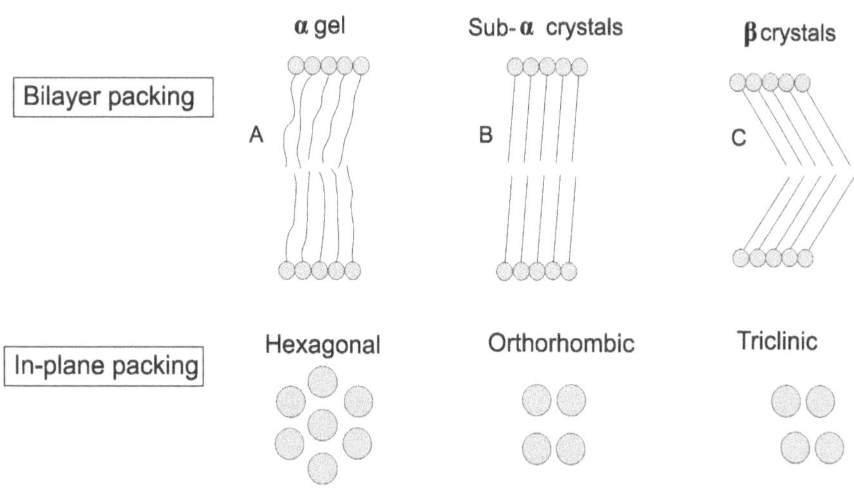

Fig. 8.4. Three basic molecular packings of lipids below the demixing point. (A) α-gel, which contains the hexagonal surfactant packing in bilayer lamellae, (B) sub-α crystals, which show the orthorhombic crystalline chain packing, (C) β-crystals with the triclinic chain packing.

been examined over the past 40 years, but many questions remain. Among the three basic polymorphic forms, the β–crystalline state is most stable, while sub-α crystal and α–gel forms are metastable.

When MG are mixed with water, with its strong affinity to the polar groups, lyotropic liquid crystalline mesophases are promoted, as extensively discussed in several classic reviews (Small, 1986; Larsson, 1994, Garti & Sato, 2001). Liquid crystalline structures in aqueous systems are important for the interaction of carbohydrates or proteins and the ability to emulsify the solvents (Marangoni & Narine, 2001; Duynhoven et al., 2005; Mezzenga et al., 2005; Sari et al., 2003). Concentrated mixtures of MG in water form cream-like materials, which are widely used in the personal products and food industry (Sagalowicz et al., 2006; Sein et al., 2002). The phase diagram of a pure 1-MG in water, as described in Larsson (1966) is sketched in Fig. 8.5. In the concentrated region, the prominent liquid crystalline phase is the

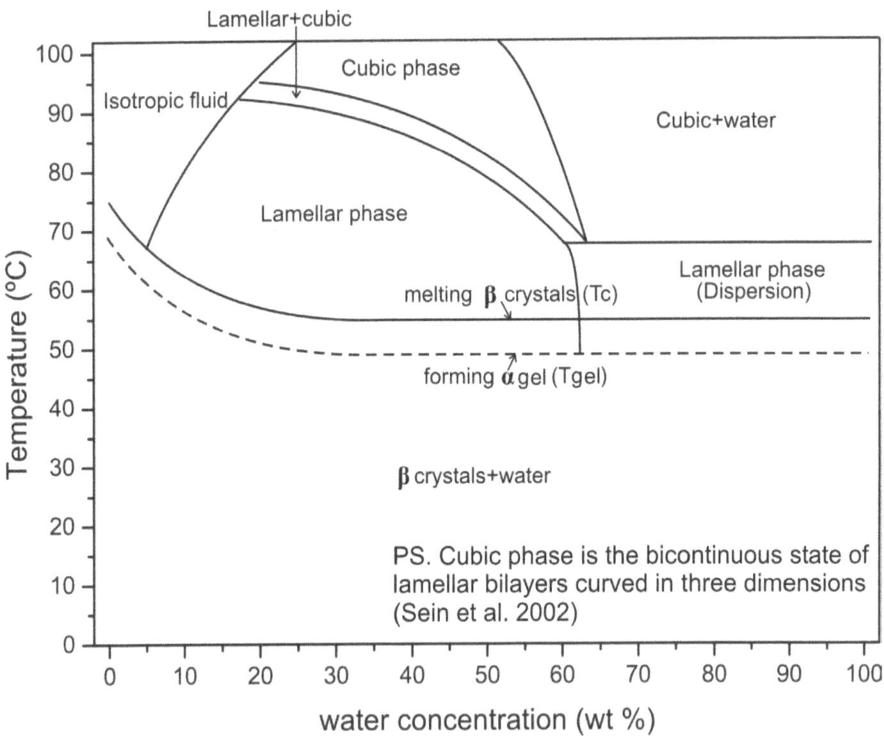

Fig. 8.5. Phase diagram of 1-MG (C16, palmitic acid) in water at pH ~7.0 to approach the optimal solubility of MG in water. Abbreviations in the plot are TC, the melting temperature of β-crystals, and T_{gel}, the gelation temperature to form α-gel. This phase diagram is determined by the X-ray scattering patterns at different MG concentrations and temperatures (figure adapted from Krog & Larsson, 1968).

L_α lamellar phase, where disordered bimolecular lipid layers are separated by water. On cooling the L_α lamellar phase the mobility of surfactant in bilayers is reduced, which leads to its dense packing, forming the α–gel, in which the hexagonal surfactant packing is found in the layers, characterized by a single spacing of 4.18Å (Larsson, 1966; Marangoni & Narine, 2001; Sein et al., 2002; Batte et al., 2007a). The structure is still lamellar, but the dense packing of polar heads gives the short-distance ordering. Although the aliphatic chains are still melted, they are forced to be extended by dense lateral constraints. This leads to the gel-like nature of the material (Chen & Terentjev, 2009). Recent studies have shown that the α–gel is metastable and gradually transforms into an anhydrous MG crystalline form, the β–crystal with triclinic chain packing that is also frequently called the coagel (Cassin et al., 1998; Gehlert et al., 1996). We shall see from the detailed investigations described below that the β–crystalline state of MG is due to hydrogen bonds established within head groups in bilayers, which in turn leads to further crystallization of aliphatic tails. On a long timescale of aging, the chiral *D*- and *L*- isomers of MG gradually separate within crystalline bilayers, leading to more dense packing and the expulsion of water (Cassin et al., 1998; Morley & Tiddy, 1993).

Bicontinuous cubic phases are also formed in aqueous solutions of MG (Sagalowicz et al., 2006), characterized by a face-centered lattice, with MG molecules exposing their polar groups toward the water-side of the bilayers (Morley & Tiddy, 1993; Pezron et al., 1990; Mcintosh et al., 1989). However, they have little impact on applications of MG as creaming and rheology-controlling agents, and so are not discussed further here.

From early studies (Krog & Larsson, 1968), it was found that the phase behavior of mixed MG with a broad distribution of aliphatic chains was similar to that of pure MG in water. The isotropic phase occupies a larger proportion than in pure MG and the melting point of β-crystals shifts to a higher temperature.

In many practical cases, the presence of water in a surfactant mixture is undesirable. In the food industry, in particular, water is always associated with growing bacteria and the corresponding products having limited lifetime. Completely hydrophobic mixtures, or the ones with water "pockets" not exceeding several microns in size, are required. For example, spreads, margarines, and shortenings with reduced and saturated fat content are increasingly being demanded by consumers (Chen et al., 2009; Larson, 1999). Traditional ways of fabricating these products by hydrogenation and addition of large amounts of saturated fat (triglyceride) are becoming unpopular (Batte et al., 2007a; Batte et al., 2007b; Marangoni et al., 2007). Therefore an alternative way to structure liquid oils into a gel-like materials must be sought. One of the promising strategies to achieve the structuring of oil without fat is to exploit the aggregations of long-chain saturated MG in hydrophobic solvents (Chen et al., 2009; Chen & Terentjev, 2009; Ojijo et al., 2004a; Ojijo et al., 2004b; Kesselman & Shimoni, 2007). A particular feature is the use of MG oil mixtures as healthy substitute for butter (Ojijos et al., 2004a; Marangoni et al., 2007).

Compared with pure MG and aqueous MG mixtures, the phase behavior of MG in a fully hydrophobic solvent and MG-water-oil systems have only recently been studied. One of the first systematic studies of rheological and storage properties has been reported by the group of Shimoni (Ojijo et al., 2004a; Ojijo et al., 2004b; Kesselman & Shimoni, 2007), but due to the absence of the phase diagram and confident structure description, the connection between the molecular arrangement and macroscopic observations was incomplete at the time. In fact, many authors in this field have continued to assume that the phase diagrams of MG-oil and MG-water-oil are the same (Batte et al., 2007a; Batte et al., 2007b; Marangoni et al., 2007), or similar, as was indeed the case of MG in water, whereas it is clear that hydrogen-bonding patterns of aggregated glycerol groups would be different. Systematic analysis of MG-oil system is the subject of this chapter.

Phase Behavior

In a nutshell, the two fundamental transitions on cooling from the isotropic phase, the gelation and the crystallization, are still observed in different types of oil. Below the gelation temperature, the inverse lamellar phase of MG in oil is formed, as expected in a hydrophobic system. This inverse lamellar phase has been revealed to be the same phase as the α–gel that was well-known in MG systems, but unlike in the presence of water, this phase contained an additional in-plane hexagonal head ordering within the bilayers. Further cooling below the crystallization temperature of MG, the low-temperature phase emerges. The lateral hexagonal packing transformed into the sub-α crystalline phase, also well-known in a pure MG melt.

Phase Diagram

Before discussing the completely water-free oil solution, MG-oil mixing with small quantities of water should be investigated to quantify the meaning of "dry" in this context (since some degree of spontaneous hydration will always occur in an assembly of molecules capable of hydrogen bond formation). MG-water-oil systems can be produced by vigorously mixing a hot oil-MG solution with distilled water at a temperature around 80C. Fig. 8.6 shows the characteristic thermal signature of the phase transition sequence in mixtures containing 10 wt% MG in hazelnut oil, but with different concentrations of water. Below 0.5 wt% of water there are no significant changes in the transition sequence. Above the water content of ~0.5 wt% the lower temperature transition shifted significantly, from 36°C to 18°C, and the shape of the peaks becomes sharper. To understand these data, the molar concentrations should be identified: 10 wt% MG corresponds to 0.279 M(mol/L) and 0.5 wt% water corresponds to 0.277 M(mol/L). It is now clear that when the molar ratio water/MG increases above one, the changes of phase behavior occur. Water has a strong affinity to the glycerol groups, forming a thin layer inside the inverse lamellar

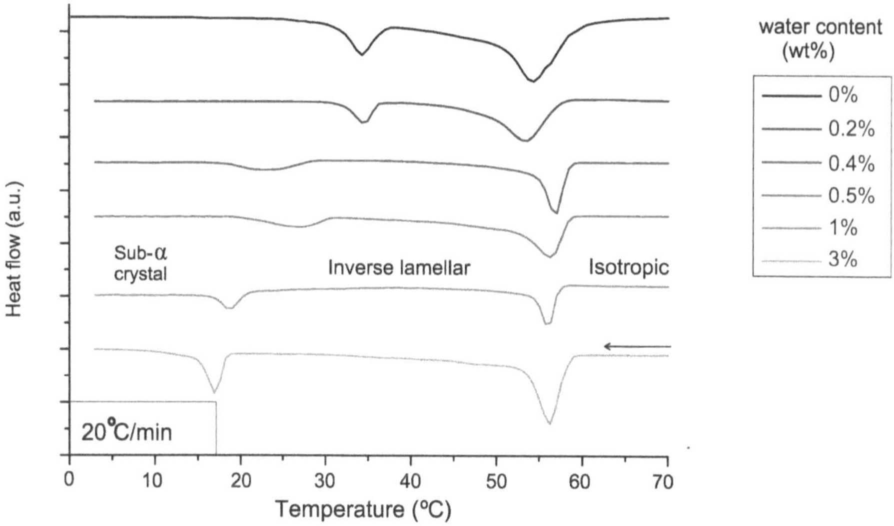

Fig. 8.6. Calorimetric scans of transitions in a mixture of 10 wt% MG in hazelnut oil, with varying amounts of added water, indicated in wt%. Above a critical concentration of water, ~0.5 wt%, the crystallization temperature shifts downward.

bilayer. With sufficient water this layer disturbs the packing of MG heads in the middle of the bilayer and causes the transition point to shift to a lower temperature. MG in dry oil, therefore, should be distinguished from any water-containing systems, and it occurs when there is fewer than one molecule of water per one molecule of MG.

Based on similar calorimetric scans performed at different mixture compositions, a phase diagram of the dry MG in hazelnut oil was constructed (Fig. 8.7). The high-temperature transition line corresponds to the formation of the inverse lamellar phase. The second (low-temperature) transition line represents the crystallization into sub-α crystalline phase with orthorhombic chain packing. From the phase diagram, we find that increasing the MG concentration shifts the gelation transition upwards. However, the crystallization remains essentially independent of the concentration of the MG. This is a characteristic feature of the Krafft point: the demixing of surfactant from the solvent, which in this case results in consolidating of inverse lamellar bilayers into the dense crystalline regions, expels the oil to their periphery.

Based on many industrial applications, hazelnut oil was a natural choice for a research target. However due to a complicated mixture of different fatty acids in hazelnut oil, it is necessary to test a pure hydrocarbon solvent to confirm that the results from hazelnut oil can be generalized for a wide range of hydrophobic solvents, which turns out to indeed be the case (Chen et al., 2009; Chen & Terentjev, 2010).

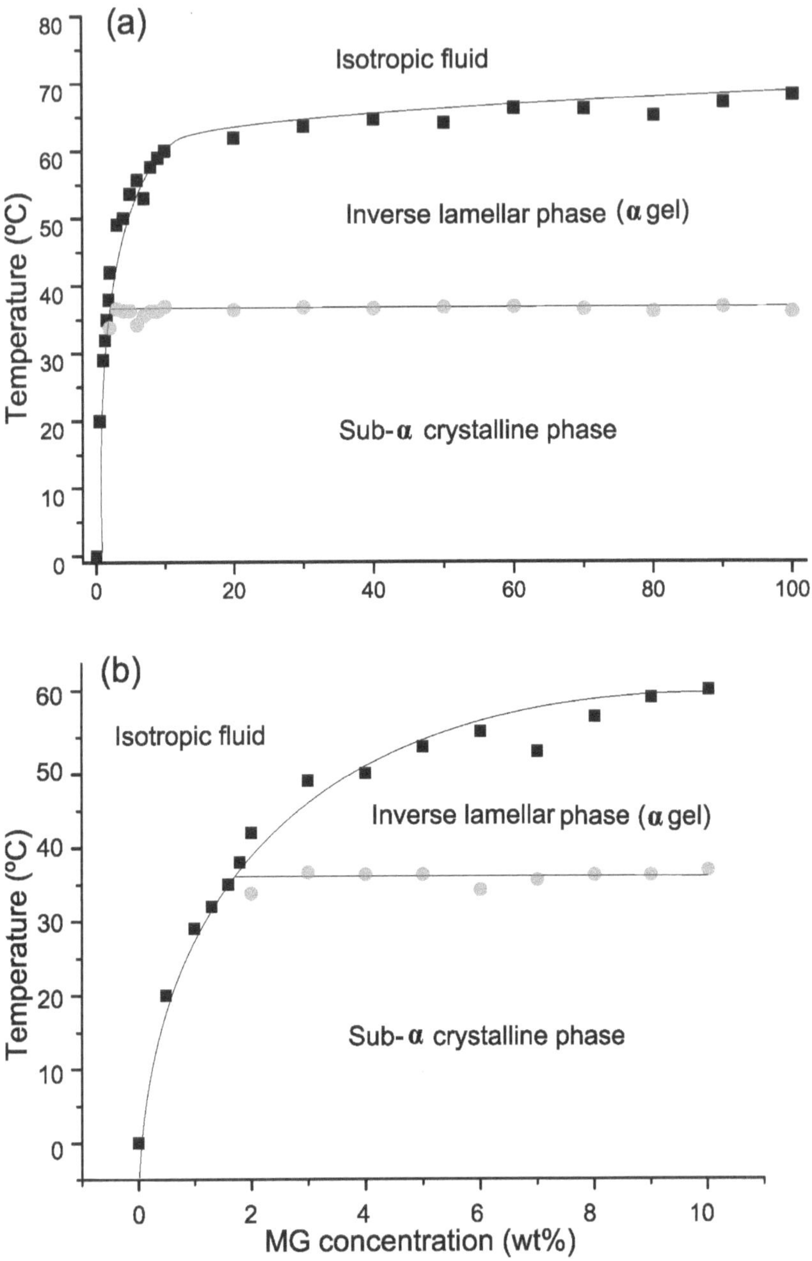

Fig. 8.7. The phase diagram of dry MG (C18) in hazelnut oil, from the summary of calorimetric data. The full range of concentrations is shown in (a), while a detailed study at low MG concentrations is shown in (b). One finds clear boundaries of the three phases: isotropic fluid, inverse lamellar phase, and the sub-α crystalline phase. Below 2 wt% the carbon chains of MG crystallize directly.

Structure and Rheology

The structure of aggregated phases of MG in oil was fingerprinted using X-ray diffraction patterns, as illustrated in Fig. 8.8. To compare with several important studies in the aqueous systems and pure MG, we particularly highlighted the difference between oil-based systems and aqueous systems by comparing the characteristic X-ray

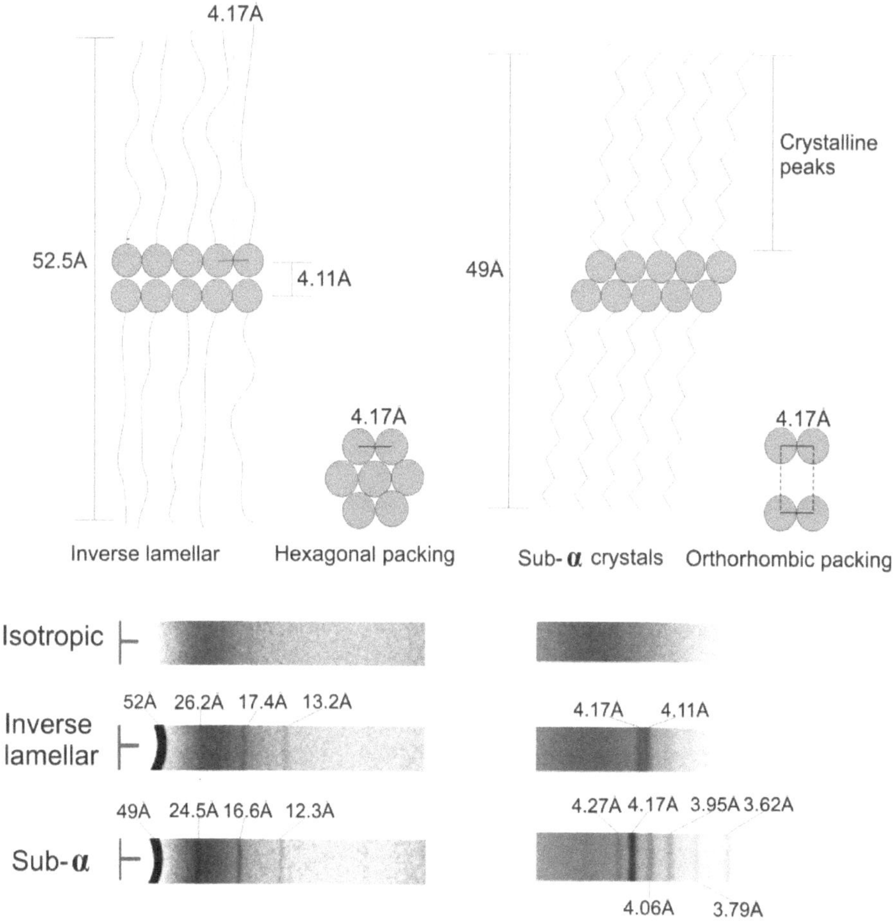

Fig. 8.8. The X-ray diffraction patterns for 10 wt% MG in hazelnut oil. In the isotropic phase, there are no significant aggregates to give the periodical spacing. In the inverse lamellar and sub-α crystalline phases, the X-ray diffractions in the small-angle region (left side) describe the lamellar ordering. The wide-angle region (right side) shows the differences in molecular packing. Twin wide-angle peaks could be observed in the inverse lamellar phase of the dry MG-oil. They correspond to the periodic spacing between neighboring glycerol heads inside the inverse lamellar bilayers, which could not be found in the aqueous systems. In the sub-α crystalline phase a series of wide-angle X-ray peaks indicate the orthorhombic chain packing.

Table 8.B. X-ray Diffraction Data of 10 wt% MG C18 in Oil and Water.

Systems	Long Spacing (Å)	Short Spacings (Å)
MG-oil (inverse lamellar)	52	4.17 - 4.11
MG-oil (sub-a crystal)	49	4.27 - 4.17 - 4.06 - 3.95 - 3.79 - 3.62
MG-water (L_α lamellar)	64.7	-
MG-water (α-gel)	54.3	4.18
β-crystal	48.9	4.55 - 4.51 - 4.38 - 4.26 - 4.09 - 3.94 - 3.84 - 3.78

peak positions in Table 8.B (Garti & Sato, 2001). Comparing these dimensions, one has to conclude that the inverse lamellar phase of MG aggregation in oil is the same structure as the α–gel in MG-water. The key difference is the twin peak at wide angles (4.11–4.17Å), reflecting the two-dimensional, hexagonal, dense packing of glycerol heads, which are tightly compressed in the mid-plane of the inverse lamellar bilayer in the oil matrix (Chen et al., 2009). In contrast, the single layer of glycerol heads exposed to water on the outside of the classical water-borne L_α lamellar bilayer do not have the constraints to pack into an ordered structure. Only the aliphatic tails carry the sequence of transitions. In the aqueous α-gel phase one again recovers the 2D hexagonal (4.18A) packing, but no twin peaks from the neighboring glycerol plane.

It is important to note that the inverse lamellar phase is not a truly crystalline state, as is its equivalent α-gel. Due to the unique size ratio between the glycerol head and the lateral area of the aliphatic chain in the fully extended state of a dense brush, as illustrated in Fig. 8.8, the inverse bilayers have a definite hexagonal in-plane ordering (Kampf et al., 2005; Kaganer et al., 1999). One might assume that the hexagonal order arises from laterally densely packed aliphatic tails. However, if we consider other densely grafted polymer brushes, and more importantly, other single-aliphatic tail surfactants, none of which have such a hexagonal packing. The more likely scenario is that the glycerol heads are densely packed in 2D, in mid-planes, and then the tails (in the fully extended due to the high brush density, but nevertheless molten amorphous state) simply follow that effective hexagonal grafting pattern (Kaganer et al., 1999; Neveu et al., 1977).

Below the crystallization temperature the sub-α crystalline phase appears. In this phase the large-scale structure is still of the inverse lamellar nature, but with the thickness of each bilayer slightly reduced to 50Å. Here a sharp peak at 4.17Å was observed to reflect the lateral spacing between neighboring glycerol heads. However, the crystallized carbon chains yield a pronounced sequence of wide-angle diffraction peaks at spacing between 4.0 and 3.6Å, characteristic of the orthorhombic chain packing.

It is interesting to examine how the macroscopic rheological response of this system evolves following the underlying microscopic phase changes. Rheological experiments in lyotropic liquid crystalline phases are non-trivial because of the easy onset of strain non-linearity (caused by shear-alignment of mobile anisotropic structures). Also, for qualitative characterization of what an end-user might perceive, the measurements ought to be carried out at low frequencies of oscillation, away from complications of viscoelastic stress relaxation in the complex medium. A typical result of such low-amplitude and low-frequency measurements is shown in Fig. 8.9, presenting how the linear shear modulus changes as a function of temperature as the system evolves from the isotropic phase to sub-α crystalline phase. The plot shows three phases. In the high temperature region, the sample is in isotropic fluid phase. The loss modulus G″ is higher than the storage modulus G′ and both are in the range of 0.1 Pa; the system is essentially a simple liquid. When the temperature dropped below the gelation point, the rapid increase in mechanical rigidity was immediately expressed by the modulus. The system acquired mechanical characteristics of a gel, which is the reason why this transition into the inverse lamellar phase is called the gelation point (in contrast to aqueous MG systems where the gelation occurs at the transition from L_α lamellar to the α-gel phase). On further cooling towards the crystallization temperature, a region of pre-transitional effects is observed, and then the sample transforms into the sub-α crystalline phase. Because the network structure of lamellar domains has kept its shape, the mechanical response in this phase does not show significant differences with the inverse lamellar phase. This is an important point and, again, a difference with aqueous MG systems. The ordinary L_α lamellar phase has thermally fluctuating bilayers that do not carry rigidity; as a result, the bulk sample remains fluid (albeit with a much higher viscosity and complex non-linear viscoelastic effects). The true gel nature of the material arises when the bilayer planes acquire rigidity. The signature characteristic is

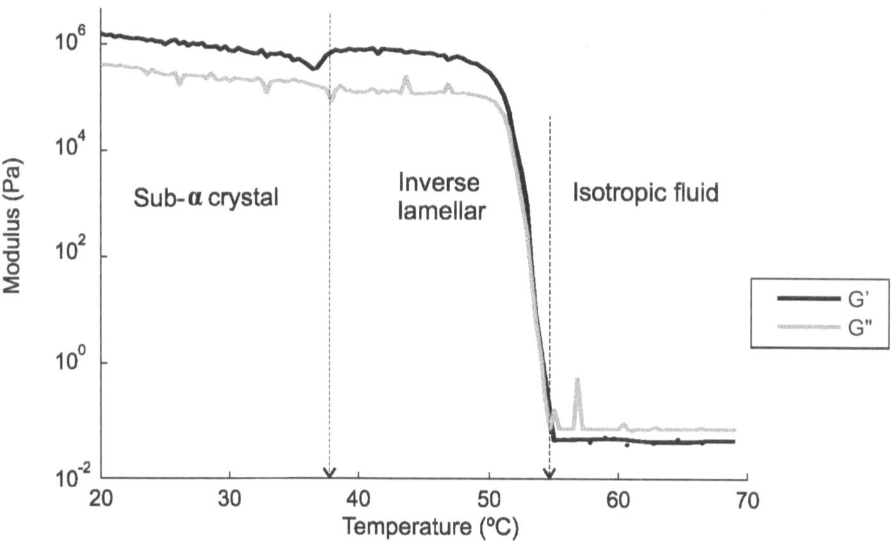

Fig. 8.9. The three regions in a low-frequency (1 Hz) linear rheological response of 10 wt% MG in hazelnut oil corresponded to the three basic phases. In the inverse lamellar phase the shear modulus increases to 10^3 Pa. After the pre-transitional drop, the sub-α crystalline phase is reached, but the modulus remains the same as in the inverse lamellar phase.

the dense hexagonal packing of glycerol heads, which happens in different regions of corresponding phase diagrams in water- and oil-based systems, in the phase that we identify as the inverse lamellar α-gel.

Monoglycerides with Mixed Fatty Acids

Mixtures of MG with fatty acid tails of C16 and C18 are the most common lipid used in the food industry (Garti & Sato, 2001). Having studied properties of pure MG-oil systems, the effect of mixing MG with different fatty acid tails can be examined and compared with our knowledge of molecular arrangements in the phases.

First, we examine the phase diagram of the industry-standard distilled MG mixture called Dimodan® HP, supplied by Danisco. This material has approximately equal proportions of stearic acid (C18, 50 wt%) and palmitic acid (C16, 50 wt%) MG. Following the same calorimetric procedures, the phase diagram of this mixture is shown in Fig. 8.10. It has a similar topology, with the high-temperature isotropic fluid mixture, followed on cooling by the inverse lamellar phase (the α gel), and by the crystalline sub-α phase. The depression of aliphatic chain crystallization (from 36°C to 12°C) is an expected consequence of a minor discrepancy in chain length. This mixing of chemically identical compounds with only minor differences in molecular shape is a standard way to manipulate transition points of materials; in contrast, adding random impurities, although they would also depress the crystallization point, may cause phase separation, as the main phase is more likely to crystallize.

The transitions in MG with different proportions of mixed fatty acid C16 and C18 in oils are characterized in Fig. 8.11. It is clear that the transition to the inverse

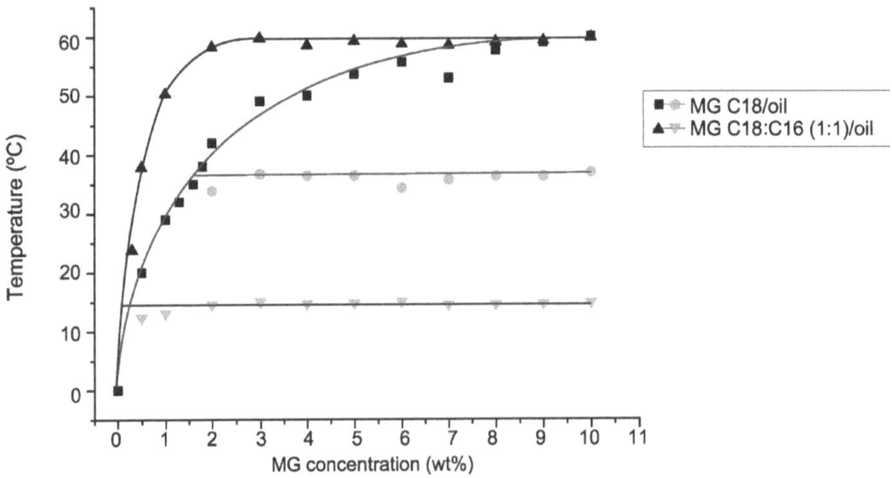

Fig. 8.10. Phase diagram of Dimodan® HP, the mixed MG (C16/C18 1:1) in hazelnut oil. This plot focuses on the relevant region of low MG content (below 10 wt%) and compares the transition lines with those of pure MG C18 (see Fig. 8b). The crystallization point of the mixture decreases from 36°C to 12°C, and the concentration region of inverse lamellar phase extends to below 1 wt%.

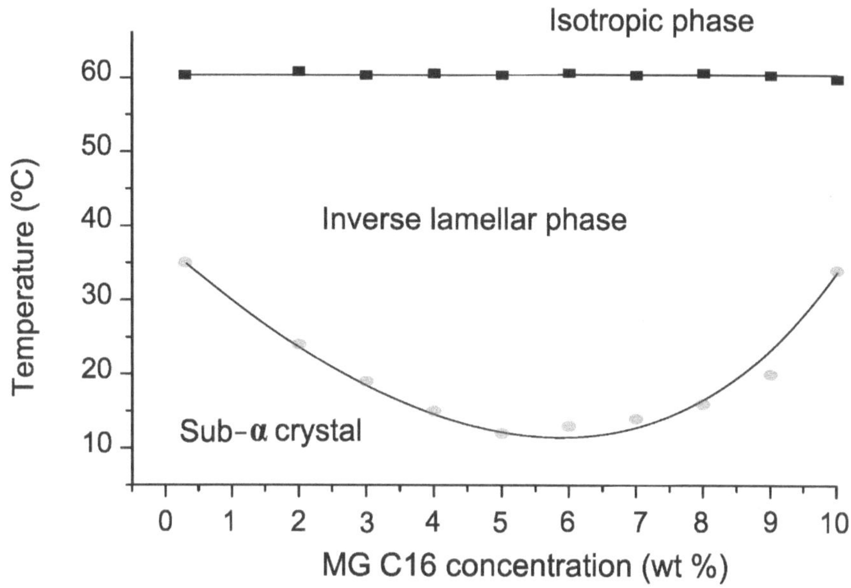

Fig. 8.11. At a fixed 10 wt% MG in hazelnut oil, this diagram maps the transition points as a function of the proportion of C16 and C18 in the mixture. The points on the left and right correspond to the pure C18 and pure C16, respectively. Dimodan® HP is the 1:1 mixture corresponding to the point at 5 wt%.

lamellar phase is not affected by this mixing, which is again expected since the aliphatic chains are molten in lamellae and the 2-carbon difference in their length has no significant effect on their formation. This again confirms the conclusion that the characteristic wide-angle diffraction peaks found in this inverse lamellar phase (and in the α gel traditionally found in hydrated systems) are due to the regular packing of glycerol heads rather than the ordering of aliphatic tails in this phase. In contrast, the crystallization point of transition into the sub-α phase has a strong variation with the mixture composition. The lens-shape of such a phase diagram is characteristic of generic mixtures of chemically similar compounds in many areas of chemical physics. It reflects the entropy increase that opposes the potential energy gain on crystallization, which reaches the maximum at or near the 1:1 fraction.

Effects of Water

Having seen the similarity and contrasts of MG-oil and MG-water systems, it is interesting to examine how these effects compete in the MG-water-oil ternary mixture. A number of studies have addressed this problem. At high water content the ternary system is an oil-in-water emulsion in which lamellar MG walls surround and stabilize oil droplets (Sari et al., 2003; Marangoni et al., 2007). At sufficient MG concentration these walls are continuous from one droplet to the next, forming a continuous solid network to retain the solvents inside (Batte et al., 2007a; Batte et al., 2007b;

Guillot et al., 2006; Rush et al., 2008). The gel formed by MG-water-oil is sensitive to factors such as MG concentration, the pH, the type and the amount of co-surfactants used, and the storage time and temperature.

Before going any further it is important to mention that the solubility of MG is limited in the aqueous phase. When referring to hydrated MG gels, the literature usually reports concentrations around 15 wt% of water (Gros & Feuge, 1951; Brokaw & Lyman, 1958). No more than 25 wt% of water can be dispersed in MG; there are a few reports claiming that gels with water up to 45 wt% water could be dispersed by vigorous mechanical mixing, but in all cases these are highly unstable systems. Therefore adding low concentrations of co-surfactant (most commonly stearic acid or diacetyl tartaric acid) is necessary to stabilize the MG in water (Garti & Sato, 2001; Marangoni et al., 2007; Batte et al., 2007a; Rush et al., 2008). Because of these additional acidic co-surfactants, the swelling capacity of lamellar bilayers becomes dependent on the pH of the aqueous phase used. At higher pH, the acid co-surfactant is ionized, which ensures the effective repulsion between the bilayers and enhanced stability of lamellar gels.

We saw in the previous section that the presence of small amounts of water (over 0.5 wt% but less than 3 wt%) shifts the crystallization to lower temperatures (Chen et al., 2009). Once the amount of water increases further, the effects of pH become significant. In effect, MG does not dissolve in water with too low pH. Two examples are discussed here, with pH/co-surfactant combination just below and just above the swelling boundary. The fully hydrated ternary system with a sufficient proportion of water does not crystallize into the sub-α crystalline phase at low temperature, as oil-dominated systems do, while at lower pH the sub-α crystalline phase is found at any oil-water combinations. As before, to confirm that the conclusions are general, different kinds of oil have been compared: the food-grade hazelnut oil and the pure hydrocarbon (decane). The structures and trends of transitions on increasing the water content are the same in complex and pure oils. Although their gelation and crystallization points are not exactly the same, the phase behavior is similar and clearly driven by the same forces.

The incorporation of acidic co-surfactant into MG is necessary to ensure the solubility in aqueous matrix. Typically, the stearic acid is mixed with MG in a ratio of 1:20 (Batte et al., 2007a; Marangoni et al., 2007). The co-surfactant stearic acid and a small amount of impurities from both MG itself and the oil cause the resulting hydrophobic solution to have an acidic nature. For hazelnut oil (mainly oleic acid) the pH approaches 3. In the case of decane, the pH is around 4. The swelling capacity of the hydrated L_α lamellar phase of MG is strongly dependent on pH. At pH below 6 swelling is very limited, while at pH above 7 greater swelling is obtained. However, an important study of MG in water used a pH of 4.2 and yet reported good MG stability, perhaps due to the use of a different co-surfactant (diacetyl tartaric acid) (Cassin et al., 1998).

Calorimetric cooling scans of the samples with changing proportions of hazelnut oil and water are given in Fig. 8.12. When a more alkaline water is used (addition of 0.05M-NaOH) and the concentration of this water is below 20 wt%, the pH of the solution remains below 6, which causes only a partial swelling of MG. Two transitions are then observed in cooling scans, similar to the dry hydrophobic systems.

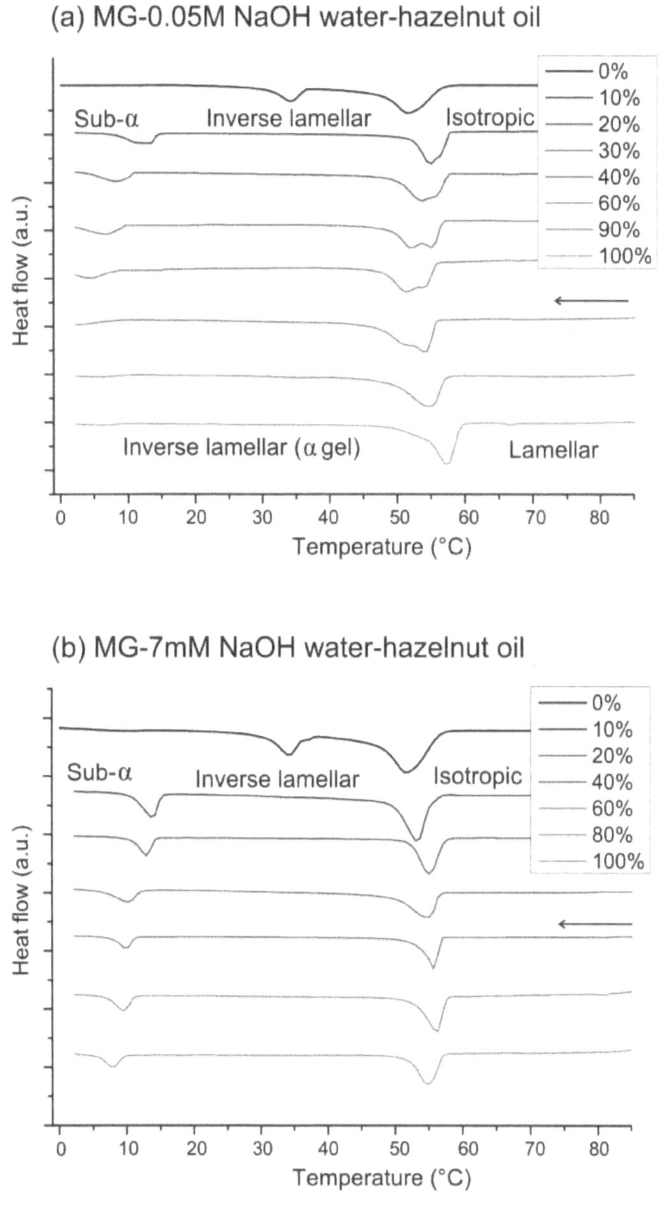

Fig. 8.12. Calorimetric scans indicate the effect of adding water (proportion in the solvent is indicated on plots) to MG-hazelnut oil mixtures at a fixed 10 wt% MG and for two different pH levels (a,b). For more alkaline water (0.05M-NaOH), in the hydrophobic region the familiar gelation and crystallizations transition are observed. At higher fractions of water, the hydrated L_α lamellar phase forms at high temperatures and the crystallization of MG at low temperatures is disturbed. For more basic water (7mM NaOH), the gelation and the crystallization transition are observed at all oil-water combinations.

The high-temperature transition corresponds to the gelation point that occurs on formation of inverse lamellar phase from the initially isotropic fluid solution. The second (low-temperature) transition represents the Krafft point of MG demixing from the dominant hazelnut oil component in the solution, which is accompanied by MG crystallization to form the sub-α crystals. Let us denote this point as T_{Ko} to distinguish it from the point of demixing from water that we shall encounter shortly (T_{Kw}). Evidently, as soon as the glycerol in the mid-plane of inverse lamellar bilayers becomes hydrated, the effective miscibility of MG in oil is enhanced and the crystallization point shifts to lower temperatures and eventually disappears altogether above 40–50 wt% of water in the solution.

When the fraction of the aqueous phase is above 20 wt%, we find an additional high-temperature phase transition emerging at 52°C. This transition is not the same as the gelation point in the oil-dominated system that we discussed previously. By examining the cooling curves in Fig. 8.12.A and extensive literature reports on MG in various aqueous phases, we conclude that this is the emerging Krafft point of demixing of MG from water, T_{Kw}, which is the only transition remaining in water-dominated ternary systems (Larsson, 1966; Krog & Larsson, 1968). On increasing the amount of water continuously, the intensity of this additional peak increases, in contrast to the low-temperature crystallization transition at T_{Ko}, which in turn becomes weaker. Above 50 wt% of water in solution, the full swelling is reached. In this case, MG is dissolved in the aqueous phase and shows the clear lamellar L_α ordering in the high-temperature region. Note the difference: at low water content the high-temperature phase was an isotropic fluid (see the subsequent discussion of gelation rheology). The association of water molecules with glycerol heads of MG suppresses the crystallization (however, the equilibrium β crystalline phase gradually emerges in all cases. See the next section on aging). Moreover the water crystallization temperature in the solutions also shifts to −20°C as the formation of ice is suppressed in the mixed systems.

It is interesting to compare the above phase sequence with the results for the analogous systems only differing water pH (Fig. 8.12.B). In this case NaOH cannot totally neutralize the acidic MG-hazelnut oil mixture to meet the required swelling conditions. The presence of water continues to disturb the crystallization process by hydrating the glycerol heads in bilayers, the same as in the alkaline water case. However due to the incomplete swelling of MG, water molecules cannot completely prevent crystallization of carbon chains, so the sub-α phase could always be observed. Two transitions, the isotropic to the inverse lamellar, and the subsequent sub-α crystallization, are seen at all oil-water combinations. Exactly the same conclusions were obtained by replacing the natural hazelnut oil with decane (Ojijo et al., 2004), confirming that the described picture is a universal feature of MG behavior.

Phase Diagrams

By summing up the calorimetric data, the phase diagrams with different oil-water combinations could be obtained. The phase map in MG-hazelnut oil-0.05M-NaOH water and MG-decane-0.05M-NaOH water is presented in Fig. 8.13.A. MG-hazelnut

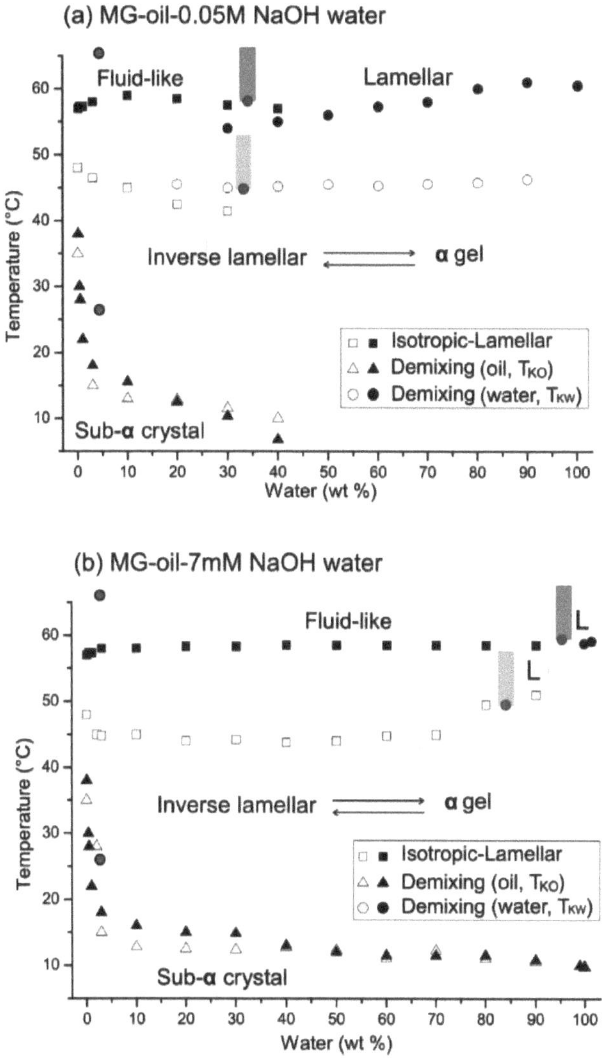

Fig. 8.13. Phase diagrams of MG-oil-0.05M-NaOH water and MG-oil-7mM-NaOH water are presented separately (a,b). Both hazelnut oil (solid data points) and decane (open data points) results are plotted on the same maps, showing the same qualitative features. (a) In the mixtures with alkaline 0.05M-NaOH water, three phases appear in the oil-dominated environment: isotropic fluid, inverse lamellar, and sub-α crystal. Above 50 wt% of water in oil there is only one transition to show the formation of α–gel from the high-temperature L_α lamellar phase. The formation of L_α lamellar phase was determined from rheological measurements and its boundary is indicated by a gray box in the plot. The red circles indicate the points (at 3 wt% water) where X-ray diffraction data was collected for Fig. 16. (b) In the samples containing the more acidic 7mM-NaOH water, the gelation and crystallization transitions are observed in whole range of oil-water concentrations, except at over 99% water where the high-temperature L_α lamellar phase is also present. The concentration of MG is always 10 wt% in all solutions.

oil-7mM-NaOH water and MG-decane-7mM-NaOH water systems are shown separately in Fig. 8.13.B. Both hazelnut oil and decane systems show the same features in phase behavior upon increasing the water concentration. Below 20 wt% of water they remain in the oil-dominated conditions and there are three phases found on decreasing temperature: isotropic fluid, inverse lamellar, and sub-α crystals. Above 20 wt% water two things happen in the more alkaline mixture. First, we find that the high temperature phase is no longer an isotropic fluid but rather is a hydrated lamellar gel. Secondly, the principle transition splits into two, with the extra transition (at 52°C in hazelnut oil and at 48°C in decane) corresponding to the demixing of MG from water, which remains for the rest of the aqueous systems. Above 50 wt% of water the low-temperature transition into the sub-α crystalline phase (MG-oil demixing) disappears completely. For the water-dominated mixtures, with more than 60 wt% of 0.05M-NaOH water well-mixed with MG-oil, the strong interaction between MG head-groups and water disturbs the crystallization process.

Compared with 0.05M-NaOH water, the 7mM-NaOH water was not able to neutralize the naturally acidic MG-oil to achieve the full swelling. Therefore the effect of water was not strong enough to completely prevent the crystallization of aliphatic chains, and the formation of sub-α crystals was observed for all oil-water combinations (Fig. 8.13.B). Similarly, in the high-temperature region the isotropic fluid-like state extends over almost the whole range of water content, the ordinary hydrated L_α lamellar phase was only found with well over 95 wt% of water in solution, as determined by the rheological measurements in the next section.

Another slice of the phase diagram of the ternary system is presented in Fig. 8.14 to illustrate the effect a small amount of water has on the phases. At low water content the mixtures do not need co-surfactant and NaOH to enhance their stability. Keeping a constant 3 wt% water and changing the concentration of MG in oils gives the transition lines that have the same topology as in the dry system; the comparison is highlighted in Fig. 8.14. An important difference is the more pronounced lack of concentration dependence of both transitions, which hints that it is the formation (and later crystalline structuring) of individual MG bilayers that is responsible for these phases, as opposed to the inter-layer interaction playing a significant role in ordinary smectic and lamellar liquid crystals.

Structure and Rheology

X-ray diffraction was used to analyze the molecular arrangement of MG-oil with 3 wt% water and 10 wt% MG; the compositions and temperatures are labeled on Fig. 8.13. This composition was chosen because it is well above the "dry" oil boundary (~0.5 wt%), is far from the triple point where the three key phases meet at low MG concentration (see Fig. 8.14), and these three phases are well-formed and not obscured by other phase transitions nearby.

The wide-angle diffraction shows the difference between dry and hydrated systems. In the dry MG-oil system the characteristic twin wide-angle X-ray diffraction peaks are observed, representing the two-dimensional packing of two layers of glycerol

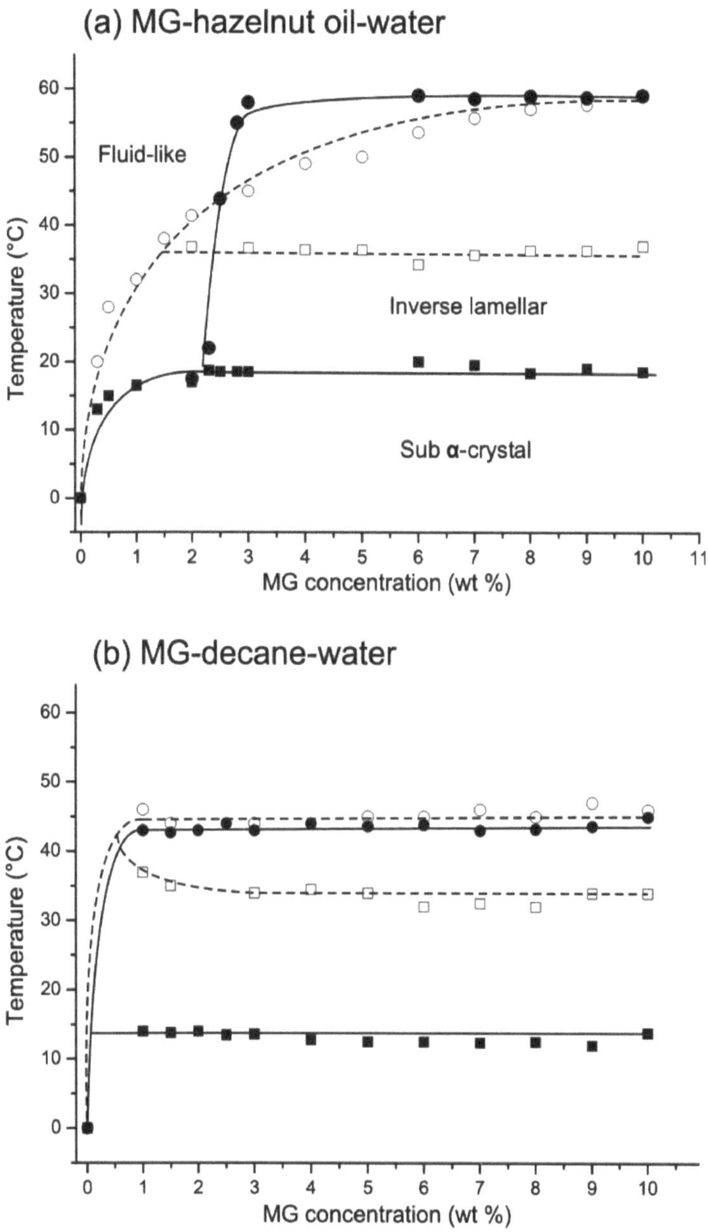

Fig. 8.14. (a) The phase diagram of MG-hazelnut oil with a constant 3 wt% water (solid lines) is presented for comparison with the completely dry MG-oil system (dashed lines). (b) The phase diagram of MG-decane with the same concentration of water (3 wt%, solid lines) is similarly compared with the completely water-free system (dashed lines). Both systems show the same trends upon addition of a small quantity of water to the MG-oil matrix.

in the bilayer midplane, which occurs in hydrophobic environment and also below the water-demixing Krafft point T_{Kw} when it exists. There is a remarkable coincidence between the size of the glycerol group and the cross-sectional size of the extended C16-C18 alkyl chain when it is laterally packed in a dense polymer brush (Milner, 1991). In contrast, there is only one wide-angle X-ray diffraction peak at 4.17Å in the MG-water-oil system. This suggests that the water molecules are absorbed in the middle of inverse lamellar bilayers to disrupt the regular spacing between the two hexagonal planes. The in-plane head arrangement is also somewhat disturbed by the presence of water, which shows in a much broader diffraction peak at 4.17Å. See Fig. 8.15.

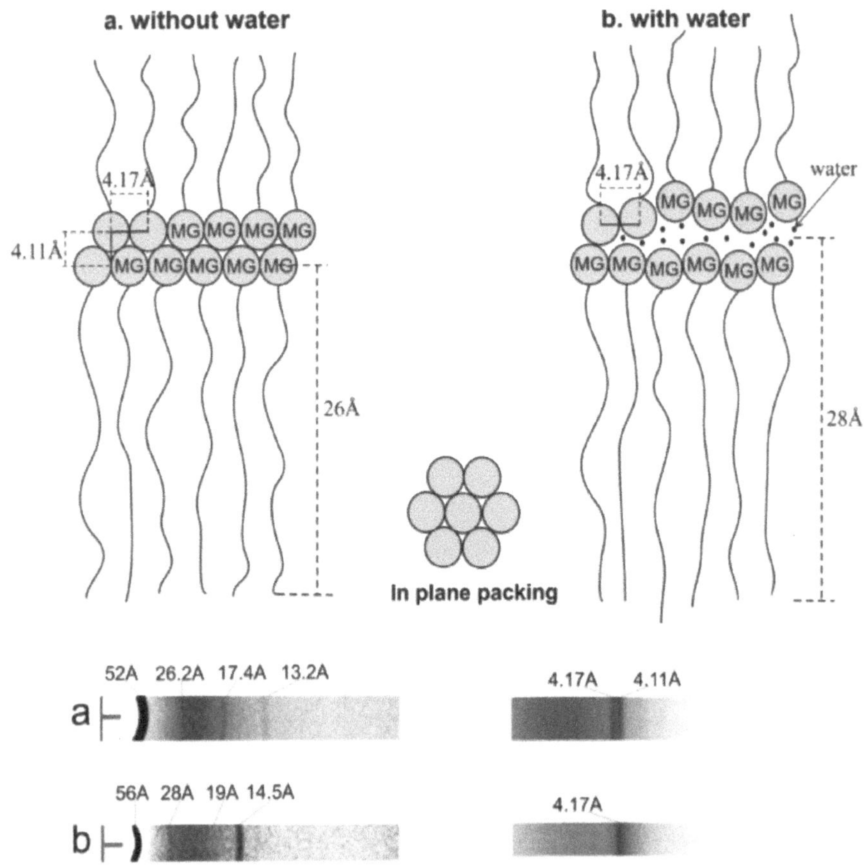

Fig. 8.15. Structures of (a) dry 10 wt% MG in hazelnut oil and (b) 10 wt% MG in hazelnut oil with 3 wt% water are described by X-ray diffraction. The small-angle peaks (left side) identify the lamellar ordering following the characteristic sequence of 1, 1/2, 1/3, and 1/4. The thickness of inverse bilayers increases from 52 Å to 56 Å on adding 3 wt% of water. The wide-angle scattering (right side) and illustrations show how the presence of water disturbs the dense packing between layers so the spacing of 4.11Å is absent in the ternary system.

Rheological experiments were carried out in the linear viscoelastic region. The measurement involved observing the low-frequency shear modulus changing as a function of temperature in samples with different water fraction and pH as the systems evolve from fluid-like phase to the gel-like materials. The typical results for MG-hazelnut oil are shown in Fig. 8.16 as 0% curves. Without water, the modulus at high temperature is close to zero (~10^{-2} Pa at this frequency), and there are two subsequent regions corresponding to the key phases: isotropic fluid, inverse lamellae phase, and sub-α crystals. When the temperature drops down below the gelation transition point T_{gel}, the rapid increase in mechanical rigidity is immediately expressed by the storage modulus G′. The system acquires mechanical characteristics of a gel (Petrov et al., 2002). On continuously cooling towards the crystallization temperature T_{Ko}, the sample enters the sub-α crystalline phase; the storage modulus continuously increases to reach the saturated level, but no significant transition is found, indicating that mechanical stability of the polydomain lamellar structure changes little.

In a ternary MG-hazelnut oil-water system, the two cases of MG-hazelnut oil-0.05M-NaOH and MG-hazelnut oil-7mM-NaOH have to be discussed separately. Cooling down from the isotropic fluid phase, the MG would aggregate under the gelation temperature, which occurs on formation of lamellar domains, and the mechanical rigidity immediately increases. The subsequent crystallization transitions could be discerned in the rheological signature of MG-oil with low concentrations of both 0.05M-NaOH water and 7mM-NaOH water, labeled by vertical arrows on plots. Above 50 wt% 0.05M-NaOH water, the lower-temperature change in the storage modulus G′ is not associated with any phase transition in the α–gel (since we have verified by X-ray and DSC that the microscopic structure remains the same). The most likely reason for this remaining feature in G′(T) is the kinetic effect due to the stress-relaxation rate of lamellar domains approaching and then exceeding the applied rate of cooling in the rheometer.

The rheological measurements at room temperature indicate that the samples are gel-like over a wide range of water dilution, with the relatively high storage modulus G′, although there is a systematic decrease in the modulus on increasing water concentration. The high-temperature region is sensitive to pH and the amount of the aqueous phase in the mixture. Above 30 wt% of 0.05M-NaOH water in MG-oil, the storage modulus at high temperature is no longer zero, indicating that the lamellar structure exists well above 70°C (see Fig. 8.16.A). Since this structure is most prominent at higher water fraction, and we do not find any wide-angle X-ray reflections here, one has to conclude that this is an ordinary (hydrated) L_α lamellar phase, which then transforms into the inverted lamellar (or α–gel) phase below the water-demixing line T_{Kw}, just below 60°C. Unfortunately, since water evaporation is involved, we could not establish the upper-temperature phase boundary of this L_α phase (see phase diagrams in Fig. 8.13).

The mechanical strength of this high-temperature L_α lamellar phase varies: in the sample with 50 wt% 0.05M-NaOH water, the storage modulus at high temperature was around 100 Pa, and at 70 wt% water it was up to 300 Pa. In a 100%

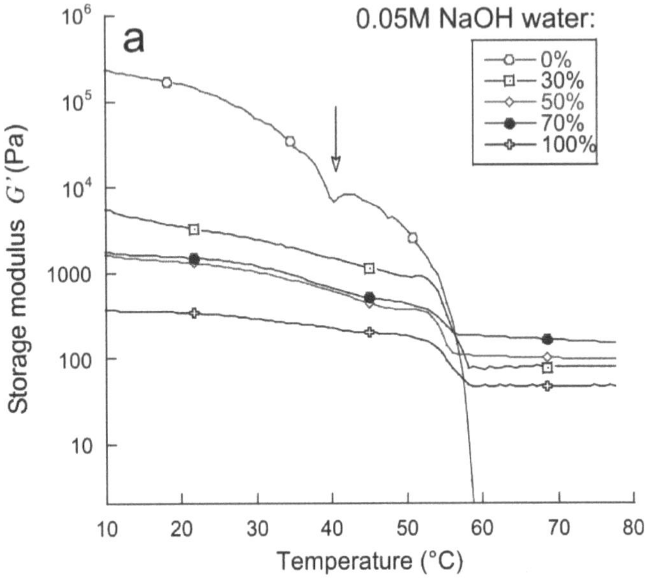

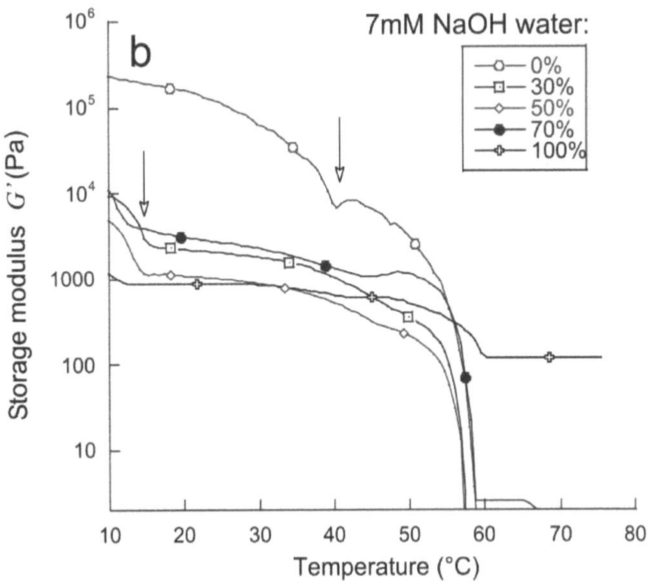

Fig. 8.16. Rheological responses of MG-hazelnut oil-0.05M-NaOH water and MG-hazelnut oil-7mM-NaOH water are illustrated in plots (a) and (b), respectively. Generally the samples lost their gel-like character upon increasing the concentration of water. The non-zero modulus at high temperature indicates the L_α lamellar phase. (a) In the sample with 0.05M-NaOH water, the high-temperature L_α lamellar phase appeared once the water concentration was above 30 wt%. (b) In the sample with 7mM-NaOH water the L_α phase was formed only in the nearly-100 wt% aqueous environment.

MG-0.05M-NaOH water system, the storage modulus at high temperature decreased to 50 Pa, and matches that of the less alkaline 100% MG-7mM-NaOH water system (see Fig. 8.16.B). Unlike the more alkaline sample containing 0.05M-NaOH water, in 7mM-NaOH water systems with any non-vanishing amount of oil the storage modulus at high temperature always remained around zero, indicating that an isotropic fluid phase has transformed into the inverse lamellar (α-gel) phase directly below the gelation line.

Conclusions

It is time to conclude this brief review and take a bird's eye view of the state of the art in this area. Research in this field was inspired by the food industry. Due to its strong emulsifying ability, MG-oil is widely used in making cream-like products. However the complex phase behavior and the unusual aging phenomena of MG-oil remains poorly understood, limiting its applications. In the literature, it was often assumed that the phase diagram in MG-water was closely related to the MG-oil systems (Ojijo et al., 2007). In fact, MG-oil shows phase behavior that is markedly different from the aqueous systems. Two phase transitions are observed on cooling MG-oil solutions from the well-mixed isotropic phase. Below the gelation transition temperature, the inverse lamellar phase is formed, as expected in a hydrophobic surfactant solution. The structure of this inverse lamellar phase turns out to be identical to the α–gel (the known phase of aqueous MG solutions). Because of the unique size ratio between the glycerol head and the lateral area of the aliphatic chain in the fully extended dense brush, the inverse lamellar bilayers have a definite hexagonal in-plane ordering. Upon continuous cooling below the crystallization point of the surfactant, T_c, the lower-temperature phase emerges. The lateral hexagonal packing transforms into a structure analogous to the sub-α crystalline phase of pure MG melt, with orthorhombic chain packing in the unit cell. The rheology shows how these changes in microstructure translate into the macroscopic mechanical features. The inverse lamellar structures form an elastic gel immediately below the gelation transition. The hexagonally structured bilayers have less flexibility and thus form a percolating scaffold that can resist the macroscopic stress applied to the system. Upon entering the lower-temperature sub-α crystalline phase, the elastic modulus of the gel was shown to not change significantly.

With this comprehensive understanding of the phase sequence in water-free MG-oil systems, we were able to generalize the results for the ternary systems. Two solutions, MG-(hazelnut oil)-water and MG-(decane)-water, were reviewed in parallel to indicate the universality of the physical properties. The solubility of MG is naturally limited in the aqueous phase. To obtain a proper MG-water-oil gel, adding a low concentration of co-surfactant (most commonly stearic acid or diacetyl tartaric acid) is necessary. Because of these acid co-surfactants in the solution, the phase ordering is highly dependent on the amount and the pH of the additional aqueous component. In samples with small quantities of water, the phase behavior is almost independent on pH, but once the proportion of water increases, the effects of pH become significant. At low pH the solubility of MG in water is limited and the ternary system retains most features of the oil-dominated environment, such as the sequence of two transitions on cooling, with the low-temperature sub-α crystalline phase. At high pH and sufficient water, the MG layers remain properly swollen and the sub-α crystalline phase disappears from the phase diagram. In rheology, the storage modulus generally decreased upon increasing the fraction of water, but the weak gel-like response is found in the high-temperature L_α lamellar phase over a wide range of water dilution.

With the passage of time, the microscopic structure of MG aggregations in different phases changes and this aging leads to a dramatic change of physical properties. The emulsifying ability of aggregated MG reduces and the solution separates from the more densely packed β–crystals. It was observed that both inverse lamellar phase and sub-α crystalline phase are metastable. After 5 to 7 days of aging they rearrange their molecular ordering to form the stable β–crystals. To establish the relationship between aging and hydrogen bonding, the time evolution of infrared spectra was examined. The coexistence of 3- and 2-OH hydrogen bonding was observed in the initial stages of aging (Fig. 8.2), while the 2-OH hydrogen bonding was dominant and reflects the separation of *D*- and *L*-isomers as a result of aging. We conclude that the emulsifying ability of MG is eventually lost during aging by the rearrangement of hydrogen bonding (Chen & Terentjev, 2009). The same phenomena are also observed in the ternary system. However because of the disturbance of the hydrogen bond interactions by the water molecules, the stability of initial MG aggregations is enhanced significantly and the lamellar gel with longer life time is obtained (Chen & Terentjev, 2010).

References

Atovmyan, E.G.; Koshchii, S.V.; Fedotova, T.N. Hydrogen bonding in alkyl glycerol ethers. J. Appl. Chem.-USSR, 1988, 48, 287–290.

Batte, H.D.; Wright, A.J.; Rush, J.W.; Idziak, S.H.J.; Marangoni, A.G. Phase behavior, stability, and mesomorphism of monostearin-oil-water gels. Food Biophys. 2007, 2, 29–37.

Batte, H.D.; Wright, A.J.; Rush, J.W.; Idziak, S.H.J.; Marangoni, A.G. Effect of processing conditions on the structure of monostearin-oil-water gels. Food Res. Int. 2007, 40, 982–988.

Benitez-Sanchez, P.L.; Leon-Camacho, M.; Aparicio, R. Characterization of the alcoholic fraction of vegetable oils by derivatization with diphenic anhydride followed by high-performance liquid chromatography with spectrophotometric and mass spectrometric detection. Eur. Food. Technol. 2003, 270, 13–19.

Brokaw, G.Y.; Lyman, W.C. The behavior of distilled monoglycerides in the presence of water. J. Am. Oil Chem. Soc. 1958, 35, 49–52.

Cassin, G.; de Costa, C.; van Duynhoven, J.P.M.; Agterof, W.G.M. Investigation of the gel to coagel phase transition in monoglyceride-water systems; Langmuir 1998, 14, 5757–5763.

Chapman, D. Infrared spectroscopic characterization of glycerides. J. Am. Oil Chem. Soc. 1960, 37, 73–77.

Chen, C.H.; van Damme, I.; Terentjev, E.M. Phase behavior of C18 monoglyceride in hydrophobic solutions. Soft Matter 2009, 5, 432–439.

Chen, C.H.; Terentjev, E.M. Aging and metastability of monoglycerides in hydrophobic solutions. Langmuir 2009, 25, 6717–6724.

Chen, C.H.; Terentjev, E.M. Effects of water on aggregation and stability of monoglycerides in hydrophobic solutions. Langmuir 2010, 26, 3095–3105.

van Duynhoven, J.P.M.; Broekmann, I.; Sein, A.; van Kempen, G.M.P.; Goudappel, G.-J.W.; Veeman, W.S. Microstructural investigation of monoglyceride-water coagel systems by NMR and CryoSEM. Coll. Int. Sci. 2005, 285, 703–710.

Garti, N.; Sato, K. Crystallization Processes in Fats and Lipid Systems. Marcel Dekker: New York, 2001; 53–60, 505–518.

Gehlert, U.; Vollhardt, D.; Brezesinski, G.; Mohwald, H. Chiral discrimination in monolayers of monoglycerides. Langmuir 1996, 12, 4892–4896.

Gros, A.T.; Feuge, R.O.; Modification of vegetable oils: Effect of monoglycerides on the interfacial tension of oil-water systems. J. Am. Oil Chem. Soc. 1951, 28, 1–4.

Guillot, S.; Moitzi, C.; Salentnig, S.; Sagalowicz, L.; Leser, M.E.; Glatter, O. Direct and indirect thermal transitions from hexosomes to emulsified micro-emulsions in oil-loaded monoglyceride-based particles. Colloids Surf. A 2006, 291, 78–84.

Heertje, I.; Roijers, E.C.; Hendrickx. H.A.C.M. Liquid crystalline phases in the structuring of food products. Lebensm. Wiss. u. Technol. 1998, 387–396.

Holmgren, A.; Lindblom. G.; Johansson, L.B.A. Intramolecular hydrogen bonding in a monoglyceride lipid studied by Fourier transform infrared spectroscopy. J. Phys. Chem. 1988, 92, 5639–5644.

Jones, R.A.L. Soft Condensed Matter. Oxford University Press: New York, 2002; 136–150.

Kaganer, V.; Mohwald, H.; Dutta, P. Structure and phase transitions in Langmuir monolayers. Rev. Mod. Phys. 1999, 71, 779–819.

Kampf, N.; Gohy, J.F.; Jerome, R.; Klein, J. Normal and shear forces between a polyelectrolyte brush and a solid surface. J. Polym. Sci. B 2005, 43, 193–204.

Kesselman, E.; Shimoni, E. Imaging of oil/monoglyceride networks by polarizing near-field scanning optical microscopy. Food Biophys. 2007, 18, 209–230.

Krog, N.; Larsson, K. Phase behaviour and rheological properties of aqueous systems of industrial distilled monoglycerides. Chem. Phys. Lipids 1968, 2, 129–143.

Krog, N.; Friberg, I.S.E.; Larsson, K. Food Emulsions 3rd ed.; Marcel Dekker: New York, 1997; 141–188.

Krog, N.; Borup, A.P. Swelling behaviour of lamellar phases of saturated monoglycerides in aqueous systems. J. Sci. Fd. Agric. 1973, 24, 691–701.

Larsson, K. The crystal structure of the 1-monoglyceride of 11-bromoundecanoic acid. Acta Cryst. 1966, 21, 267–272.

Larsson, K. Lipids—molecular organisation, physical functions and technical applications; Oily Press: Dundee, 1994.

Larson, R.G. The structure and rheology of complex fluids; Oxford University Press: New York, 1999; 551–562.

Larsson, K.; Krog, N. Structural properties of lipid-water gel phase. Chem. Phys. Lipids 1973, 10, 177–180.

Laughlin, R.G. The aqueous phase behavior of surfactants; Academic Press: New York, 1994.

Lutton, E.S. Phase behavior of aqueous systems of monoglycerides. J. Am. Oil Chem. Soc. 1965, 42, 1068–1070.

Lutton, E.S.; Jackson, F.L. Binary systems with monoglyceride. J. Am. Oil Chem. Soc. 1966, 44, 357–358.

Marangoni, A.G.; Narine, S.S. Physical properties of lipids; Marcel Dekker: New York, 2001; 85–98, 292–307, 324–336.

Marangoni, A.G.; Idziak, S.H.J.; Vega, C.; Batte, H.; Ollivon, M.; Jantzi, P.S.; Rush, W.E. Encapsulation-structuring of edible oil attenuates acute elevation of blood lipids and insulin in humans. Soft Matter 2007, 3, 183–187.

Martin, G. Rheological Measurement. Collyer, A.A., Clegg, D.W., Eds.; Chapman and Hall: London, 1998.

Mcintosh, T.J.; Magid, A.D.; Simon, S.A. Repulsive interactions between uncharged bilayers. Hydration and fluctuation pressures for monoglycerides. Biophys. J. 1989, 55, 897–904.

Mezzenga, R.; Schurtenberger, P.; Burbidge, A.; Michel, M. Understanding foods as soft materials. Nature Mater. 2005, 4, 728–740.

Milner, S.T.; Polymer brushes. Science 1991, 251, 905–914.

Morley, W.G.; Tiddy, G.J.T. Phase behaviour of monoglyceride/water systems. J. Chem. Soc. Faraday 1993, 89, 2823–2831.

Neveu, D.M.L.; Rand, R.P.; Parsegian, V.A.; Gingell, D. Measurement and modification of forces between lecithin bilayers. J. Sci. Food Agric. 1977, 18, 209–230.

Ojijo, N.K.O.; Neeman, I.; Eger, S.; Shimoni, E. Effects of monoglyceride content, cooling rate and shear on the rheological properties of olive oil/monoglyceride gel networks. J. Sci. Food Agric. 2004a, 84, 1585–1593.

Ojijo, N.K.O.; Kesselman, E.; Shuster, V.; Eichler, S.; Eger, S.; Neeman, I.; Shimoni, E. Changes in microstructural, thermal, and rheological properties of olive oil/monoglyceride networks during storage. Food Res. Int. 2004b, 37, 385–393.

Pecharsky, V.; Zavalij, P. Fundamentals of powder diffraction and structural characterization of materials, 2nd ed.; Springer: New York, 2009.

Petrov, P.G.; Ahir, S.V.; Terentjev, E.M. Rheology at the phase transition boundary: Lamellar L_α phase of AOT surfactant solution. Langmuir 2002, 18, 9133–9139.

Pezron, I.; Pezron, E.; Bergenstahl, B.A.; Claesson, P.M. Repulsive pressure between monoglyceride bilayers in the lamellar and gel states. J. Phys. Chem. 1990, 94, 8255–8261.

Pitzalis, P.; Monduzzi, M.; Krog, N.; Larsson, H.; Wahren, H.L.; Nylander, T. Characterization of the liquid-crystalline phases in the glycerol monooleate/diglycerol monooleate/water system. Langmuir 2000, 16, 6358–6365.

Rush, J.W.E.; Jantzi, P.S.; Dupak, K.; Idziak, S.H.J.; Marangoni, A.G. Effect of food preparation on the structure and metabolic responses to a monostearin-oil-water gel based spread. Food Res. Int. 2008, 41, 1065–1071.

Sagalowicz, L.; Leser, M.E.; Watzke, H.J.; Michel, M. Monoglyceride self-assembly structures as delivery vehicles. Trends Food Sci. Technol. 2006, 17, 204–214.

Sari, P.; Razzak, M.; Tucker, I.G. Isotropic medium chain mono-diglyceride/oil/water formulations for solubilization of lipophilic and hydrophilic drugs. Int. J. Pharm. 2003, 270, 287–296.

Sein, A.; Verheij, J.A.; Agterof, W.G.M. Rheological characterization, crystallization, and gelation behavior of monoglyceride gels. J. Coll. Int. Sci. 2002, 249, 412–422.

Silver, B.L. The physical chemistry of membranes; Solomon: New York, 1985.

Small, D.M. The physical chemistry of lipids; Marcel Dekker: New York, 1986; 386–392, 475–492.

9

Physical Properties of β-fat Gel Made of Fully-hydrogenated Rapeseed Oil and Vegetable Oils

Kiyotaka Sato and Satoru Ueno
Laboratory of Food Biophysics, Graduate School of Biosphere Science, Hiroshima University, 739–8528, Japan

Introduction

Fats and oils are generally employed in industrial applications in solid (crystalline), semi-solid, emulsion (oil-in-water and water-in-oil), gel, and liquid states. A gel state is defined as a two-phase colloidal system consisting of a solid and a liquid in a more solid form than sol, although there are various definitions representing the complicated nature of the gel states (Ziegler & Foegeding, 1990). Because of their smooth texture, visco-elasticity, appearance, handling ease, and comfortable mouth-feel, gel materials have attracted considerable attention in food, cosmetic, and pharmaceutical sciences and technology (Clark, 1992; Gallegos & Franco, 1999; Eccleston et al., 2000; Jeong et al., 2002; Vintiloiu & Leroux, 2008). Therefore, a wide variety of research of the gel phases made of polysaccharides, proteins, and other biopolymer materials has been performed (Ould Eleya & Turgeon, 2000; De Bont et al., 2002).

In the gel phase made of lipids, emulsifiers, and water, a lamellar-type lyotropic liquid crystal phase (LC phase) is formed at elevated temperatures. In the LC phase, a water phase is swelled into a continuous lamellar LC phase. When this LC phase is cooled to undergo the transformation from the LC phase to a crystallization phase, the lamellar structure involving swollen water phase is maintained, forming a highly viscous phase called an α-gel phase (Shah et al., 2001; Aoki et al., 2001; Exerowa, 2002; Chronakis & Kasapis, 1995; Chronakis, 1997). Further cooling forms a rheologically hard phase called a coagel (Ojijo et al., 2004; Chen & Terentjev, 2010; Chen et al., 2009).

In this chapter we present the physical properties of organogels using high-melting (HM) and low-melting (LM) fat mixtures that exhibit the gelating behavior, and to clarify the formation mechanism of the gelating fat (Higaki et al., 2003a; 2003b; 2004). No other ingredients, such as emulsifiers, water or waxes, were added to the mixtures. Throughout the present work, we have set the highest concentration of the HM-fat as 4.0 wt %. This is due to the presence of large amounts of HM-fats, which often results in a fat-blend mixture that causes granular crystal formation and sandy taste. These properties are not favored for food uses such as confections and creams.

In the present work, we examined five kinds of HM-fats and four kinds of LM-oils, and various HM-fat and LM-fat mixtures were subjected to different modes of cooling and heating treatment (tempering). We found that particular combinations of the HM-fats and LM-oils formed the gelating fats when the liquid mixture was rapidly cooled from high-temperature and a subsequent heating procedure was applied. Solid fat content (SFC) and X-ray diffraction analyses showed that the transformation from α to β of the HM-fat fraction was a prerequisite for the formation of the gelating fat mixtures. Eventually, the β crystals of fully-hydrogenated rapeseed containing a high concentration of behenic acid moiety exhibited the most functional properties of an organogel.

Materials and Methods

Table 9.A shows the melting point and weight percent fatty acid components and melting points (T_m) of fully-hydrogenated rapeseed oil with a high amount of behenic acid (FHR-B), fully-hydrogenated rapeseed oil with a high amount of stearic acid (FHR-S), sal fat olein (SFO), palm super olein (PSO), olive oil, and cocoa butter (CB). Multiple DSC melting peaks were observed for LM-oils because the oils contain a variety of triacylglycerols, each having slightly different high and low

Table 9.A. Melting Point and Weight Percent Fatty Acid Composition of the High-melting Fats and Low-melting Fats.

	High-melting Fats		Low-melting Fats			
	FHR-B	FHR-S	SFO	PSO	Olive Oil	CB
Melting Point (°)	65.8°	68.0°	15°	7°	5°	34°
C12:0	—	—	—	0.3	13.1	—
C14:0	0.1	—	0.1	1.0	0.5	—
C16:0	3.2	4.9	8.6	32.0	—	24.4
C16:1	—	—	—	0.4	—	—
C18:0	37.3	91.3	24.6	3.0	2.1	35.4
C18:1	0.2	—	56.7	49.0	79.3	38.1
C18:2	—	—	4.0	13.5	4.7	2.1
C18:3	—	—	0.7	0.8	0.2	—
C20:0	9.8	2.5	4.9	—	0.1	—
C22:0	47.8	1.3	0.4	—	—	—
C24:0	1.6	—	—	—	—	—

FHR-B: Fully-hydrogenated rapeseed oil with high amount of behenic acid

FHR-S: Fully-hydrogenated rapeseed oil with high amount of stearic acid

SFO: sal fat olein

PSO: palm super olein

CB: cocoa butter

melting points. As a result the melting points shown in Table 9.A correspond to those of the major fat components with the highest DSC melting peaks. Minor HM-components are also present in the LM-oils. For example, SFO contains a few percent of the HM-fraction, which contributes to the SFC values of the mixtures. SSS (T_m = 73.1°C), AAA (T_m = 78.0°C), and BBB (T_m = 82.5°C) were purchased (Tokyo Kasei Co. Tokyo), each having a purity above 99%.

The HM-fats and LM-oils were mixed at 70°C and subjected to the tempering process illustrated in Fig. 9.1. We varied three factors of the tempering process to search for the optimal thermal treatments for inducing gelating behavior: (i) Cooling rate (V_c) is a rate of cooling from 70°C to crystallization temperature (T_c) and V_c=10°C/min and V_c=15°C/min were examined; (ii) T_c is the first crystallized structure that determines the polymorphic forms of the HM-fats in the HM-fat and LM-oil mixtures. The T_c values examined were 20, 25, 30, 35, and 38°C. The crystallized samples were kept at T_c for 10 minutes and subjected to heating at a rate of 2°C/min.; and (iii) T_a is the temperature at which isothermal treatment was applied to cause the melting of the crystal forms, which occurred at T_c, and transformation to more stable forms. The T_a values examined were 30, 35, 38, 41, 43, and 45°C.

We observed the gelation behavior visually by putting the fat mixtures in test tubes and tilting them by 45° (visual observation). The gel point was defined as the temperature/time point when no flow was evident upon tilting of the test tube and the liquid and solid phases did not separate in the mixture.

A Rigaku RINT-TTR powder X-ray diffractometer (Rigaku, Tokyo) was used in this study to determine the polymorphic forms of the crystallized HM-fats. This machine uses a copper rotating-anode running at 8.4kW; 42kV, 200mA to generate Kα (wave length, 0.154 nm) radiation, and was equipped with a nickel filter.

SFC values were measured by pulse-NMR (Bruker. PC120, Germany). The sample (1.5 g) was placed in a glass cell (diameter: 7 mm) and a tempering process was performed using the temperature-control system of the NMR equipment.

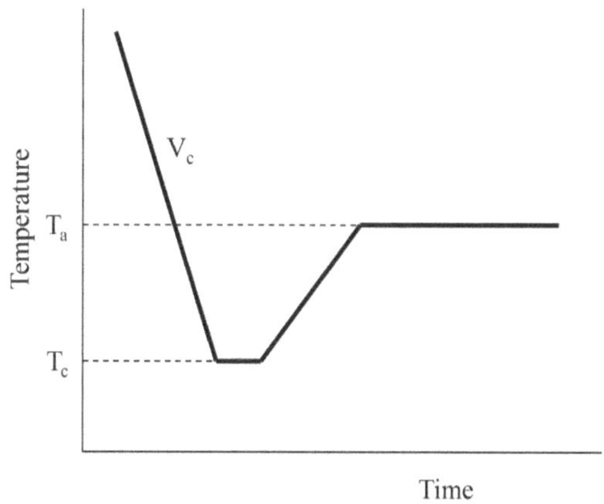

Fig. 9.1. Temperature variation profile.

A digital high-fidelity microscope (model VH-8000, Keyence Co., Ltd., Osaka, Japan) with a controlled-temperature sample holder was used for the in-situ optical observation. A thermal device (LINKAM TH-600RMS, Japan High Tech Co., Ltd., Fukuoka, Japan) controlled the temperature variation with circulating water connected to a thermostat. Samples weighing 0.5 mg were placed on a glass and then covered using a thin 30 μm glass. All optical observations were performed at 2000× magnification.

A cone-plate viscometer (Tokyo Keiki Ltd., Tokyo, Japan) was used for measuring flow properties of the gelled fat mixture. This experiment was performed using a cone-plate geometry of 24mm radius (R) with conic angle of 2.34×10^{-2} (rad). The molten sample (1.0 g) was placed on the measuring plate of a viscometer, and the sample temperature was varied with a temperature-controlled sample holder. The measurement was performed at 38°C while varying the rotor revolutions (N) from 0.5 to 100 rpm.

Storage module (G′), loss module (G″), and strain (γ) were measured by small-amplitude dynamic measurements using a RheoStress RS100 (Gebr. HAAKE Gmblt, Karlsruhe, Germany) dynamic rheometer with application software V2 for RheoStress. G′ describes the elastic, solid-like, or gel behavior, and G″ is the viscous behavior of the materials. This experiment was performed in a cone-plate system of 17.5mm radius, with a conic angle of 6.98×10^{-2} (rad). The thermal treatment was performed in RS100 circulating coolant connected to a thermostat. The molten sample (1.0 g) was mounted on the temperature-controlled measuring plate. The height, force, gap, and temperature of the rheometer were calibrated automatically. To determine the linear viscoelastic range for the sample, a dynamic stress-sweep measurement at a frequency of 1.0 Hz from 0.01 to 10 Pa was conducted at 38°C after proper thermal treatment for exhibiting the gel behavior. The dynamic force per area selected for the linear viscoelastic range was lower than the stress corresponding to the yield value, which was 6 Pa. G′ and G″ moduli were obtained for a frequency scan from 0.1 to 10 Hz using a fixed stress value of 2 Pa.

Thermal, Structural, and Morphological Properties

Visual Observation

We evaluated the visual observation of the fat mixtures' gelling behavior for different combinations of HM-fats (FHR-B, FHR-S, SSS, AAA, and BBB, 2.0 wt %) and LM-oils (SFO, PSO, olive oil, and CB, 98.0 wt %) by applying very specific thermal treatments for organogel formation (see the following). The conclusion was that only two mixtures of FHR-B/SFO and FHR-B/CB exhibited organogelling behavior, and all other mixtures were in a sol state. Therefore, we present here the results for only the mixtures using FHR-B as the HM-fat.

Table 9.B shows the results of visual evaluation of the fat mixtures of FHR-B/SFO at different concentrations of FHR-B, when T_c was varied from 20 to 38°C while the tempered fats were kept at T_a =38°C for 60 minutes. At all T_c values examined, the sol mixtures were formed when the concentrations of FHR-B did not exceed 1.5 wt %. At the 1.5 wt % concentration of FHR-B, gelling behavior was observed at T_c =20°C, whereas the mixture became sol when T_c increased above 20°C. At the concentrations of FHR-B of 2.0 wt % and 4.0 wt %, the mixtures

Table 9.B. Effects of T_c and T_a on Gelating Behavior of FHR-B/SFO Mixtures. B-1: Effects of T_c at T_a=38°C (for 60 Minutes). B-2: Effects of T_a (for 60 Minutes) at V_c=10°C/min and T_c=20°C.

B-1			T_c (°C)			
Concentration of FHR-B(wt%)	20	25	30	35	38	
0	—	—	—	—	—	
0.5	—	—	—	—	—	
1.0	—	—	—	—	—	
1.5	▲	—	—	—	—	
2.0	●	●	—	—	—	
3.0	●	●	▲	—	—	
4.0	●	●	●	—	—	
B-2			T_a (°C)			
Concentration of FHR-B(wt%)	30	35	38	41	43	45
0	—	—	—	—	—	—
0.5	—	—	—	—	—	—
1.0	—	—	—	—	—	—
1.5	●	●	▲	—	—	—
2.0	●	●	●	—	—	—
3.0	●	●	●	●	●	—
4.0	●	●	●	●	●	●

●: Gel, ▲: Partly-gel, —: Sol.

showed gelating behavior below T_c = 25°C (FHR-B: 2.0 wt %) and below T_c = 30°C (FHR-B: 4.0 wt %). This result shows the importance of the crystallization of FHR-B in the FHR-B/SFO liquid mixture at T_c = 20~25°C by chilling from 70°C at a rate of 10°C/min at the concentrations of FHR-B above 2.0 wt %. The FHR-B crystals thus formed were α form, which converted to β crystals through α-melt mediation after subsequent rapid heating, as shown below.

Table 9.B.2 shows the results of the visual evaluation of the FHR-B/SFO fat mixtures that were kept at different T_a values for 60 minutes, after crystallization at T_c = 20°C with V_c = 10°C/min. Enhance gelation behavior was exhibited with increasing amounts of FHR-B at the all T_a values, whereas the sol mixture was formed at T_a above 41°C at FHR-B concentrations of 1.5 and 2.0 wt %. As for the FHR-B concentrations of 3.0 and 4.0 wt %, the gelating behavior was revealed by the fluidity evaluation.

Solid Fat Content Measurements

To in-situ analyze the gelation processes, SFC values were measured during the isothermal treatment of the FHR-B fat mixture (2.0 wt %)/SFO at T_a = 38°C for 60 minutes after cooling to T_c = 20°C at V_c = 10°C/min and V_c = 1.5°C/min as shown in Fig. 9.2. The SFC value was about 3.0 wt % before the temperature changed from 20°C. This

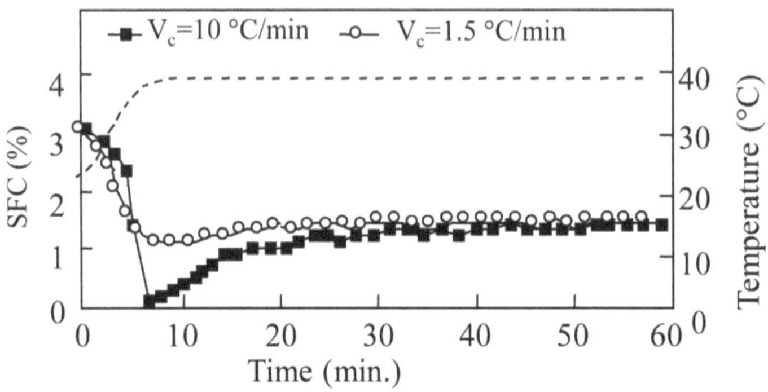

Fig. 9.2. SFC values of FHR-B (2.0 wt %)/SFO mixture during the isothermal treatment at $T_a = 38°C$ for 60 minutes, after tempering with $T_c = 20°C$ at $V_c = 10°C/min$ and $V_c = 1.5°C/min$.

value increased to more than the initial concentration of the HM-fat in the mixture, due to crystallization of the high-melting fraction in SFO at 20°C. About 6 minutes after the isothermal treatment started, the SFC value decreased to 1.2 wt %, and did not change within 60 minutes. By contrast, the SFC value of the fat mixture that was cooled with $V_c = 10°C/min$ to 20°C decreased to 0% at 8 minutes after the isothermal treatment started. Then, the SFC value gradually increased to 1.4 wt % at around 30 minutes. This result showed that the crystals, which were formed by rapid cooling to 20°C in the FHR-B/SFO mixture, completely melted at $T_a = 38°C$, and recrystallized during the isothermal treatment. Therefore, we concluded that melt-mediated transformations occurred during the heating from 20°C to 38°C and isothermal treatment at 38°C.

Fig. 9.3 shows the effects of time during isothermal treatment at $T_a = 38°C$ on the conversion from sol to gel states of FHR-B (2.0 wt %)/SFO, which was tempered with $V_c = 10°C/min$ and $T_c = 20°C$. Ten minutes after T_a was reached (Fig. 9.3.A), the fat mixture was sol-like. However, an enhanced gelation behavior was observed as the duration of the isothermal treatment was extended, and the complete gel state was achieved only after 40 minutes (Fig. 9.3.D). This result showed good agreement with the results of the SFC measurement shown in Fig. 9.2.

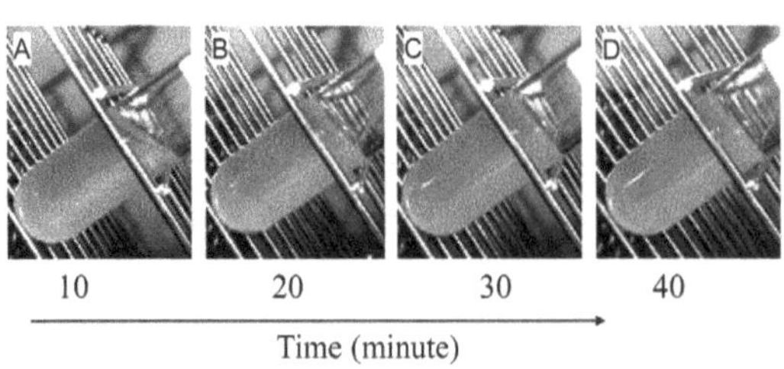

Fig. 9.3. Transformation from sol to gel states of FHR-B (2.0 wt %)/SFO mixture during isothermal treatment at $T_a = 38°C$ after tempering with $V_c = 10°C/min$ and $T_c = 20°C$.

X-ray Diffraction Measurements

Fig. 9.4 shows the in-situ XRD experiments of the FHR-B fat mixture (4.0 wt %)/SFO that was subjected to the tempering procedures of T_c = 20°C, T_a = 38°C at V_c = 10°C/min (Fig.9.4.A), and V_c=1.5°C/min (Fig. 9.4.B). The peaks in the XRD spectra were broad since the concentration of fat crystals was low and liquid content was high. However, it was confirmed that the α form was crystallized at 20°C when the mixture was cooled at V_c = 10°C/min, as revealed in a long-spacing spectrum of 1.96 nm (003 reflection) and a short-spacing spectrum of 0.41 nm (Higaki et al., 2004) (see arrows in Fig. 9.4.A). After heating to T_a=38°C, the XRD peaks of the α form disappeared and those of the β form having a long-spacing spectrum of 1.73 nm (003 reflection) and a short-spacing spectrum of 0.46 nm appeared. By contrast, the β′ and β forms were formed at 20°C when the liquid fat mixture was cooled to 20°C at V_c=1.5°C/min (Fig. 9.4.B), and the β′ form transformed into the β form during the heating to 38°C.

From these results, we may conclude that the melt-mediated transformation of the FHR-B/SFO fat mixture during isothermal treatment after tempering of V_c=10°C/min and T_c=20°C was due to the melt-mediated transformation from the α form to the β form of FHR-B crystal fractions in the mixtures.

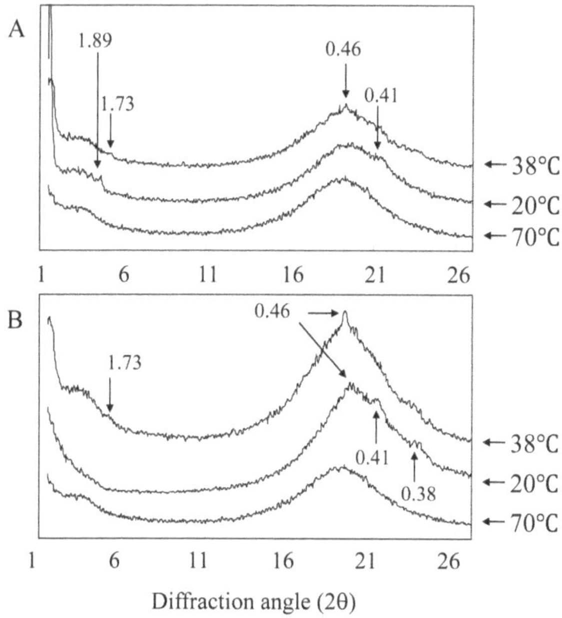

Fig. 9.4. Temperature variations in X-ray diffraction patterns of a FHR-B (4.0 wt %)/SFO mixture that was subjected to the tempering procedures of T_c = 20°C, T_a = 38°C at V_c = 10°C/min (Fig. 9.4.A), and V_c=1.5°C/min (Fig. 9.4.B).

Optical Microscopic Observation

Effect of V_c

Figs. 9.5 and 9.6 show optical images of the FHR-B crystals (concentration of 2.0 wt %) in FHR-B/SFO (98.0 wt %) during the first cooling at rates of V_c=10°C/min and V_c=1.5°C/min. For the rapid cooling shown in Fig. 9.5, small FHR-B crystals with dimensions of 10 μm were formed in the FHR-B/SFO liquid mixture at T_c=20°C (Fig. 9.5.A). After heating to T_a=38°C, the crystals melted (Fig. 9.5.B). After further heating at 38°C, new small crystals with dimensions of 2 μm re-crystallized.

By contrast, large FHR-B crystals with dimensions of 15 μm formed in the FHR-B/SFO liquid mixture with T_c=20°C and V_c=1.5°C/min (Fig. 9.6.A). These crystals did not melt completely during heating from T_c=20°C to T_a=38°C (Fig. 9.6.B), and most crystals grew into larger crystals with dimensions of 20 μm (Fig. 9.6.C). The crystals denoted by circles in Fig. 9.6 possess these properties. In this case, no gelating behavior was observed. Our previous XRD study showed that the β′ and β forms developed in the FHR-B/SFO liquid mixture at T_c=20°C and V_c=1.5°C/min. In addition, the β′ form transformed to the β form by solid-state transformation after subsequent heating from 20°C to 38°C, not by melt-mediation when V_c was 1.5°C/min. The crystals shown in Fig. 9.6 support this result, which demonstrates that the crystals indicated by circles might be β form, and the other crystals might be β′ form.

Effect of T_c

Fig. 9.7 shows the difference in the FHR-B crystal morphology of the FHR-B (2.0 wt %)/SFO (98.0 wt %) mixture that was subjected to thermal treatment with T_c = 30 and 35°C, T_a =38°C, and V_c=10°C/min. Large crystals were seen after cooling to

Fig. 9.5. Optical images of the crystallization behavior of FHR-B in the FHR-B (2.0 wt %)/SFO mixtures during rapid-cooling of V_c=10°C/min. A) is at T_c=20°C, B) is soon after heating from T_c to T_a=38°C, and C) is at T_a=38°C. Scale bar = 20 μm.

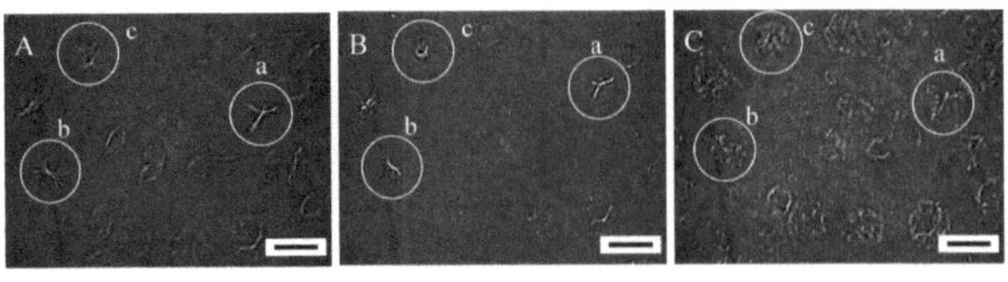

Fig. 9.6. Optical images of crystallization behavior of FHR-B in the FHR-B (2.0 wt %)/SFO mixtures during slow-cooling of V_c=1.5°C/min. A) is at T_c=20°C, B) is soon after heating from T_c to T_a=38°C, and C) is at T_a=38°C. Scale bar = 20 μm.

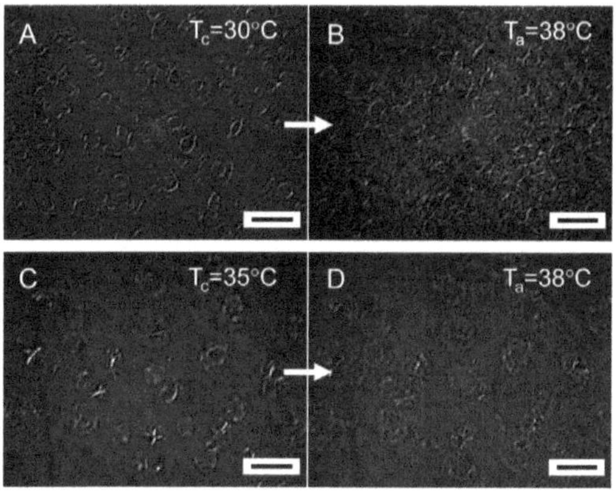

Fig. 9.7. Optical images of the crystallization behavior of FHR-B in the FHR-B (2.0 wt %)/SFO mixtures when T_c was varied from 30 to 35°C. Upper portion: T_c=30°C (A) → T_a=38°C (B). Lower portion: T_c=35°C (C) → T_a=38°C (D), and V_c =10°C/min. Scale bar = 20 μm.

30°C and 35°C, and these crystals grew by heating to 38°C. At T_c=38°C, the crystal morphology was the same as that of the simple cooling from 70°C to 38°C (T_c=T_a). The morphology and crystallization behavior of FHR-B were similar to the results after slow cooling (V_c=1.5°C/min) to T_c=20°C, as shown in Figs. 9.6.A to 9.6.C.

Effect of LM-oils

Fig. 9.8 presents optical images of the crystal morphology of two fat mixtures containing FHR-B (2.0 wt %), and CB and olive oil (98.0 wt %), respectively, which

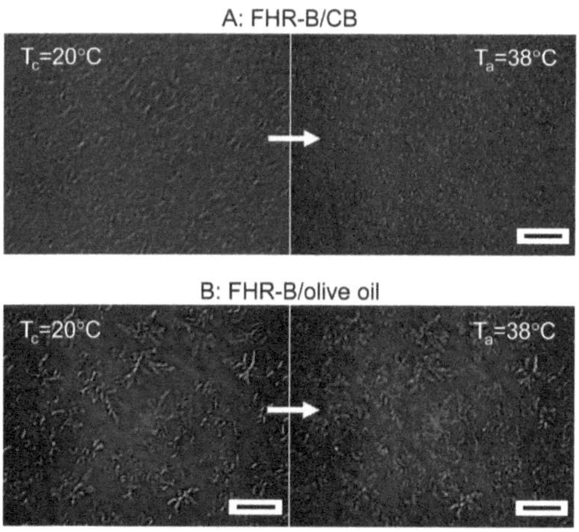

Fig. 9.8. Optical images of the crystallization behavior of FHR-B in the mixtures of FHR-B (2.0 wt %) and cocoa butter (CB) and olive oil. A) is FHR-B/CB at T_c=20°C, B) is FHR-B/CB at T_a=38°C, C) is FHR-B/olive oil at T_c=20°C, and D) is FHR-B/olive oil at T_a=38°C. Tempering conditions were T_c=20°C, T_a=38°C, and V_c=10°C/min. Scale bar = 20 μm.

were subjected to thermal treatment at T_c=20°C, T_a=38°C, and V_c=10°C/min. The former mixture displayed gelating behavior, whereas the latter did not.

The FHR-B crystals in the mixture with CB (Fig. 9.8.A) exhibited the same morphological behavior as that of the FHR-B/SFO mixture during the temperature variation from T_c=20°C to T_a=38°C. In particular, the β crystals of FHR-B in these fat mixtures at T_a=38°C were small, with dimensions of 2 μm (Fig. 9.8.A). In contrast, the FHR-B crystals in olive oil were significantly larger, with dimensions of 20 μm at T_c=20°C, and grew into large crystals at T_a=38°C (Fig. 9.8.B). Although not shown, the FHR-B crystals in PSO possessed the same morphological properties as those of Fig. 9.8.B.

Effect of HM-fats

Fig. 9.9 illustrates the crystal morphology of the FHR-S and SSS in the FHR-S (2.0 wt %)/SFO (98.0 wt %) and SSS (2.0 wt %)/SFO (98.0 wt %) mixtures, which were subjected to thermal treatment of T_c=20°C, T_a=38°C, and V_c=10°C/min. FHR-S/SFO and SSS/SFO fat mixtures did not exhibit gel behavior. The crystals of FHR-S in FHR-S/SFO and SSS in SSS/SFO at T_c=20°C and T_a=38°C were larger than those in FHR-B/SFO, and both crystals were dispersed non-uniformly, resulting in crystal aggregates at T_a=38°C.

Fig. 9.10 shows the SFC values of the FHR-B/SFO, FHR-S/SFO, and SSS/SFO fat mixtures. The SFCs of FHR-S/SFO and SSS/SFO (1.9 wt %) were larger than that of FHR-B/SFO (1.4 wt %) at T_a=38°C. This result demonstrates that the crystal content of FHR-S and SSS was higher than that of FHR-B. This suggests that the development of gelating behavior does not depend merely on the crystal contents of HM-fats but, more importantly, on the crystal morphology, as presented above.

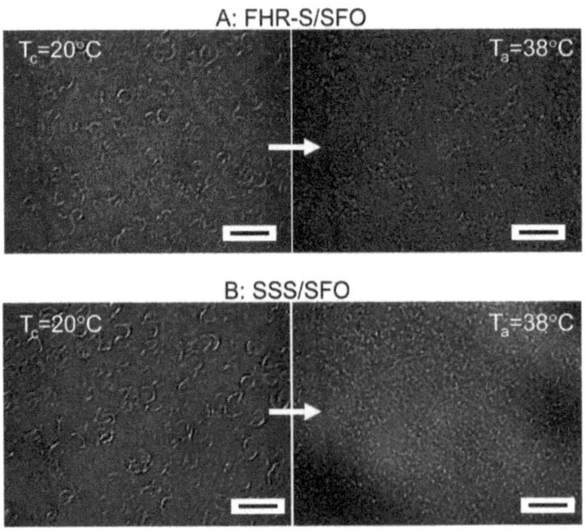

Fig. 9.9. Optical images of the crystallization behavior in mixtures of two HM-fats (2.0 wt %, FHR-S and SSS) with SFO (98.0 wt %). A) is FHR-S/SFO at T_c=20°C, B) is FHR-S/SFO at T_a =38°C, C) is SSS/SFO at T_c=20°C, and D) is SSS/SFO at T_a =38°C. Tempering conditions were T_c=20°C, T_a =38°C, and V_c=10°C/min. Scale bar = 20 μm.

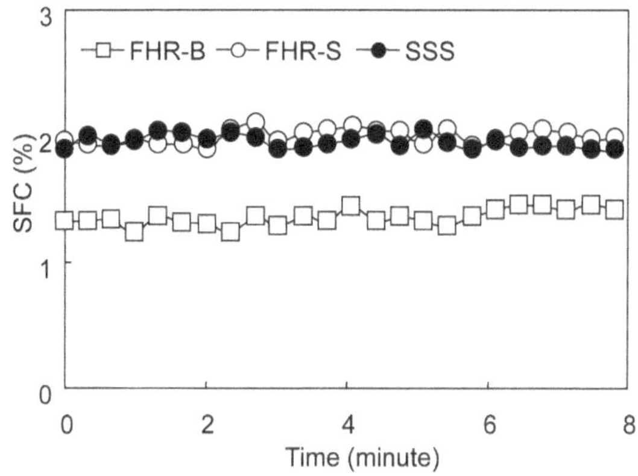

Fig. 9.10. SFC values of three HM-fats (2.0 wt %) with SFO (98.0 wt %) after tempering of T_c=20°C, T_a=38°C for 60 min, and V_c=10°C/min.

The results of the structural and morphological observation can be summarized as follows.

(1) The FHR-B crystals of FHR-B/SFO and FHR-B/CB fat mixtures that exhibited gelating behavior possessed unique microstructures, such as large numbers, minimal size, and uniform distribution. These microstructures were observed at high-cooling rates during the first crystallization, but not during the slow-cooling process.
(2) XRD analysis showed that this unique crystallization behavior is associated with the α-melt mediated transformation to β form.
(3) The mixtures of FHR-S/SFO and SSS/SFO also showed the melt-mediated transformation from α to β form. However, the crystal morphology of the HM-fat fractions was quite different from that of FHR-B/SFO in terms of size and aggregation of the β crystals.
(4) While the mixtures of FHR-B/olive oil and other LM-fats did not exhibit gel behavior, crystal morphology similar to FHR-S/SFO was observed. Morphological properties of the HM-fats play decisive roles in exhibiting the gel behavior of the HM and LM-fat mixtures.

Rheological Properties

The rheological properties of crystal networks formed by fat crystals are important in food products containing significant amount of fats and oils. Many sensory characteristics, such as appearance, texture, spreadability, and hardness, depend on the mechanical strength of the underlying network of fat crystals (DeMan & Beers, 1987; Heertje, 1993; Juriaanse & Heertje, 1988; Marangoni & Hartel, 1998; Narine & Marangoni, 1999). Also, fat crystal networks form a particular class of soft materials that demonstrate yield stress and viscoelastic properties. The rheological behavior of fats is mainly determined by the amount of fat crystals and the types of crystals present in the fat mixtures (Herrera & Hartel, 2000; Litwinenko et al., 2002; Marangoni & Narine, 2002).

In this section, we present the results of rheological analysis of the gel fat mixture of FHR-B/SFO produced by proper thermal treatment.

Flow Properties of the Gelated Fat Mixtures

Fig. 9.11 illustrates the flow properties of FHR-B (concentration of 2.0 wt %)/SFO (98.0 wt %) mixture and SFO liquid at 38°C that was subjected to thermal treatment with V_c=10°C/min, T_c=20°C, and T_a=38°C held for 60 min. SFO exhibited the straight line of a Newtonian fluid because SFO melted completely. The consistency factor k and flow behavior index n of SFO were 4.57×10^{-2} Pa•s and 1.0. The FHR-B/SFO fat mixture exhibiting gelating behavior is shown as pseudo-plastic fluid curves with the apparent yield value of 6 Pa with k and n values of 1.38 Pa•s^n and n=0.65.

Influence of V_c on Viscoelasticity

Fig. 9.12 illustrates the change of storage moduli (G′) and loss moduli (G″) of FHR-B (2.0 wt %)/SFO (98.0 wt %) mixture with the first cooling at rates of V_c=10°C/min and 1.5°C/min. The G′ and G″ of SFO liquid are also shown as the control. The measurement was performed at 38°C after thermal treatment with T_c=20°C and T_a=38°C held for 60min.

In the rapid cooling of V_c=10°C/min shown in Fig. 9.12.A, G′ approximated the same values in the scanned frequency. G″ values were smaller than G′ values and decreased with increasing frequency in the measured range. The behavior of G′ and G″ in Fig. 9.12.A was the same as the typical gel state.

In general, G′ and G″ of the gel state are larger than those of the liquid state, and G′ is larger than G″. G′ and G″ of the liquid state increased linearly with increasing frequency, with G″ dominating the viscoelasticity of the liquid state.

At the slow cooling rate of V_c=1.5°C/min, the gelating fat mixture was not formed, as shown in Fig. 9.12.B. G′ and G″ rose with increasing frequency, and these

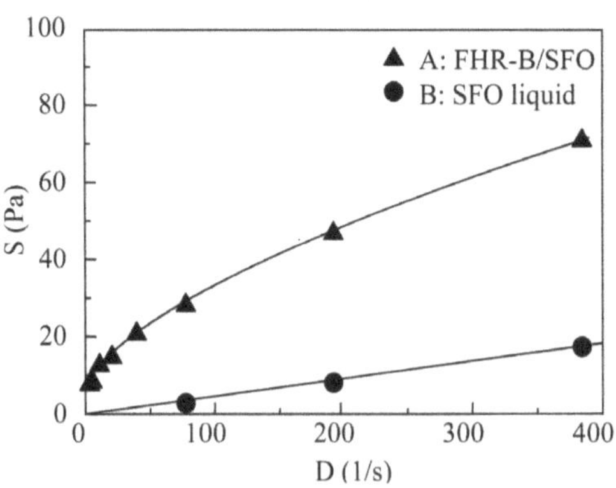

Fig. 9.11. Shear stress-shear rate curve showing flow properties of FHR-B (2.0 wt %)/SFO (98.0 wt %) fat mixture and SFO liquid at 38°C (T_c=20°C, V_c=10°C/min, and T_a=38°C held for 60 min).

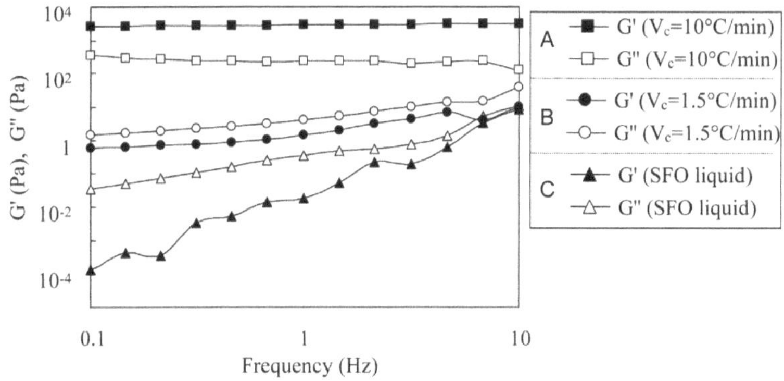

Fig. 9.12. Influence of V_c on storage (G′, black symbol) and loss (G″, white symbol) moduli values measured for FHR-B (2.0 wt %)/SFO (98.0 wt %) fat mixture and SFO liquid at 38°C during a frequency scan from 0.1 to 10 Hz with 2 Pa (T_c=20°C and T_a=38°C held for 60 min).

values were smaller than those in rapid cooling. Furthermore, G″ dominated the viscoelasticity, corresponding to sol behavior. Finally, G′ and G″ of the SFO liquid exhibited liquid behavior because there were no crystals in the SFO oil (Fig. 9.12.C).

Influence of FHR-B Concentration on Viscoelasticity

Table 9.C shows the conditions of FHR-B/SFO fat mixtures and the values of G′ and G″ at a frequency of 1.0Hz. Fig. 9.13 shows the G′ and G″ values summarized in Table 9.C and the corresponding SFC values. The thermal treatment was V_c=10°C/min, T_c=20°C, and T_a=38°C held for 60 min. For fat mixtures containing a FHR-B concentration of less than 1.5 wt %, G″ dominated the viscoelasticity, exhibiting sol behavior. However, for FHR-B concentrations exceeding 1.5 wt % exhibited gel behavior in which the G′ values were larger than the G″ values. Furthermore, G′ had almost constant values throughout the scanned frequencies. In addition, G′ and G″ increased with increasing FHR-B concentration. This means that the SFC values representing the

Table 9.C. States of FHR-B/SFO Fat Mixture, G′, and G″ at a Frequency of 1.0 Hz when the FHR-B Concentration was Varied from 0 to 4.0 wt % (V_c=10°C/min, T_c=20°C, and T_a=38°C Held for 60 min).

Concentration (wt.%)		State of Fat Mixture		
FHR-B	SFO	at T_a = 38°C	G′ (Pa)	G″(Pa)
0.0	100.0	sol	2.04×10^{-2}	3.53×10^{-1}
1.0	99.0	sol	6.75×10^{-1}	2.07
1.5	98.5	gel	7.13×10^{2}	8.50×10
2.0	98.0	gel	2.85×10^{3}	2.37×10^{2}
3.0	97.0	gel	3.37×10^{4}	2.19×10^{3}
4.0	96.0	gel	8.95×10^{4}	7.47×10^{3}

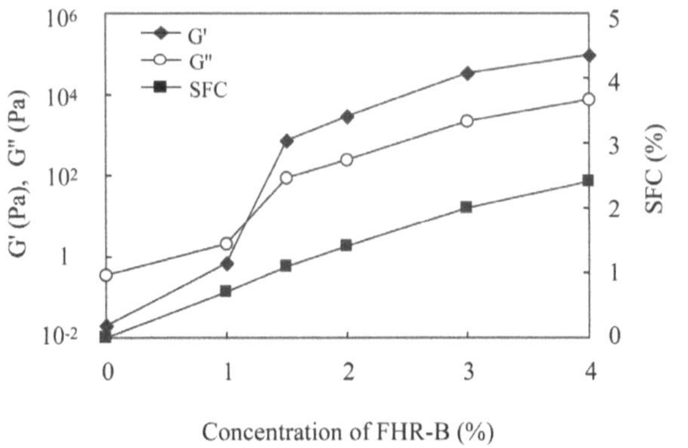

Fig. 9.13. Influence of initial concentrations of FHR-B crystals on storage (G′, black symbol) and loss (G″, white symbol) moduli values shown in Table 9.C and the corresponding SFC values.

crystal amount of FHR-B rose with increasing FHR-B concentration. This indicates that induction of gelating behavior was necessary for crystallizing the tiny β crystals of FHR-B above the 1.5 wt % FHR-B concentrations with proper thermal treatment.

Influence of T_a on Viscoelasticity

Fig. 9.14.A depicts the change of G′, G″, and oscillatory shear strain (γ), at a frequency of 1.0 Hz with 2 Pa stress during the heating process from 25 to 45°C at a rate of 2.0°C/min. The measurement was performed after the gelated mixture of FHR-B (2.0 wt %)/SFO (98.0 wt %) cooled from 38 to 20°C. G′ and G″ decreased linearly with increasing temperature until 38°C. Above 38°C, G′ and G″ decreased drastically. Around 40.5°C, G′ and G″ had the same values as the FHR-B (1.5 wt %)/SFO (98.5 wt %) fat mixture that exhibited gel behavior at 38°C. It is worthy to note that G″ became larger than G′ at 43°C. At 44.5°C, G′ and G″ were almost the same values as in the FHR-B (1.0 wt %)/SFO (99.0 wt %) fat mixture that did not exhibit gelated behavior at 38°C.

γ increased linearly with temperature until 38°C and then increased sharply above 38°C. This result indicates that the structure of the gel fat mixture of FHR-B (2.0 wt %)/SFO (98.0 wt %) weakened and gradually converted to the sol state. This means that the amount of β crystals of FHR-B formed by proper thermal treatment started to decrease.

Fig. 9.14.B illustrates the reduction of the SFC values of β crystals of FHR-B, which gradually decreased at temperature from 25 to 45°C. Around 38°C where G′ and G″ started to decrease rapidly, the SFC value was 1.5 wt %, and around 43°C where G′ became larger than G′ the SFC value was 1.0 wt %.

We can summarize the results of the rheological analysis as follows.

(1) The gel-mixture of FHR-B (2.0 wt %)/SFO (98.0 wt %) formed by proper thermal treatment with V_c=10°C/min, T_c=20°C, and T_a=38°C exhibited pseudo-plastic fluid properties with an apparent yield value of 2 Pa. By

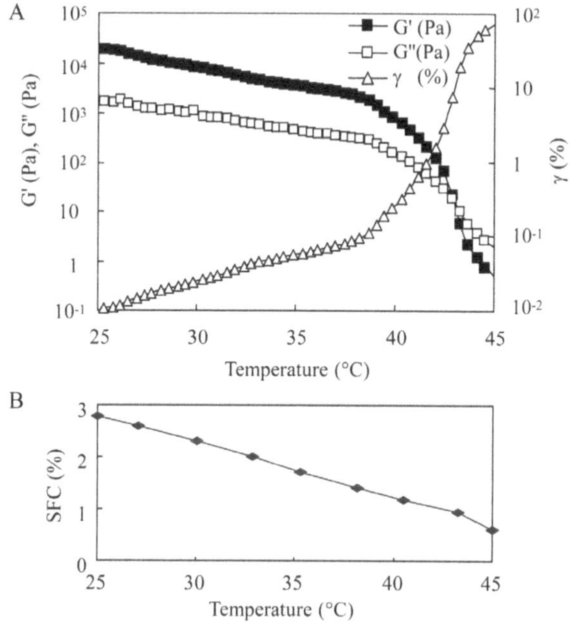

Fig. 9.14. (A) Change of G′ (black squares), G″ (white squares), and oscillatory shear strain γ (white triangles) at a frequency of 1.0 Hz with 2 Pa during heating from 25 to 45°C at 2.0°C/min after the FHR-B (2.0 wt %)/SFO (98.0 wt %) mixture was re-cooled from 38 to 20°C (V_c=10°C/min, T_c=20°C, and T_a=38°C held for 60min), and (B) the corresponding SFC values.

contrast, G′ and G″ of this fat mixture exhibited sol-mixture behavior at a slow cooling rate of V_c=1.5°C/min.

(2) For FHR-B concentrations above 1.5 wt %, G′ and G″ of FHR-B/SFO fat mixtures exhibited gel behavior when proper thermal treatment was applied. However, G′ and G″ of the fat mixtures containing FHR-B at less than 1.5 wt % concentrations showed sol behavior even when proper thermal treatment was applied.

(3) The gel state of the FHR-B (2.0 wt %)/SFO (98.0 wt %) fat mixture started to change to a sol state as T_a increased from 38 to 45°C. G′ dominated the viscoelasticity of this fat mixture until 43°C, whereas G″ dominated the viscoelasticity above 43°C.

(4) The SFC value of 1.1% at 40.5°C was the minimum FHR-B crystal amount to exhibit gel behavior. The gel state was not maintained below 1.1%, although tiny β crystals of FHR-B were formed by proper thermal treatment.

The rheological analysis of the gelated mixtures of FHR-B/SFO formed at different cooling rates (V_c) and different FHR-B concentrations indicated that the FHR-B crystals present in the fat mixture play decisive roles for exhibiting gel behavior. G′ and G″ values typical of the gel state changed to those typical of the sol state with decreasing V_c and FHR-B concentration; 1.5 wt % was the minimum concentration of FHR-B crystals. This concentration was worth the SFC value of 1.1%, the minimum crystal amount of FHR-B for exhibiting gel behavior. For FHR-B concentrations less than 1.5 wt %, the crystal amount of FHR-B was insufficient for forming the gel state.

Gelation Mechanisms

Based on the results of presented in previous sections, we explored the formation mechanisms of the gel fat mixture containing HM (2.0 wt %) and LM-fats (98.0 wt %), as shown in Fig. 9.15. For mixtures that exhibit gel behavior (Fig. 9.15.A), the β crystals of FHR-B were small and uniformly distributed, formed only through the α-melt-mediated transformation. These fine β crystals may form the crystal network, incorporating the liquid phase of the LM-fats. By contrast, the β crystals formed by thermal treatment with slow cooling (Fig. 9.15.B), or by rapid cooling to T_c=20°C in the LM-fats of PSO and olive oil (Fig. 9.15.C), were fewer in number and larger, and thus did not form a gel fat mixture. The β crystals of FHR-S and SSS formed by thermal treatment even with rapid cooling to T_c=20°C were large and non-uniformly dispersed at T_a=38°C through the α-melt-mediated transformation (Fig. 9.15.D). These β crystals might be precipitated and aggregated because liquid and solid fractions are separated and form the sol phase.

The mechanisms displayed in Fig. 9.15 are justified as follows. First, we assumed that the viscosity of the LM-fats during the rapid cooling to T_c from 70°C influenced

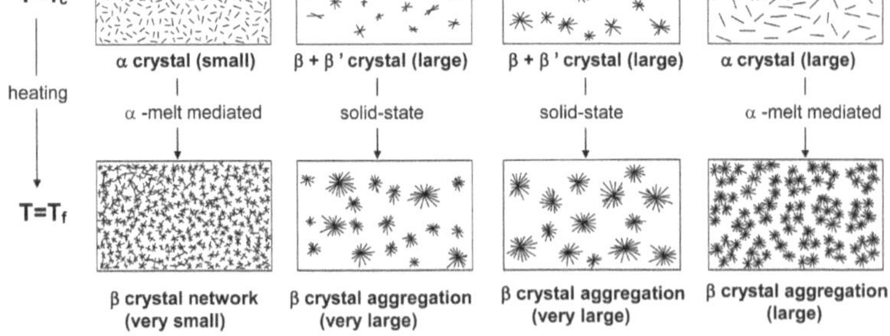

Fig. 9.15. Models for composing gel fat mixtures. (A) FHR-B/SFO and FHR-B/CB at a rapid cooling rate formed the gel fat mixture. (B) FHR-B/SFO and FHR-B/CB at a slow cooling rate. (C) FHR-B/PSO and FHR-B/olive oil at a rapid cooling rate. (D) FHR-S/SFO and SSS/SFO at a rapid cooling rate. (B)-(D) did not form gel fat.

the polymorphic crystallization behavior of FHR-B (Higaki et al., 2003a; 2003b; Toro-Vazquez et al., 2002). The α form crystallized directly from fat mixtures with increasing viscosity of the liquid phase and the cooling rate. Therefore, the α form was crystallized when FHR-B was mixed with SFO and CB that contained the middle-melting TAGs such as POP. On the other hand, PSO and olive oil contain lower-melting TAGs such as POO. Therefore, the β′ and β crystals of FHR-B might be crystallized.

Second, the TAG compositions of HM-fats may influence the crystallization behavior of three HM-fats during the thermal treatment. In general, crystal morphology is influenced by the combination of TAG and fatty acid, cooling rate (rapid/slow) from the melt, and degree of super-cooling (Martini et al., 2002). FHR-B consists mainly of stearic acid, arachidonic acid, and behenic acid (Higaki et al., 2003a; 2003b), whereas FHR-S consists mainly of the stearic moiety, which is similar to SSS. The difference in the TAG compositions might affect the crystallization (and morphology) of the HM-fat crystals in highly viscous LM-fats such as SFO and CB. As a result, the β crystals of FHR-B became smaller and thinner than those of FHR-S and SSS, incorporating the liquid fraction of the LM-fats.

Acknowledgments

The authors are deeply indebted to Dr. Tetsuo Koyano and Dr. Kaoru Higaki, Meiji Seika Kaihsa Co. Ltd., since this chapter is based on a collaborative work between Meiji Seika Kaisha Co. Ltd. and Hiroshima University.

References

Aoki, H.; Koto, T.; Kodama, M. Calorimetric investigation of conversion to the most stable subgel phase of phosphatidylethanolamine-water system. J. Therm. Anal. Calorim. 2001, 64, 299–306.

Chen, C-H.; Terentjev, E.M. Aging and Metastability of Monoglycerides in Hydrophobic Solutions. Langmuir 2010, 26, 3095–3105.

Chen, C-H.; van Damme, I.; Terentjev E.M. Phase behavior of C18 monoglyceride in hydrophobic solutions. Soft Matter 2009, 5, 432–439.

Chronakis I.S.; Kasapis, S. Preparation and analysis of water continuous very low fat spreads. Lebensmi.-Wissenschaft und-Technol. 1995, 28, 488–494.

Chronakis, I.S. Structural-functional and water-holding studies of biopolymers in low fat content spreads. Lebensmi.-Wissenschaft und-Technol. 1997, 30, 36–44.

Clark, A.H. Gels and gelling. Physical Chemistry of Foods, Schwartzberg, H.G., Hartel, R.W., Eds.; Marcel Dekker Inc.: New York, 1992; 263–305.

De Bont, P.W.; van Kempen, G.M.P.; Vreeker, R. Phase separation in milk protein and amylopection mixtures. Food Hydrocolloid 2002, 16, 127–138.

DeMan, J.M.; Beers, A.M. Fat crystal networks: structure and rheological properties. J. Texture Stud. 1987, 18, 303–318.

Eccleston, G.M.; Behan-Martin, M.K.; Jones G.R.; Towns-Andrews, E. Synchrotron X-ray investigations into the lamellar gel phase formed in pharmaceutical creams prepared with cetrimide and fatty alcohols. Int. J. Pharm. 2000, 203, 127–139.

Exerowa, D. Chain-melting phase transition and short-range molecular interactions in phospholipid foam bilayers. Adv. Colloid Interf. Sci. 2002, 96, 75–100.

Gallegos, C.; Franco, J.M. Rheology of food, cosmetics and pharmaceuticals. Curr. Opin. Colloids Interf. Sci. 1999, 4, 288–293.

Heertje, I. Microstructural studies in fat research. Food Structure 1993, 12, 77–94.

Herrera, M.L.; Hartel, R.W. Effect of processing conditions on physical properties of milk fat model system: rheology. J. Am. Oil Chem. Soc. 2000, 77, 1189–1195.

Higaki, K.; Sasakura, Y.; Koyano, T.; Hachiya, I.; Sato, K. Physical properties of gel-like behavior of binary mixtures of high-melting and low-melting fats. J. Am. Oil Chem. Soc. 2003a, 80, 263–270.

Higaki, K.; Sasakura, Y.; Koyano, T.; Hachiya, I.; Sato, K. In-situ optical observation of microstructures made of β-fat gel made of binary mixtures of high-melting and low-melting fats. Food Res. Intern. 2003b, 37, 2–10.

Higaki, K.; Koyano, T.; Hachiya, I.; Sato, K.; Suzuki, K. Rheological properties of β-fat gel made of binary mixtures of high-melting and low-melting fats. Food Res. Intern. 2004, 37, 799–804.

Jeong, B.; Kim, S.W.; Bae, S.W. Thermosensitive sol-gel reversible hydrogels. Adv. Drug Deliv. Rev. 2002, 54, 149–161.

Juriaanse, A.C.; Heertje, I. Microstructure of shortenings, margarine and butter—a review. Food Microstructure 1988, 7, 181–188.

Litwinenko, J.W.; Rojas, A.M.; Gerschenson, L.N.; Marangoni, A.G. Relationship between crystallization behavior, microstructure, and mechanical properties in a palm oil-based shortening. J. Am. Oil Chem. Soc. 2002, 79, 647–654.

Marangoni, A.G.; Hartel, R.W. Visualization and structural analysis of fat crystal networks. Food Technol. 1998, 52, 46–51.

Marangoni, A.G.; Narine, S.S. Identifying key structural indicators of mechanical strength in networks of fat crystal. Food Res. Intern. 2002, 35, 957–969.

Martini, S.; Herrera, M.L.; Hartel, R.W. Effect of processing conditions on microstructure of milk fat fraction/sunflower oil blends. J. Am. Oil Chem. Soc. 2002, 79, 1063–1068.

Narine, S.S.; Marangoni, A. G. Structure and mechanical properties of fat crystal networks. Adv. Food Nut. Res. 2002, 44, 33–145.

Ojijo, N.K.O.; Neeman, I.; Eger, S.; Shimoni, E. Effects of monoglyceride content, cooling rate and shear on the rheological properties of olive oil/monoglyceride gel networks. J. Sci. Food Agric. 2004, 84, 1585–1593.

Ould Eleya, M.M.; Turgeon, S.L. Rheology of κ-carrageenan and β-lactoglobulin mixed gels. Food Hydrocolloids 2000, 14, 29–40.

Shah, J.C.; Sadhale, Y.; Chilukuri, D.M. Cubic phase gels as drug delivery systems. Adv. Drug Deliv. Rev. 2001, 47, 229–250.

Toro-Vazquez, J.F.; Herrera-Coronado, V.; Dibildax-Alvarado, E.; Charo Alonso, M.; Gomez-Aldapa, C.A. Induction time of crystallization in vegetable oils. Comparative measurements by differential scanning calorimetry and diffuse light scattering. J. Food Sci. 2002, 67, 1057–1065.

Vintiloiu, A.; Leroux, J-C. Organogels and their use in drug delivery—a review. J. Controlled Release 2008, 125, 179–192.

Ziegler, G.R.; Foegeding, E.A. The gelation of proteins. Advances in Food and Nutrition Research; Kinsella, J.E., Ed.; Academic Press: San Diego, 1990; 34, 203–298.

10

Ceramide Oleogels

Michael A. Rogers[1], Amanda J. Wright[2], and Alejandro G. Marangoni[3]
[1]School of Environmental and Biological Sciences, Department of Food Science; Rutgers University; The State University of New Jersey, New Brunswick, New Jersey; [2]Department of Human Health and Nutritional Sciences, College of Biological Sciences, University of Guelph, Guelph, Ontario, Canada; and [3]Department of Food Science, Ontario Agricultural College, University of Guelph, Guelph, Ontario, Canada

Lipid Structuring

The texture of lipid-based foods is commonly provided by the presence of crystalline triacylglycerols (TAGs). Upon cooling, the hardstock TAGs become supercooled and aggregate, eventually forming a fat crystal network. This network, able to entrain liquid oil, exhibits both viscous and elastic components (Tang & Marangoni, 2006). While hardstock fats contribute plasticity and stability to foods, they are comprised of predominately saturated- and/or *trans*-fatty acids, which have deleterious implications for cardiovascular health (Keys et al., 1965). Excessive consumption of *trans* and saturated fats has been shown to have deleterious health effects, including negative effects on lipoprotein (cholesterol) profiles and increases in the incidence of heart disease and metabolic syndrome (Keys et al., 1965). It is estimated that 230 million people suffer from metabolic syndrome, which is associated with an increased risk of developing type 2 diabetes mellitus, cardiovascular disease, and premature death. As of July 1, 2006 labeling of *trans* fats became mandatory in the United States as required by the Food and Drug Administration (FDA), causing food manufactures to rethink the use of these heart-unhealthy fats to replace saturated fats. The American Heart Institute believes that even more action is required and recommends that no more than 10% of daily energy should come from a combination of *trans* and saturated fats. The recent deadline for the total ban of *trans* fats in New York City has just passed and it is more evident than ever that new technologies need to be implemented to curb the associated epidemics.

A lipid-based product may be structured by dispersing inert solids (suspension) or liquids (emulsion) in the continuous lipid phase. The inert particles must be concentrated to exceed the close packing fraction, allowing mechanical contact between particles. Mechanical contact between particles forms a three-dimensional network that entrains the liquid phase. The limitation to producing this dispersed system is the large amount of structurant required to immobilize the liquid phase, which makes it unsuitable for most food products (Pernetti et al., 2007).

Alternatively, liquid oil may be immobilized using a foreign phase dispersion such as small inert particles, crystallized solids at low concentrations, and separated droplets, or by using "surfactant-like" building blocks with a specific molecular mechanism of assembly (i.e., low-molecular-weight-organogelators) (Pernetti et al., 2007). Structuring a lipid material involves the use of "surfactant-like" molecules with limited solubility that precipitate and induce gel formation. In terms of a molecular gel, the sol is defined as a dispersion of solid particles in a liquid colloidal solution. "Surfactant-like" molecules may include macromolecules (polymers or proteins) or low molecular weight compounds (fatty acids, fatty alcohols, or TAGs). The functionality depends on solvent properties, processing conditions, and the presence of other minor components. An excellent review on alternatives to hardstock fats has been written by Pernetti et al. (2007).

In the food industry, there is great potential for organogels to be utilized to reduce oil migration in multi-component foods, such as cream-filled chocolates. Organogels may also provide structure to edible unsaturated oils, reducing the need for saturated and *trans* fatty acids. With such diverse applications, an intimate understanding of the physical chemistry of these systems is of paramount importance. The potential of organogels as soft materials for the food industry is further impeded due to a lack of recognized food grade organogelators. One solution may be ceramides derived from food sources. Further, sphingomyelins, which exist in greater quantities than ceramides, may be extracted from foods and converted to ceramides using phospholipase C (Zhang et al., 2006). The amounts of sphingolipids in foods vary from a few μmol per kg (in fruits) to several mmol per kg in dairy, eggs, and soybeans (Vesper et al., 1999). Sphingolipids derived from animal sources (i.e., dairy, eggs, and meat) contain a large variety of complex sphingolipids. The great diversity of ceramides differing with chain length and substituted sugar groups may change the functionality of the ceramide organogel.

Benefits of dietary ceramides are that they are able to pass though the gut, where minimal hydrolysis occurs, and then are enzymatically broken down into sphingosine in the intestinal tract (Nilsson, 1968; 1969; Schmelz et al., 1994; Schmelz, 2000; Kobayashi, 1997). Since these enzymes are found in the small and large intestine and not the stomach, this technology may be of interest for protecting nutraceuticals, allowing them to pass though the gut (Schmelz et al., 1994). Sphingolipids have several physiological effects, which include influencing cell growth, differentiation, and cell death (Schmelz, 2000), as well as modifying cholesterol profiles (Kobayashi et al., 1997; Gellen et al., 1995).

Typically ceramides are located in cellular membranes, exhibiting a stabilizing effect for the structure (Schmelz, 2000). Ceramides have several physiological effects. They are involved in interactions between cells as well as between cells and the underlying matrix, and also serve as binding sites for bacteria, toxins, and viruses (Kolesnick & Kroncke, 1998).

Ceramides and Cancer

Sphingolipid metabolites, including ceramides, are lipid "secondary messengers" in the signal transduction pathway for growth factors (e.g., platelet-derived growth factor) and cytokines [e.g., tumor necrosis factor α (TNF-α] (Schmelz, 2000; Cremesti & Fischl, 2000). Platelet-derived growth factor induces sphingomyelin hydrolysis to ceramide (using sphingomyelinase), which is then further metabolized to sphingosine and sphingosine 1-phosphate (using ceramidase and sphingosine kinase) (Vesper et al., 1999). TNF-α, on the other hand, only activates sphingomyelinase, which results in the accumulation of ceramides. During platelet-derived growth factor activity there is an induction of ceramide, sphingosine, and sphingosine 1-phosphate, allowing for sphingosine 1-phosphate to act as a potent mutagen and an inhibitor of apoptosis (Cuviller et al., 1998; Olivera & Spiegel, 1993). Conversely, when TNF-α is active, it results in elevated ceramide levels, leading to inhibition of growth and/or apoptosis (Hannun, 1994; Jayadev et al., 1995). Sphingolipids and ceramides are referred to as tumor suppressing lipids (Parodi, 2001). The presence of TNF-α, produced and excreted in the white blood cells and in the endothelium, kills tumors and is promoted by the presence of ceramides (Cermesti & Fischl, 2000). Ceramides also promote the production of interlukin 1B, which causes apoptosis in tumor cells (Cermesti & Fischl, 2000). There are numerous possible health implications for the development of a ceramide-based spreadable product. Possibly, the most exciting is the role of ceramides as inhibitors of colon carcinogenesis (Parodi, 1996; 1999). Schmelz and Merrill (1998) postulated that they may exert their effect on developing malignant adenocarcinomas in the colon.

Colon cancer was chemically induced using 1,2-dimethylhydrazine on CF1 mice, followed by treatment with and without the ingestion of sphingomyelin (0.025-0.1 wt %). During the first 4-week study a 70% reduction of aberrant crypt foci (ACF) was observed. ACF are one of the earliest morphological changes in the colon and it is thought that adenomas and adenocarcinomas develop from these early lesions. After 40 weeks, it was found that the mice that were fed the sphingomyelin-enriched diet had a reduction in malignant adenocarcinomas versus benign adenomas (Schmelz & Merrill, 1998). It was proposed by Schmelz and Merrill (1998) that ceramides and sphingosine have anticarcinogenic effects while sphingolipids and dihydroceramide do not. Ceramides inhibit growth and induce differentiation and apoptosis, while sphingosine may inhibit or stimulate growth as well as induce apoptosis (Schmelz & Merrill, 1998).

Ceramides and Cholesterol

Feeding sphingolipids to rats has been shown to reduce the total cholesterol by 30% (Koybayashi et al., 1997). Cholesterol in turn increased the very low density lipoprotein (VLDL) sphingomyelin content (Gellen et al., 1995). Sphingomyelin is able to quench free radicals inhibiting lipid peroxidation in VLDL (Subbaiah et al., 1999). This suggests that sphingomyelin may directly and/or indirectly reduce the risk factors for atherosclerosis (Schmelz, 2000).

Sphingomyelin is involved in many aspects of cholesterol transport and metabolism (Vesper et al., 1999). There are several studies that report that sphingomyelin may aid in the reduction of atherosclerosis. Chatterjee (1999) found that cultured cells have different LDL binding and utilization capabilities if sphingomyelin is present in the culture. Another study found that when LDL with sphingomyelin is hydrolyzed by an extracellular sphingomyelinase enriched in atherosclerotic lesions, there is an alteration in the aggregation state of the particle that promotes the activity of macrophages (Marathe et al., 1998). Furthermore, oxidized lipoproteins have been reported to stimulate the growth factor of vascular smooth muscle cells and human blood monocytes via the sphingomyelin signal pathway (Kinscherf et al., 1997). Sphingomyelin is converted to ceramide so that it may be then converted to sphingosine 1-phosphate, which mediates growth stimulation (Vesper et al., 1999). Elevated levels of sphingomyelin occur in aortic lesions (Barenholz & Gatt, 1982), which is attributed to the synthesis in the arterial tissue accompanied by decreased cellular turnover (Eisenberg et al., 1969).

Gelation Ability of Differing Ceramide Chain Length

C2 ceramide/canola oil gels were formed at concentrations as low as 2.0 wt % (Fig. 10.1) and C24 ceramide formed a gel at 5 wt %. More gelator molecules were required to immobilize the oil as the fatty acid sidechain increased in length. At low concentrations, with a synthetically pure gelator (synthetic C2, C24), the gel was translucent. Fig. 10.1 also shows the potential to produce a gel using the enzymatically

Fig. 10.1. Gels formed from 2% synthetic C2 ceramide and 10 wt % egg sphingomyelin converted using phospholipase C (Adapted from Rogers et al., 2009).

converted egg sphingomyelin to ceramide at 7 wt % crude extract. Sphingomyelins were converted to ceramides using phospholipase C derived from *Clostridium perfringens* (Sigma Aldrich, catalogue number P-7633, Lot number 108H8638, Mississauga, ON, Canada).

This enzyme was selected because of its high conversion rate (Zhang et al., 2006). The enzyme contained 9.9 U/mg, where one unit liberates 1 μmol of water-soluble organic phosphorus from egg yolk at pH 7.3 at 37°C (Sigma-Aldrich, 2010). Sphingomyelin is hydrolyzed between the primary hydroxyl group of ceramide and the choline phosphate ester. The phosphate ester bond cannot be cleaved by chemical hydrolysis (Susuma et al., 1999). Since the sphingomyelin is water insoluble and phospholipase C is not soluble in organic solvents, a complex media of 3 U/mg enzyme, 6% water, 1.8% ethanol, and 46% hexane in ethyl acetate (ethyl acetate:hexane 50:50) at room temperature for 72 hours was required to ensure complete hydrolysis (Zhang, 2006). Following the enzymatic conversion, the sample was dried at 60°C under nitrogen.

In contrast to the gels prepared using pure ceramides, these gels were opaque. It was determined that the enzymatic conversion was 100% (Fig. 10.2). Samples of enzymatically converted ceramide were prepared in a 1 wt % solution of 61:41:4 acetonitrile:methanol:water. The samples were filtered injected into the HPLC with a mobile phase 61:41:4 acetonitrile:methanol:water, run at 0.5 mL/min and the stationary phase was an Econosil C18 Column-w (250 × 4.6 mm × 5 (m), maintained at 40°C. The samples were detected using a Sedex 75 evaporative light scatter. The choline phosphate esters were not soluble in oil, and therefore solid material was suspended in the organogel. Although 7 wt % egg ceramide was added to canola oil, the ceramide content was 5.3%, based on the molecular weight of ceramides and the phosphocholine ester, due to the presence of contaminants (choline phosphate esters and sphingomyelins).

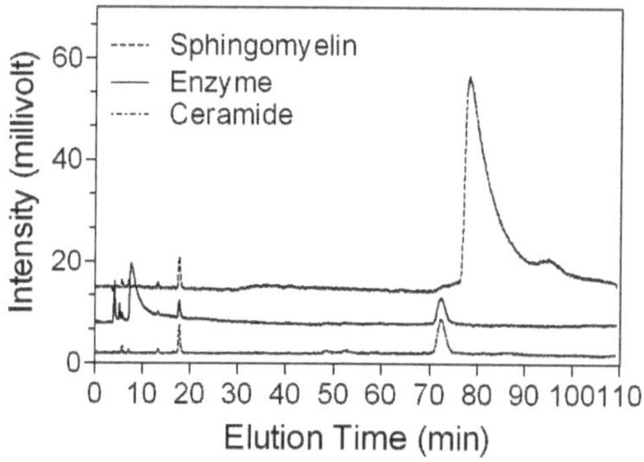

Fig. 10.2. HPLC diagram of the elution of pure sphingomyelin, pure ceramide, and enzymatically converted sphingomyelin to ceramide from egg.

Crystal Morphology of Ceramides

Fig. 10.3 shows the birefringent crystalline network of the C2 ceramide-canola oil gel. The highly branched strands suggest a tendency for crystallographic mismatches due to the different sidechain lengths on the same molecule (Fig. 10.3).

The high purity of the compound favored unidimensional growth, which allowed C2 ceramide to gel at lower concentrations than enzymatically converted sphingomyelin because the long "fiber-like" strands more readily form a continuous network. The high degree of variation in the sidechain length (Fig. 10.4) between the

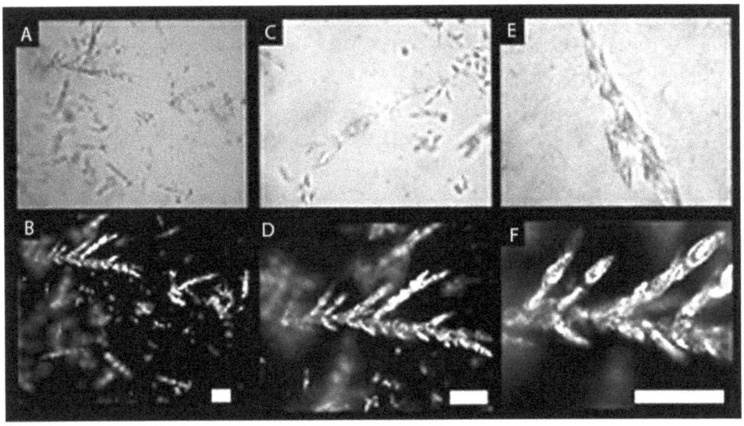

Fig. 10.3. 2 wt % C2 ceramide-canola oil gels observed *in-situ* using brightfield (A, C, D) and polarized light (B, D, F) at different magnifications (Magnification bar = 100 µm) (Adapted from Rogers et al., 2009).

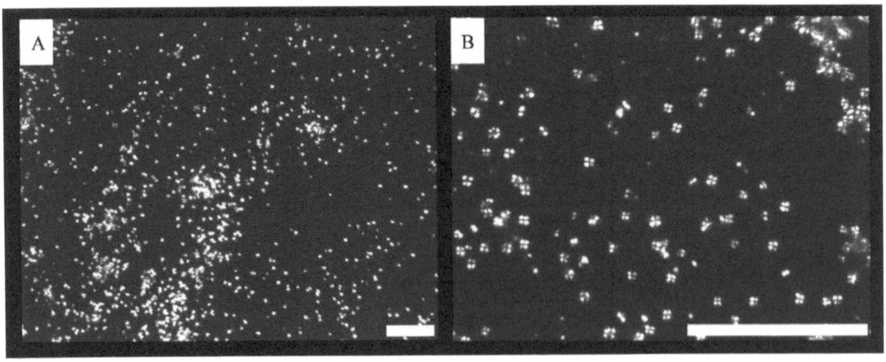

Fig. 10.4. 10 wt % milk ceramide-canola oil gels observed *in-situ* using polarized light at different magnifications (Magnification bar = 100 µm) (Adapted from Rogers et al., 2009).

sphingomyelin-derived gelator molecules promoted small dendritic crystals, which interacted to form flocs and required higher levels of ceramides to gel the canola oil. In part, this may be due to the purity of the system.

The heterogeneous nature of ceramide sidechains may disrupt the unidirectional growth of the fibrillar aggregates and promote a more three-dimensional crystallization, similar to TAG crystallization (Fig. 10.5) (Rogers et al., 2007).

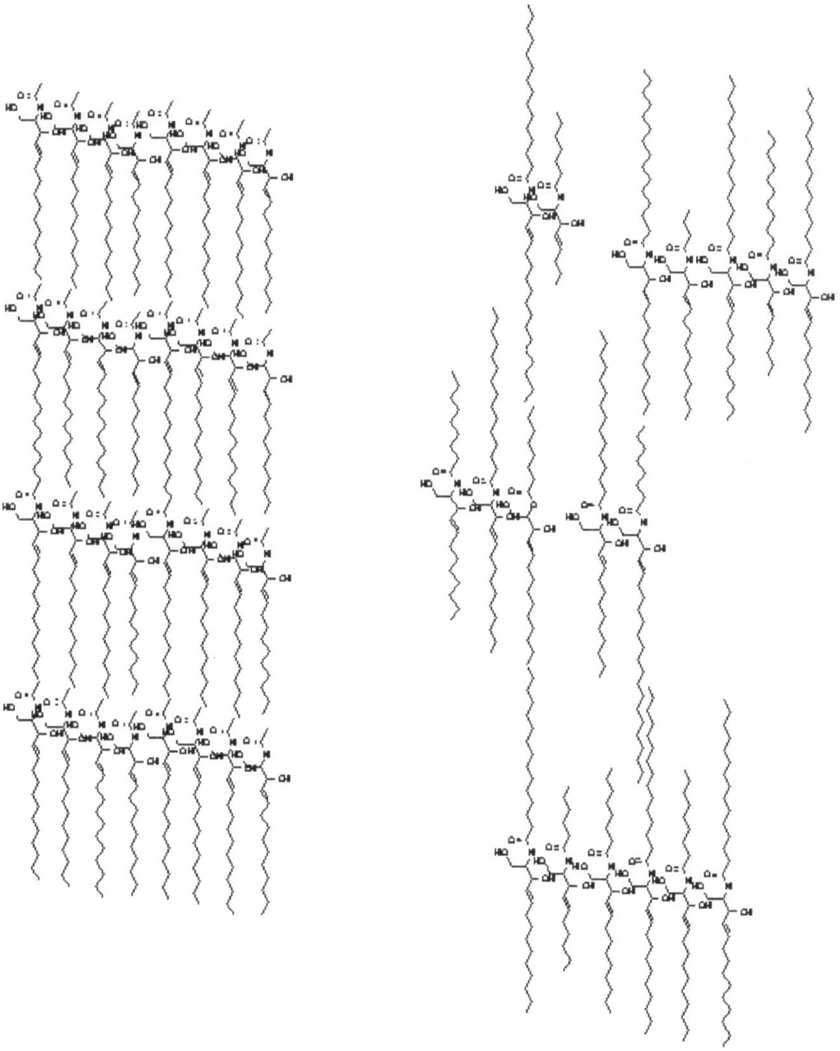

Fig. 10.5. Idealized packing of a pure C2 ceramide and a mixed system representing an enzymatically converted sphingomyelin to mixed ceramide.

Small spherulites did not seem to form a network in canola oil when imaged using polarized light microscopy (Fig. 10.4). Therefore, the ceramide obtained from enzymatically converting sphingomyelin were imaged using cryo-SEM (Fig. 10.6) (Rogers et al., 2007). Fig. 10.6 indicated a high degree of ordering of the ceramide crystallites, which aggregated into a three-dimensional network, effectively entraining the liquid oil. The microscopic structure of the ceramides gels was similar to wax esters or mixtures of long-chain fatty acids and long-chain alcohols (Gandolfo et al., 2003; 2004; Toro-Vazquez et al., 2007).

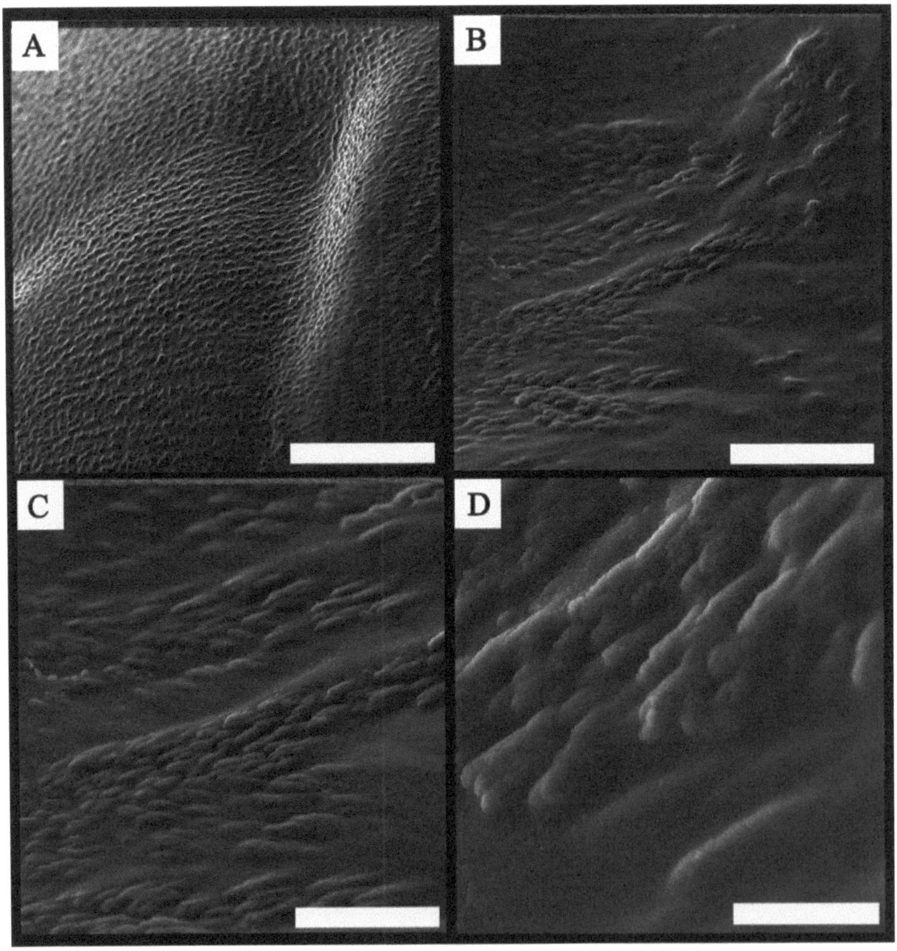

Fig. 10.6. 10 wt % milk ceramide-canola oil gels observed using cryo-SEM at different magnifications (Magnification bars A, B=3 0 μm, C=15 μm, and D=6 μm).

Crystal Structure of Ceramides Organogel Fibers

Since the gels were birefringent, X-ray diffraction was used to examine the crystalline nature of the gels (Fig. 10.7). The diffraction pattern of the enzymatically converted milk sphingomyelin displayed long spacings (001 plane) at 46.46Å, which

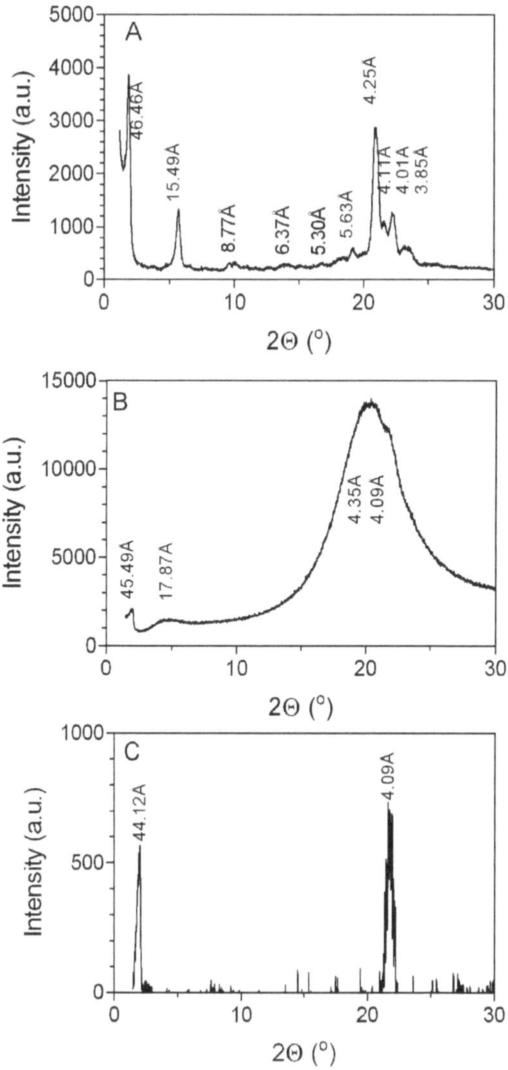

Fig. 10.7. Powder X-ray diffraction profile of milk ceramide/canola oil (10 wt %) stored at 5°C for 24 hours (A) neat spectrum, (B) 10% milk ceramide/canola oil, and (C) 10% milk ceramide/canola oil with baseline subtraction (Adapted from Rogers et al., 2009).

corresponded to ceramides orientated in bilayers (Fig. 10.7.A). This was followed by a higher order reflection from the 003 plane at 15.49Å. The presence of peaks at 3.8Å and 4.2Å indicate an orthorhombic subcell packing. The spacings for the milk ceramide gel (Figs. 10.7.B and 10.7.C) were slightly different than those of the neat sample, with a strong amorphous scattering due to the presence of canola oil at 17.87Å and 4.35Å (Fig. 10.7.B). After baseline subtraction, a sharp reflection was evident at 44.12Å, suggesting a δ configuration for ceramides (Raudenkolb et al., 2003). The δ spacing for ceramides corresponds to a molecular arrangement where the ceramides have a torsion angle of 180°. The short spacing at 4.09Å suggests a hexagonal packing for the methylene groups in the ceramide sidechains.

Their crystalline nature suggest a similar gelation mechanism as for 12HSA-canola oil (Roger et al., 2007; Rogers et al., 2008) and ricinelaidic acid-canola oil gels (Wright & Marangoni, 2006; 2007). It is postulated that the mechanism of gelation follows the nucleation-crystal growth-branching-crystal growth model. Ceramide-based gels have a wide structural diversity that may allow for specific tailoring of the physical properties of the gels. Of even greater interest is that these compounds are food grade and display nutraceutical properties. By modifying the chain length and chain variability, it may be possible to modify the network from a "fiber-like" network to a network comprised of dendritic crystals.

Physical Properties of Ceramide Organogel

The thermal properties differ for C2 and egg ceramide (Fig. 10.8). The onset of melting for the 2 wt% C2 ceramide was 37.4°C while crystallization occurred at 39.4°C as seen using the derivative of the heat capacity. The mixed ceramides derived from the conversion of egg sphingomyelin had a much higher onset of melting at 50.4°C and a higher temperature for the onset of crystallization 50.1°C. About 86% of the fatty acid sidechains in egg sphingomyelin are palmitic acid while in the C2 ceramide 100% of the fatty acid sidechains were propionic acid. It becomes obvious that the melting behavior of the organogels can be varied by altering the chain length as well as the ceramide composition (homogeneity of sidechains). As the chain length increases we expect an increase in the melting temperature of the gel.

The rheological behavior of the egg sphingomyelin was examined using small deformation oscillatory rheology with a frequency sweep and stress sweep (Fig. 10.9). Using Clark and Ross-Murphy's (1987) classical definition of gels, based on the frequency sweeps, it was determined that the 10 wt% ceramide/canola oil networks were weak gels. The gels produced from egg sphingomyelin have a 10-fold higher storage modulus (10000Pa) than loss modulus (1000Pa) (Fig. 10.9.B).

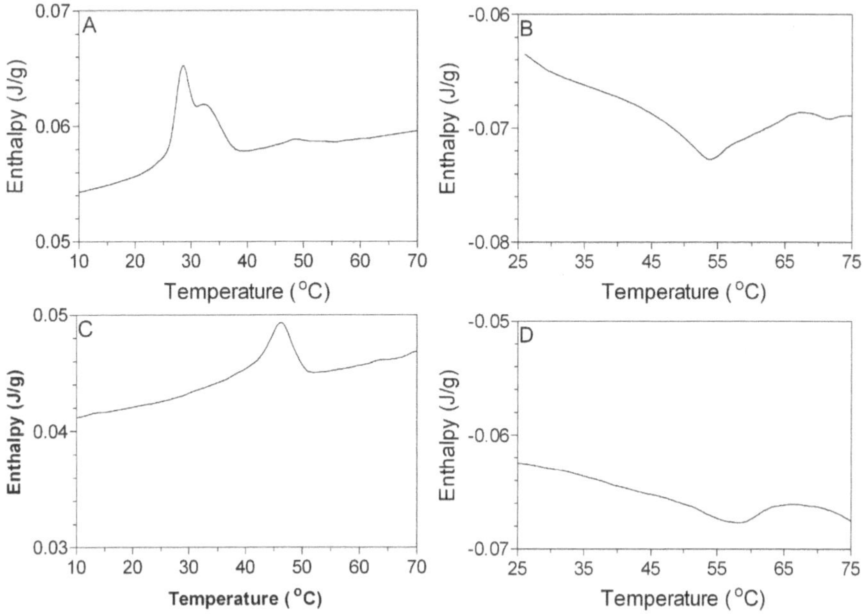

Fig. 10.8. DSC thermograms for the onset of crystallization (A, C) and melting (B, D) for 2 wt% C2 ceramide (A, B) and 10 wt% ceramide converted from egg sphingomyelin (C, D) (Adapted from Rogers et al., 2009).

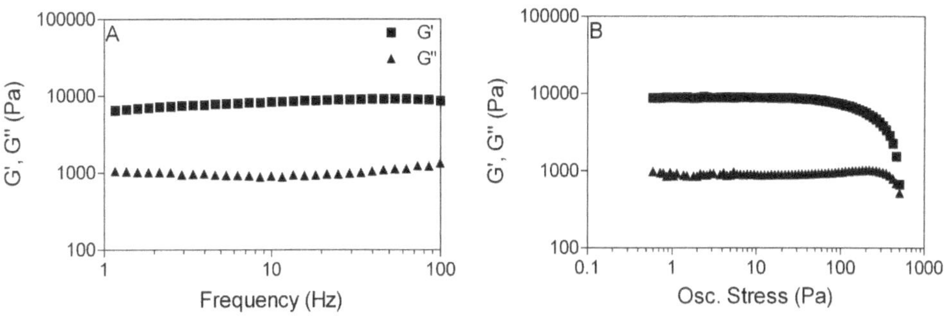

Fig. 10.9. Frequency sweep (A) and oscillatory stress sweep (B) of 10 wt% ceramide converted from egg sphingomyelin (Adapted from Rogers et al., 2009).

References

Barenholz, Y., Gatt; S. Phospholipids. Phospholipids; Hawthorne, J.N., Ansell, G.B., Eds.; Elsevier Biomedical Press: Amsterdam, The Netherlands, 1982; chapter 4.

Chatterjee, S. Neutral sphingomyelinase increase the binding, internalization, and degradation of low density lipoproteins and synthesis of cholesteryl ester in cultured human fibroblasts. J. Bio. Chem. 1999, 268, 3401–3406.

Clark, A.H.; Ross-Murphy, S.B. Structural and mechanical properties of biopolymer gels. Adv. Polymer. Sci. 1987, 83, 57–192.

Cremesti, A.E.; Fischl, A.S. Current methods for the identification and quantization of ceramides: An overview. Lipids. 2000, 35 937–945.

Cuvillier, O.; Rosenthal, D.S.; Smulson, M.E.; Spiegel, S. Sphingosine 1-phosphate inhibits activation of caspases that cleave poly(ADP-ribose) polymerase and lamins during FAS- and ceramide-mediated apoptosis in Jurkat T lymphocytes. J. Bio. Chem. 1998, 273, 2910–2916.

Eisenberg, S.; Steini, Y.; Stein, O. Phospholipases in arterial tissue. IV. The role of phosphatide acyl hydrolase, and sphingomyelin choline phosphohydrolase in the regulation of phospholipid composition in the normal human aorta with age. J. Clin. Invest. 1969, 48, 2320–2329.

Gellen, M.J.H.; Tijburg, L.M.B.; Bouma, C.J.; Beynen, A.C. Cholesterol consumption alters hepatic sphingomyelin metabolism in rats. J. Nut. 1995, 125, 2294–2300.

Gandolfo, F.G.; Bot, A.; Floter, E. Phase diagram of mixtures of stearic acid and stearyl alcohol. Thermochimica Acta. 2003, 404, 9–17.

Gandolfo, F.G.; Bot, A.; Floter, E. Structuring of edible oils by long-chain FA, fatty alcohols and their mixtures. J. Am. Oil Chem. Soc. 2004, 81, 1–5.

Hannun, Y.A. The sphingomyelin cycle and the second messenger function of ceramide. J. Bio. Chem. 1994, 269; 3125–3128.

Jayadev, S.; Liu, B.; Bielawska, A.E.; Less, J.Y.; Nazaire, F.; Pushkareva, M.Y.; Obeid, L.M.; Hannun, Y.A. Role of ceramide in cell cycle arrest. J. Bio. Chem. 1995, 270, 2047–2052.

Keys, A.; Anderson, J.T.; Grande, F. Serum cholesterol response to changes in diet, IV. Particular saturated fatty acids in the diet. Metabolism. 1965, 14, 776–687.

Kinscherf, R.; Claus, R.; Deigner, H.P.; Nauen, O.; Gehrke, C.; Hermetter, A.; Russwurm, S.; Daniel, V.; Hack, V.; Metz, J. Modified low density lipoprotein delivers substrate for ceramide formation and stimulates the sphingomyelin-ceramide pathway in human macrophages. FEBS Letters 1997, 405, 55–59.

Kobayashi, T.; Shimizugawa, T.; Osakabe, T.; Watanabe, S.; Okuyama, H. A long-term feeding of sphingolipids affected the level of plasma cholesterol and hepatic triacylglycerol but not tissue phospholipids and sphingolipids. Nut. Res. 1997, 17, 111–114.

Kolesnick, R.N.; Kroncke, M. Regulation of ceramide production and apoptosis. Ann. Rev. Phys. 1998, 60, 643–665.

Olivera, A.; Spiegel, S. Sphingosine-1-phosphate as second messenger in cell proliferation induced by PDGF and FCS mitogens. Nature 1993, 365, 557–560.

Marathe, S.; Schissel, S.L.; Yellin, M.J.; Beatini, N.; Mintzer, R.; Williams, K.J.; Tabas, I. Human vascular endothelial cells are a rich and regulatable source of secretory sphingomyelinase. Implications for early atherogenesis and ceramide-mediated cell signaling. J. Bio. Chem. 1998, 273, 4081–4088.

Nilsson, A. Metabolism of sphingomyelin in the intentional tract of the rat. Biochim. Biophys. Acta. 1968, 164, 575–580.

Nilsson, A. Metabolism of cerebrosides in the intestinal tract of the rat. Biochim. Biophys. Acta. 1969, 187, 113–117.

Parodi, P.W. Cow's milk components with anti-cancer potential. Aust. J. Dairy Tech. 2001, 56, 65–71.

Parodi, P.W. Milk fat components: Possible chemopreventive agents for cancer and other diseases. Aust. J. Dairy Tech. 1996, 51, 24–32.

Parodi, P.W. Conjugated linoleic acid and other anticarcinogenic agents of bovine milk fat. J. Dairy Sci. 1999, 82, 1339–1349.

Pernetti, M.; van Malssen, K.F.; Floter, E.; Bot, A. Structuring edible oils by alternatives to crystalline fat. Curr. Opin. Colloid Inter. Sci. 2007, 84, 989–1000.

Raudenkolb, S.; Wartewig, S.; Neubert, R.H. Polymorphism of ceramide 3. Part 2: a vibrational spectroscopic and X-ray powder diffraction investigation of N-octadecanoyl phytosphingosine and the analogous specifically deuterated d_{35} derivative. Chem. Phys. Lipids 2003, 124, 89–103.

Rogers, M.A.; Wright, A.J.; Marangoni, A.G. A novel cryo-SEM technique for imaging vegetable based organogels. J. Am. Oil Chem. Soc. 2007, 84, 899–906.

Rogers, M.A.; Wright, A.J.; Marangoni, A.G. Nanostructuring fiber morphology and solvent inclusions in 12-Hydroxystearic acid/canola oil organogels. Curr. Opin. Colloid Interface Sci. 2008, 14, 33–42.

Rogers, M.A.; Wright, A.J.; Maranoni, A.G. Oil Organogels: The Fat of the Future. Soft Matter 2009, 5, 8, 1594–1596.

Schmelz, E.M.; Crall, K.J.; Larocque, R.; Dillehay, D.L.; Merrill, A.H. Uptake and metabolism of sphingomyelin in isolated intestinal loops. J Nut. 1994, 124, 702.

Schmelz, E.M. Dietary sphingomyelin and other sphingolipids in health and disease. Nut. Bull. 2000, 25, 135–139.

Schmelz, E.M.; Merrill, A.H. Ceramides and ceramide metabolites in cell regulation: evidence for dietary sphingolipids as inhibitors of colon carcinogenesis. Nutr. 1998, 14, 717–719.

Susumu, H.; Hidehiko, T.; Yumiko, T. Process for preparing sphingomyelin and ceramide from erythrocyte as a starting material and a curing agent for cosmetic formulated with ceramide. 1999, US5912152.

Sigma-Aldrich. http://www.sigmaaldrich.com/catalog/search/ProductDetail /SIGMA/P7633. (accessed 12/20/2010).

Tang, D.; Marangoni, A.G. Quantitative study on the microstructure of colloidal fat crystal networks and fractal dimensions. Adv. Colloid Inter. Sci. 2006, 128–130, 257–265.

Toro-Vazquez, J.F.; Morales-Rueda, J.A.; Dibildox-Alvarado, E.; Charo-Alonso, M.; González-Chávez, M.; Alonzo-Macias, M.M. Thermal and textural properties of organogels developed by candelilla wax in safflower oil. J. Am. Oil Chem. Soc. 2007, 84, 989–1000.

Vesper, H.; Schmelz, E-M.; Nikolova-Karakashian, M.N.; Dillehay, D.L.; Lynch, D.V.; Merrill Jr., A.H. Sphingolipids in food and the emerging importance of sphingolipids to nutrition. Am. Soc. Nut. Sci. 1999, 129, 1239–1250.

Wright, A.J.; Marangoni, A.G. Formation, structure and rheological properties of ricinelaidic acid-vegetable oil organogels. J. Am. Oil Chem. Soc. 2006, 83, 497–503.

Wright, A.J.; Marangoni, A.G. Time, temperature and concentration dependence of ricinelaidic acid-canola oil organogelation. J. Am. Oil Chem. Soc. 2007, 84, 3-9.

Zhang, L.; Hellgren, L.I.; Xu, X. Enzymatic production of ceramide from sphingomyelin. J. Biotechno. 2006, 123, 93–105.

11

Oleogels Based on Non-lamellar Lyotropic Liquid Crystalline Structures for Food Applications

Dima Libster, Abraham Aserin, and Nissim Garti
Ratner Chair of Chemistry, Casali Institute of Applied Chemistry, The Institute of Chemistry, Hebrew University of Jerusalem, Edmond J. Safra Campus, Givat Ram, Jerusalem 91904, Israel

Introduction

Liquid crystals (LC) are self-assembled organized mesophases with properties intermediate to those of crystalline solids and isotropic liquids (Gin et al., 2008). In LC phases, long-range periodicity exists, although the molecules exhibit a dynamical disorder at atomic distances, as is the case in liquids. Accordingly, these materials can also be considered ordered fluids (Larsson, 1989). Lyotropic LC (LLC) are materials that are composed from at least two molecules: an amphiphilic molecule and its solvent. A hydrophilic solvent, such as water, hydrates the polar moieties of the amphiphiles via hydrogen bonding, while the flexible aliphatic tails of the amphiphiles aggregate into fused hydrophobic regions based on van der Waals interactions. In addition to morphologic dependence on the chemical composition, LLC are also sensitive to external parameters, such as temperature and pressure (Gin et al., 2008; Larsson, 1989; Amar-Yuli, 2008). As a function of the molecular shape of the surfactants, packing parameters, and interfacial curvature energy considerations, LLC can be formed with aqueous domains ranging from planar bilayer lamellae to extended, cylindrical channels, to three-dimensional interconnected channels and manifolds (Gruner, 1989). These mesophases are defined as lamellar (L_α), hexagonal (H), bicontinuous cubic [Q (or V)], and discontinuous cubic (I) phases, based on their symmetry (Seddon, 1990). In addition, most lyotropic mesophases exist as symmetric pairs, a "normal" (type I) oil-in-water system, consisting of lipid aggregates in a continuous water matrix, and a topologically "inverted" (type II) water-in-oil version. The headgroups hydrated by water are arranged within a continuous non-polar matrix, which is composed of the fluid hydrocarbon chains (Seddon, 1990). In addition to its biological significance, inverse lipid phases could be useful as host systems for incorporation of food additives (Sagalowicz et al., 2006b), the crystallization of membrane

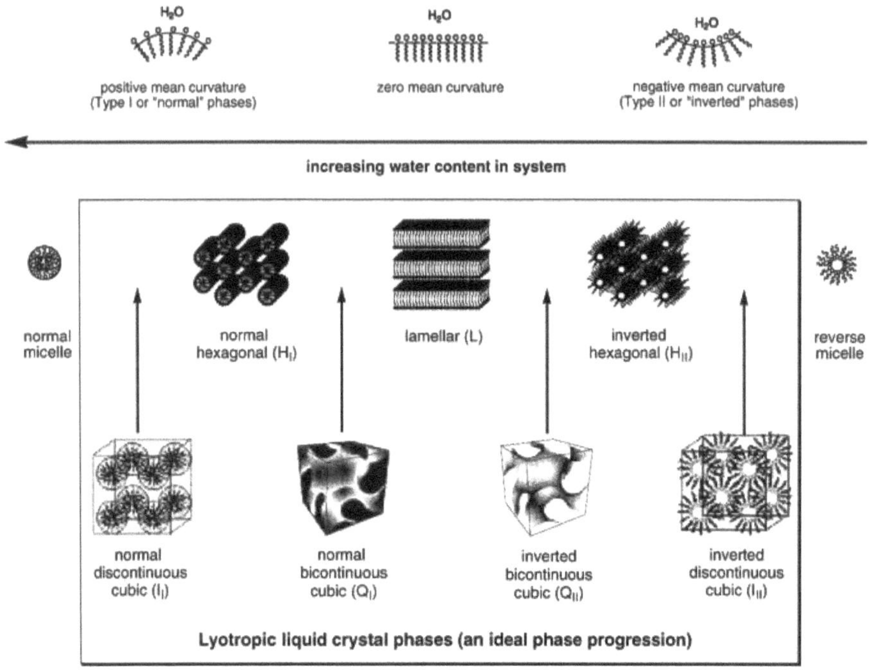

Fig. 11.1. Schematic representations of common LLC-phases formed by amphiphiles in water. *Source:* Gin et al., 2008.

proteins for drug delivery (Sagalowicz et al., 2006b; Cherezov et al., 2006), and for inorganic synthesis (Boyd et al., 2007).

The lamellar structure does not possess any intrinsic curvature and is considered as the midpoint of an ideal, symmetrical LLC phase progression (Fig. 11.1) (Gruner, 1989; Seddon, 1990). The current review is mainly focused only on the inverted (W/O) mesophases (cubic and hexagonal), representing an important class of oleogels for potential food applications.

Cubic Phases

The bicontinuous cubic phase is the most complicated among all the LLC. Here the lipids are located in a complex, optically isotropic three-dimensional lattice. Cubic structures can be either normal (type I, O/W) or inverse (type II, W/O) and their topology is differentiated as bicontinuous or micellar, resulting in seven cubic space groups. Only three space groups are both inverse and bicontinuous structures (Fig. 11.2): the gyroid (G) type (*Ia3d,* denoted Q^{230}), the diamond (D) type (primitive lattice *Pn3m,* denoted Q^{229}), and the primitive (P) type (body-centered lattice *Im3m,* denoted Q^{224}) (Gruner et al., 1988). The aqueous channels of G-surface consist of

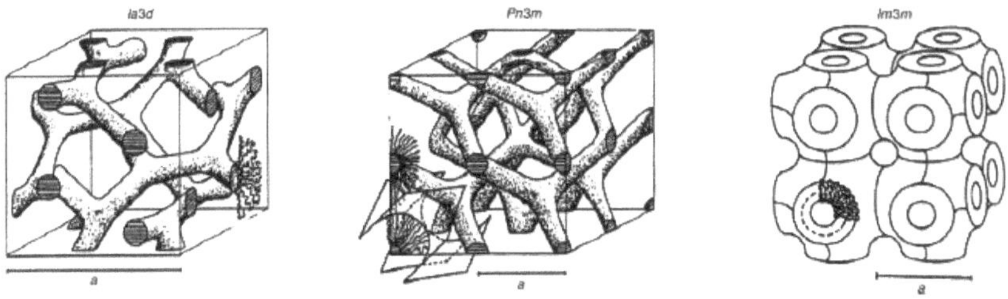

Fig. 11.2. Schematic illustrations of the inverse bicontinuous cubic phases *Ia3d* (gyroid type), *Pn3m* (diamond type), and *Im3m* (primitive type). (Seddon et al., 2000).

two separate, left-handed and right-handed helical channels. The aqueous channels can extend through the matrix, but the centers of the water channels never intersect. Hence they are connected and the structure adopts a helical arrangement. The structure D is characterized by a bilayer, which separates two interpenetrating aqueous channel systems forming a diamond lattice. In this configuration four aqueous channels of the D-surface meet at a tetrahedral angle of 109.5°. Structure P contains two aqueous channel systems that are separated by a bilayer. The unit cell possesses three mutually perpendicular aqueous channels, connected to contiguous unit cells, taking shape of a cubic array.

Cubic Micellar Mesophases

Cubic micellar structures are less widespread than bicontinuous cubic. Discrete micellar mesophases are made of micellar amphiphilic nanostructures organized in 3-dimensional (3D) structures/lattices with cubic symmetry. The monolinolein/water system, upon tetradecane addition, was shown to exhibit an additional LC phase not present in the monolinolein/water binary mixture: an inverse discrete micellar cubic phase of the *Fd3m* type (Q_{227} group), as revealed by Yaghmur et al. (Yaghmur et al., 2006). The structure of the *Fd3m* micellar cubic phase was composed of two types of micelles with different sizes organized in a double diamond network. These micelles are not spherical but showed a dodecahedron and a hexakaidecahedron shape, respectively. In this way they minimized the high packing frustration constraints on the alkyl tails associated with the *Fd3m* arrangement (Pouzot et al., 2007). This mesophase was also observed in other systems upon oil addition, such as block copolymers (Uddin et al., 2001; Alexandridis et al., 1996) or phospholipid/fatty acid systems (Luzzatti et al., 1992). It is generally agreed that formation of a *Fd3m* micellar phase is promoted in systems where lipids have a negative preferred curvature when long alkyl hydrophobic tails are associated with weakly hydrated hydrophilic heads (Seddon et al., 1996).

Efrat et al. included ethanol (EtOH) in the binary GMO/water system and obtained a micellar mesophase of cubic symmetry with significantly lower viscosity

(Efrat et al., 2007; Efrat et al., 2008). The newly designed mesophase was termed Q_L—a fluid, liquid discontinuous cubic mesophase. It was found that EtOH influences the structural parameters of the bicontinuous cubic phases at a constant GMO/water weight ratio of 0.82 by increasing the area per lipid at the headgroup (a_0). The radius of the water channels (R_{wat}) also increased significantly as the ethanol content increased, but the mean curvature (‹H›) was reduced. These changes in the structural parameters indicated that ethanol mainly affected the polar region of the surfactant since it was accommodated around the headgroups. The rheological measurements showed that the new phase is of low viscosity (36.6 Pas in comparison to around 10^5 Pas of the bicontinuous cubic phase). The intermicellar structure complied with the so-called I_2 mesophase with primitive (P) symmetry. Its lattice parameter (the d-space) is ca. ~105Å (based on synchrotron SAXS, and cryo-TEM). The new mesophase maintained its cubic order although it exhibits a low viscosity, even lower than the values that were reported for I_2. Efrat et al. also studied the solubilization effects of a variety of bioactive guest molecules on the Q_L mesophase at various compositions and loads, including coenzyme (CoQ_{10}), cholesterol, phytosterols, and lycopene (Efrat et al., 2008; Efrat et al., 2009).

The Reverse Hexagonal Mesophase

The reverse hexagonal mesophase is of a primitive type (*P6mm*) and characterized by one cylinder per unit cell corner. These densely packed, straight, water-filled cylinders exhibit two-dimensional ordering. Each cylinder is surrounded by a layer of surfactant molecules that are perpendicular to the cylinder interface such that their hydrophobic moieties point outward from the water rods (Fig. 11.3). There is a growing indication that inverse hexagonal mesophases play structural and dynamic roles in biological systems (Seddon, 1990; Gradzielski, 2004; Yang et al., 2003). These systems are assumed to be active as transient intermediates in biological phenomena that require topological rearrangements of lipid bilayers, such as membrane fusion/fission and the trans-bilayer transport of lipids and polar solutes (Seddon, 1990; Gradzielski, 2004; Yang et al., 2003). H_{II} mesophases have recently been considered promising drug delivery vehicles, mainly owing to their unique structural features (Boyd et al., 2007; Lopes et al., 2007; Boyd et al., 2006).

Phase Behavior

Most of the common surfactants form direct (type I) phases, where the interface bends away from the polar solvent, toward the tail region. However, most biological amphiphiles (such as phospholipids) form "type II" LLC, where the interface curves toward the polar region. Only a few synthetic amphiphiles can mimic the behavior of biological lipids and form inverted mesophases. The unsaturated monoglycerides (monoolein and monolinolein) belong to this category (Drummond & Fong, 2000). Glycerol monooleate (GMO, monoolein) is a polar lipid that is commonly used as

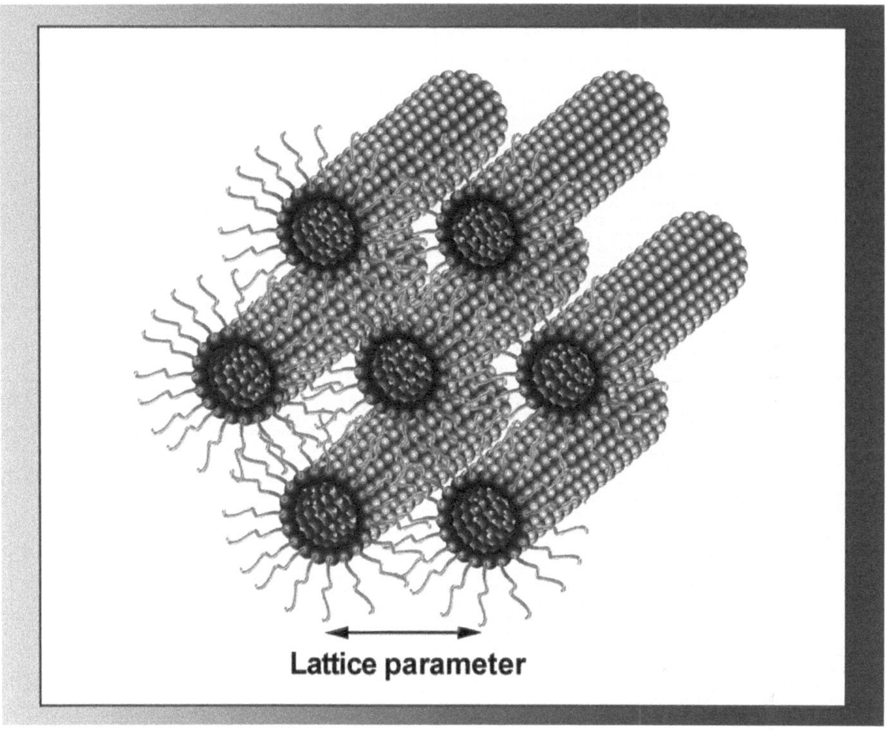

Fig. 11.3. Schematic presentation of H_{II} mesophase showing the packing of water-filled rods surrounded by lipid layers.

a food emulsifier. Presently, GMO is the preferred amphiphile for formulating LC phases for scientific research and drug delivery. This is a non-toxic, biocompatible, and biodegradable lipid, which possesses low water solubility, but in excess water it swells and forms several LC phases.

Monoolein/water is one of the most extensively investigated types of lipid-based mesophases (Drummond & Fong, 2000). The phase diagram of this system (Fig. 11.4) revealed complex structural behavior. At room temperature the following phase sequence existed upon increasing hydration: lamellar crystalline phase (L_c) in coexistence with a L_2 phase, lamellar mesophase (L_α), and the inverted bicontinuous cubic mesophases-gyroid *Ia3d* and diamond *Pn3m*. Upon heating, at about 85°C, the cubic phase is transformed into the H_{II} mesophase, followed by the micellar phase.

These concentration and temperature-dependent structural transitions can be qualitatively explained in terms of effective critical packing parameter (CPP), as developed by Israelachvili et al. (Israelachvili et al., 1976). According to this theory, amphiphiles possess geometric parameters characterized by the critical packing parameter,

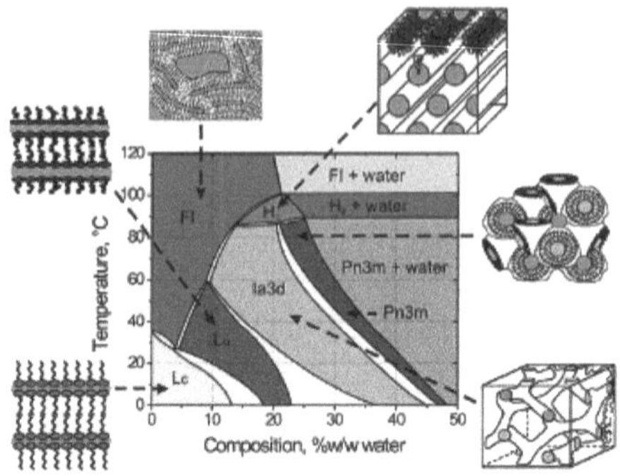

Fig. 11.4. Temperature-composition phase diagram of the monoolein/water system (up to 50 wt % water). A cartoon representation of the various phase states is included in which colored zones represent water. The mesophases are as follows: L_c-crystalline lamellar, L_α-lamellar, *Ia3d*-gyroid inverted bicontinuous cubic, *Pn3m*-primitive inverted bicontinuous cubic, H_{II}-inverted hexagonal, and F1-reverse micelles fluid phases. *Source:* Cherezov et al., 2006.

CPP = V_s/a_0l, where V_s is the hydrophobic chain volume, a_0 is the polar headgroup area, and l is the length of the chain in its molten state. The packing parameter is useful in predicting which phases can be preferentially formed by a given lipid, since it connects the molecular shape and properties to the favored curvature of the polar–apolar interface and, therefore, the topology and shape of the aggregate. The main factors responsible for alterations of the mentioned parameters are the molecular shape of the surfactant, the chemical composition, and the temperature. According to the theory developed by Israelachvili and coworkers, inverse mesophases are formed by amphiphiles with CPP >1, which adopt inversed cone shaped geometry. Such lipids should possess a small headgroup area as compared to that of the tail region. In contrast, direct mesophases are preferred when the surfactant head cross-section is larger than that of the tail, resulting in CPP <1. In the case of lamellar mesophase these parameters are equal and hence result in CPP = 1. The size of the polar headgroup area is dictated by both molecular shape and the degree of hydration.

Within the boundaries of a given mesophase the headgroups' area normally increases with increasing hydration and decreases upon temperature rise. The length and the volume of the lipophilic tails are also greatly affected by a temperature increase. Stronger thermal motion of the tails eventually decreases the length of the tails but increases their volume.

Usually the phase sequence follows a certain order; the "generic sequence" is reverse micellar → micellar cubic → reverse hexagonal → reverse bicontinuous cubic → lamellar → bicontinuous cubic → hexagonal → micellar cubic → micellar (Fig. 11.5) (Efrat, 2009).

It was shown for monoglyceride-based LLC that H_{II} mesophases are formed from amphiphiles with CPP ~ 1.7, cubic phases with CPP~1.3, and lamellar structures with CPP = 1 (Larsson, 1989). The lamellar phase is characterized by zero

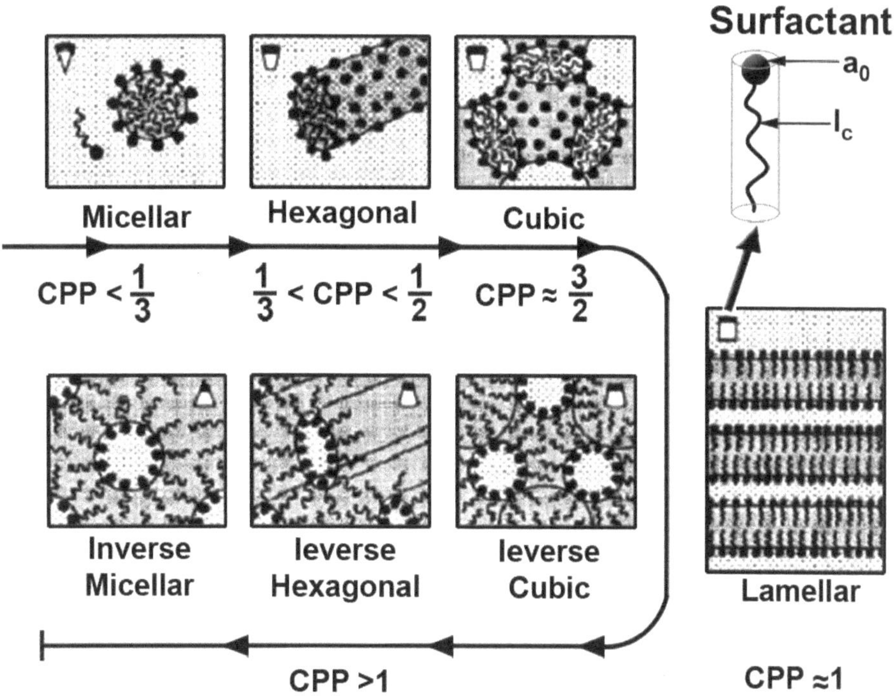

Fig. 11.5. Schematic presentation of classic mesophases, their corresponding CPP, and the 'generic sequence' with increased water content. *Source:* Efrat, 2009.

curvature since the cross sections of the polar heads and the lipophilic tails are similar. Upon increased hydration, cubic phases (*Ia3d* and *Pn3m*) with higher curvature and consequently higher CPP are preferred. The effective CPP theory can supply a reasonable explanation for the temperature-induced structural shifts from lamellar through cubic to reverse hexagonal phases, requiring higher curvature than in the lamellar phase. Increasing the thermal motion of both the hydrocarbon chains and the water molecules would increase the CPP values via expanding the volume of the lipophilic moiety but decreasing the chain length and the headgroup area. This leads to an increase in curvature and therefore induces the formation of cubic and hexagonal mesophases.

It should be noticed that the CPP concept may only be used as a "rule of thumb" to predict transitions between various forms of LLC. For example, it was shown that it fails to predict the appearance of the inverse bicontinuous phases and an intermediate phase (Shearman et al., 2006; Shah & Paradkar, 2005). This is the reason that a thermodynamic approach considering the total free energy of the LLC systems was applied by Seddon and coworkers (Seddon et al., 2000; Seddon et al., 1996; Shearman et al., 2006), however, this is beyond the scope of this chapter.

Phospholipids-based Oleogels

Reverse hexagonal mesophases based on phospholipids are also regarded as oleogels due to the low water content and macroscopic behavior similar to "oils". Most phospholipid-based oleogels are composed of phosphatidylethanolamine (PE) and water. PE is one of the most abundant lipids in eukaryotic cell membranes and is unevenly distributed between the inner and the outer leaflets of the bilayer. Hence, PEs in the membrane leaflet facing the inner media, in comparison to the external one, have been extensively studied to explore relationships between the topological properties of those surfaces and the functional roles of this lipid (Langner & Kubica, 1999; Rand, 1981; Mariani et al., 1990). Unsaturated PEs have been shown to induce nonlamellar phases formation (Seddon, 1990; Rand & Parsegian, 1989; Cullis & De Kruijff, 1979), and such structures were characterized in detail by ^{31}P-NMR, EM, AFM, and X-ray diffractions (Lewis et al., 1989; Harper et al., 2001; Zasadzinski et al., 1991).

The most common example of such a phospholipid is dioleoylphosphatidylethanolamine (DOPE), which stabilizes the H_{II} mesophase. This phenomenon is attributed to the conical shape of the lipid, in contrast with the cylindrical one of the PCs, which triggers formation of bilayers (Bouchet et al., 2009, Seddon, 1990; Rand & Parsegian, 1989; Cullis & De Kruijff, 1979). Harper et al. used X-ray diffraction to characterize fully hydrated aqueous dispersions of seven different PEs whose hydrocarbon chains have the same effective chain length but whose structures vary widely (Harper et al., 2001). These investigators found that the magnitude of the L_α and H_{II} phase d-spacings on either side of the L_α/H_{II} phase transition temperature (T_h) depends significantly on the structure of the PE hydrocarbon chains. It was shown that the L_α phase d-spacings range from 51.2 to 56.4 Å, whereas those of the H_{II} phase range from 74.9 to 82.7 Å (Fig. 11.6). Such structural dimensions allow solubilzation of active molecules for food applications.

A detailed molecular description of the L_α/H_{II} phase transition in these PEs was also elucidated (Harper et al., 2001). These authors showed that the decreasing lipid length drives the lipid–water interface to larger and larger areas in the L_α phase. Ultimately, the lamellar morphology becomes too costly energetically, and the system is forced to adopt the H_{II} phase, where the extra degree of geometric freedom allows both the hydrocarbon chains and polar headgroups to assume their desired states. In this case there are two main free energies in the system to be considered. The first one drives the hydrocarbon chains to a desired length that decreases with increasing temperature. The other one is for pushing the headgroup to an optimum area that is relatively temperature independent. Hence, these two free energies are in direct competition, with the decreasing lipid length and concomitant splaying of the hydrocarbon chains driving the lipid–water interface to larger and larger areas. Eventually, this becomes too costly and the system is forced to adopt hexagonal packing, where the extra degree of geometric freedom allows both the hydrocarbon chains and headgroups to be arranged in their desired states. The only free energy cost keeping the

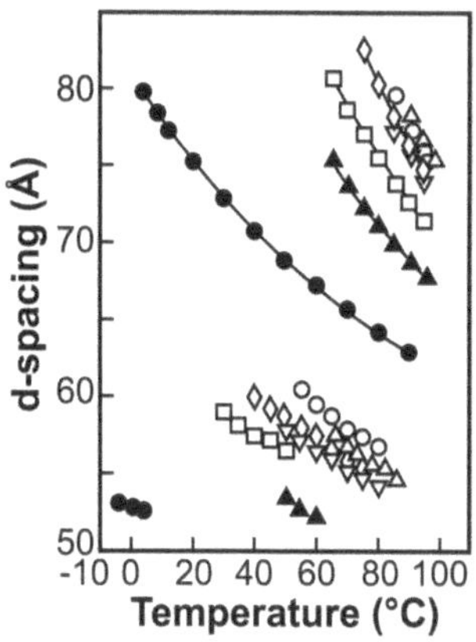

Fig. 11.6. Unit cell lengths (*d*) for some hydrocarbon chain structural variants of diacyl phosphatidylethanolamine. Solid lines connect the L_{α} data points and dotted lines connect the H_{II} data points. Representative error bars for each phase are plotted for 18:*1c*Δ*9*-PE at one data point per phase. ●, 18:*1c*Δ*9*-PE; ▲, 18:*1t*Δ*9*-PE; □, 20:0_{eai}-PE; ◇, 20:0_{dmi}-PE; △, 19:0_{i}-PE; ▽, 19:0_{ai}-PE; ○, 21:0_{ch}-PE. *Source:* Harper et al., 2001.

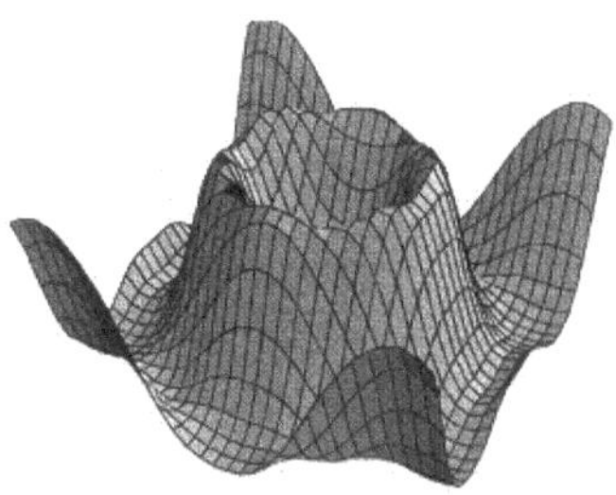

Fig. 11.7. A 3D view of the electron density reconstruction of the H_{II} phase for 18:1 *t*Δ*9*-PE at 85°C. *Source:* Harper et al., 2001.

system out of the H_{II} phase is that the lipid hydrocarbon chains are forced to assume a variety of lengths to fill the hexagon (Fig. 11.7).

On the molecular level FTIR spectroscopy studies of PEs have presented valuable information about the hydration of the polar headgroups in the interfacial water. It was indicated that the frequency of the asymmetric PO_2^- stretching vibration depends on the nature of the polar head group. It varies from 1230 cm^{-1} in fully hydrated DMPC to 1217 cm^{-1} in DMPE. The low frequency observed in DMPE was attributed to intra and intermolecular hydrogen bonding of NH_3 with the PO_2^- (Pohle et al., 1997; Hubner & Blume, 1998). In addition, it was stressed that the interaction with water would be different for PE than for PC due to the higher positive charge density of ethanolamine in comparison to cholines (Bouchet et al., 2009, Seddon, 1990; Rand & Parsegian, 1989; Cullis & De Kruijff, 1979). Hence, PEs absorb less water than PCs, resulting in higher effective CPP values and preferring stabilization of the hexagonal mesophase. While unsaturated PEs such as DOPE stabilize in the hexagonal H_{II} phase, saturated PEs stabilize in lamellar forms due to the reduced hydrophobic chain volume and lower effective CPP.

In this regard it's well-known that membranes enriched in unsaturated PEs tend to abandon the bilayer conformation. It was suggested that in mixtures with PCs, PE determines the propensity of the membrane to abandon the bilayer structure (Bouchet et al., 2009, Epand & Leon, 1992). This indicates that the dilution of PEs affects the lateral interaction. The triggering of the lamellar–non lamellar transition has been explained as a consequence of the lateral migration of PE molecules due to the strong lateral interactions. These PE properties were ascribed to the strong formation of H-bonds between the phosphate group and the amine group (Hubner & Blume, 1998; Lis et al., 1982).

Although phospholipid-based oleogels are well characterized, their main drawbacks for food applications are high viscosity and high cost of the lipids. Therefore, monoolein-based structures are more suitable for practical implementation.

General Rheological Behavior of Liquid Crystalline Mesophases

Rheological properties of soft materials are of practical importance for food applications. In this context shear rheology is emerging as an important technique for the determination of both structural and dynamical aspects of liquid crystalline mesophases. Investigation of the storage, G', and loss moduli, G'', on the linear-viscoelastic regime as a function of temperature and composition have proved to be an effective tool in determination of structural transitions between liquid crystalline structures (Shah & Paradkar, 2005; Sagalowicz et al., 2006a). Among the liquid crystalline phases, the bicontinuous cubic phases (*Ia3d, Pn3m,* and *Im3m*) are regarded as the most rigid structures, followed by the reverse hexagonal phase, which possesses intermediate viscoelastic properties, and the lamellar phase, which is characterized by a plastic fluid properties (Sagalowicz et al., 2006a). Based on these differences in viscoelastic behavior, structural transition between the mesophases was detected in several experimental systems, including cubic (space-group) to cubic, cubic to hexagonal, and hexagonal to isotropic fluid transitions (Jones & McLeish, 1999; Bonacucina et al., 2005; Shui et al., 2005a; Shui et al., 2005b; Montvalvo et al., 1996; Rodriguez-Abreu et al., 2005; Siddig et al., 2006). For example, Pouzot et al. demonstrated that an inverse micellar cubic phase of *Fd3m* structure can be obtained by adding food-grade limonene to the binary system monolinolein/water (Pouzot et al., 2007). The *Fd3m* structure formed at a ratio of limonene oil to total lipids (α) of 0.4 was stable in bulk up to a maximum hydration of 12.68 wt % water, beyond which it coexisted with water. The evolution of G', G'', and tan (δ) of the *Fd3m* phase with slow heating from 20 to 80°C revealed two major phase transitions (Fig. 11.8.A). The first was an increase in the slope of tan (δ) at *ca.* 29°C. It referred to the onset of melting of the highly rigid *Fd3m* phase to yield a coexistence region of L_2 + water + *Fd3m*. The second transition was monitored by the sharp drop in G' and G'' between 40 and 45°C. This was explained by complete loss of the rigidity associated with the viscoelastic *Fd3m* phase to yield a viscous L_2 + water mixture.

The study of variations of the longest relaxation time (τ_{max}) with composition has been shown to be the most reliable rheological method to detect order–disorder transitions in liquid crystalline phases. τ_{max} can be calculated as the inverse of the frequency at which the crossover of the storage and loss moduli takes place at $\tau_{max} = 1/\omega$. In liquid crystals this parameter is regarded as the timescale for relaxation to the equilibrium configuration of the water–surfactant interface, following perturbation by shear deformations (Sagalowicz et al., 2006a). However, it is well-recognized that a single relaxation time cannot reflect the complex viscoelastic behavior of the liquid crystalline phases. It was shown that H_{II} phases comply with the single Maxwell model only at low frequencies, but at higher frequencies the systems deviated from this model, which should display a single relaxation time (Libster et al., 2007). Hence, multiple Maxwell models with several relaxation times were used to successfully fit the experimental data of storage and loss moduli of the cubic phase as a function of frequency. However, the reverse hexagonal phase rheological spectra still could not be resolved by this method. Therefore, Mezzenga et al. suggested a more general approach by introducing the time Laplace-transform of G' (ω), G'' (ω) (Mezzenga et al., 2005a). A multitude of relaxation times was obtained, representing an individual rheological signature for each mesophase (Fig. 11.8.B). Using this fitting method, the

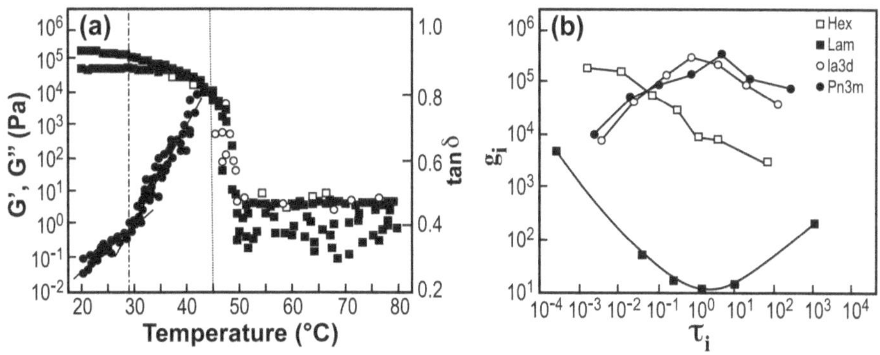

Fig. 11.8. (a) Temperature evolution of the storage modulus (*G'*, filled squares), loss modulus (*G''*, open squares), and tan (δ) (filled circles) measured for an *Fd3m* phase for ratio of limonene oil to total lipids α = 0.4 and 11.2 wt % water after 5 days equilibration at 23°C. The heating rate was 0.2°C/min. The solid lines given for tan (δ) are visual guides and denote a change in the slope at *ca.* 29.5°C. Dotted and dashed lines give phase transitions from *Fd3m* to *Fd3m* + L_2 + water and from *Fd3m* + L_2 + water to L_2 + water (Pouzot et al., 2007). (b) Discrete relaxation spectra pairs (g_i, τ_i) for the liquid crystalline phases: lamellar phase (10 wt % water at 40°C), hexagonal phase (20 wt % water at 70°C), *Ia3d* cubic phase (20 wt % at 40°C), and *Pn3m* cubic phase (20 wt % at 55°C). *Source:* Mezzenga et al., 2005a.

dominating relaxation time was assigned to τ_{max} while the residual times were attributed to secondary relaxation mechanisms that refer to confined hydrophobic and hydrophilic phases. The lamellar phase exhibited two dominating relaxation times. The first was assigned to the relaxation of the water–surfactant interface (of the order of 10^3 s), and the other (of the order of 10^{-3} s) was assigned to the relaxation of water in confined lamellae. The cubic phases demonstrated one dominating relaxation time and a secondary multiple relaxation mechanism. In contrast, the H_{II} phase revealed a monotonic distribution of relaxation times. Therefore, these kinds of measurements would allow following the structural changes induced by solubilization of guest molecules to LLC.

Solubilization of Biopolymers into LLC

The impact of hydrophilic guest molecules, such as sugars, on water domains in liquid crystalline phases has been studied by several authors. In general, kosmotropic (water-structure makers) and chaotropic (water-structure breakers) solutes have a significant impact on the properties of liquid crystalline mesophases by indirect (Hofmeister) interactions with these structures. It was demonstrated that kosmotropic solutes stabilize the structure of bulk water (Koynova et al., 1997; Tsvetkova et al., 1991; Saturni et al., 2001). They are likely to be solubilized within the bulk water and be excluded from the interfacial areas. Water-structures (kosmotropes) interfere with the tetrahedral network of water and, as a result, dehydration of the surfactant polar heads takes place ("salting-out" effect) along with the quantity of interfacial water being decreased. As a result, higher curvature is promoted and formation of L_α and H_{II} phases at the expense of L_α phase is preferred. Alternatively, chaotropes tend to destabilize the structure of bulk water and lead to an increase of the area at the water/surfactants interface and hence induce formation of the L_α phase in place of the L_α and H_{II} phases.

Mezzenga et al. investigated the effects of confining polymeric forms of glucose in the water domains of liquid crystals composed of monoglyceride and water solutions (Mezzenga et al., 2005b). The variations in the phase behavior of water–GMO as a result of solubilization of 5 wt % glucose are demonstrated in Fig. 11.9. Noticeable changes were detected on the boundaries of liquid crystalline regions. It was observed that both the lamellar (L_α) and the cubic phase regions shrank, and the transition corresponding to cubic-to-H_{II} occurred at lower temperatures. Moreover, the onset of coexistence of H_{II} mesophase and water, as well as the double diamond cubic phase (*Pn*3*m*) and water, was shifted toward lower water concentrations. The obtained results were interpreted in terms of increase of the CPP, leading to the favorable L_α–cubic–H_{II} transitions. According to the authors' interpretations, glucose, as a highly hydrophilic compound, is supposed to be located almost solely in water channels. It was suggested that the cross-sectional surfactant head area can be reduced by the presence of the solute by introducing stronger hydrogen bond acceptors ("water pumps"). The kosmotropic effect caused by glucose induced a decrease

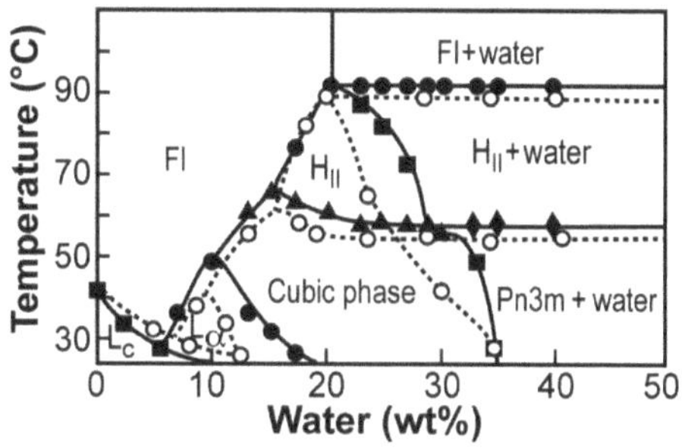

Fig. 11.9. Phase diagram of the GMO and water solution of 5 wt % glucose (dashed lines, open symbols), compared to that of GMO with pure water (continuous lines, full symbols). *Source:* Mezzenga et al., 2005b.

in the curvature of L_α GMO–water interfaces and resulted in the formation of the cubic phase or further reduced the latter to yield H_{II} phase. Using rheological and SAXS methods the authors studied the effect of sugar concentration and molecular weight on cubic-to-H_{II} transitions by further solubilization of maltose and dextran I in the system. It was found that higher concentrations of sugars resulted in a decrease of the *Pn*3*m*–H_{II} transition temperature, as a consequence of the above-mentioned Hofmeister effect. However, no dependence on the molecular weight of sugar was observed. In the case of a polymeric sugar, as soon as the end-to-end distance of the polymer approached the diameter of water channels, its molecular weight made it more difficult for the guest molecule to fit within narrow water channels. Dextrans of larger molecular weights than dextran I did not fit into the water channels of the cubic phases, inducing a phase separation. Moreover, it was noticed that dextran I, when solubilized into the *Ia*3*d* phase channels, induced the appearance of a *Pn*3*m* phase, which possesses larger water channels. Hence, when the chain end-to-end distance of the polysaccharide approached the diameter of the water channels, it induced phase transitions toward structures with different topologies, thus enhancing the solubilization capacity of large sugars.

Saturni et al. explored the molecular mechanism of the protective function of sugars on monoolein-based cubic systems that were dehydrated by osmotic stress experiments in the presence of a series of sugars (Saturni et al., 2001). Trehalose, maltose, sucrose, glucose, and maltotriose concentrations up to 3 M were tested. It was found that in sugar solutions, the *Pn*3*m* phase demonstrated high stability. Although a drastic decrease of lattice parameter was monitored as a function of sugar concentration (from 97 Å in water-based systems to 82 Å in the presence of glucose and maltotriose), no phase transitions occurred. However, the *Pn*3*m* to *Ia*3*d* phase transition took place by equilibrating the lipid phase with aqueous polymer solutions of increasing osmotic pressure. Hence, the possibility of a simple osmotic mechanism in sugar-containing systems was ruled out. Instead, it was proposed that sugars can change the chain packing by modulating the interface properties, as demonstrated by the appearance of the pivotal surface in the osmotic stress experiments. Thus, the packing stresses in the *Pn3m* were probably reduced to levels where this phase is at lower free energy than the *Ia3d*. In this case no phase transition occurred upon dehydration. Furthermore, at high maltotriose concentrations (1.22–1.39 M), the *Pn3m*–H_{II} phase transition was detected. The investigators assumed, based on the fact that packing frustration in the H_{II} phase is usually larger than that of the *Pn3m,* that a high temperature or incorporation of another hydrophobic molecule is needed to cause transition to H_{II} structure by packing stress relief (Mariani et al., 1988; Briggs et al., 1996). It was suggested that the presence of the sugar induced formation of the H_{II} phase by relieving most of the packing stress in the *Pn3m* phase.

H_{II} Mesophase Composed of GMO/Triglyceride/Water

It was shown that in the monoolein-based system, the cubic phase is transformed into an H_{II} mesophase upon heating at ca. 85°C (Drummond & Fong, 2000). CPP theory can supply a reasonable explanation to the temperature-induced structural shifts from lamellar through cubic to reverse hexagonal phases, requiring greater curvature than in the lamellar phase. Increasing the thermal motion of both the hydrocarbon chains and the water molecules would increase the CPP values via expanding the volume of the lipophilic moiety but decrease the chain length and the headgroup area. This leads to an increase in curvature and therefore induces the formation of cubic and hexagonal mesophases. In addition, the hexagonal mesophase is characterized by greater packing cost than the cubic phase, but the opposite is true for curvature elastic energy. Therefore, elevated temperatures induced the tendency for interfacial curvature, which increased the curvature elastic costs of the bicontinuous cubic phase, stabilizing hexagonal symmetry (Gin et al., 2008; Israelachvili et al., 1976).

Bearing in mind the thermodynamic and structural considerations noted above, systematic research was conducted in our lab to decrease the cubic to hexagonal temperature transition and stabilize the GMO-based H_{II} mesophase at room temperature (Amar-Yuli & Garti, 2005). To achieve this goal, experiments relative to incorporation of triglycerides (TAGs) with various chain lengths to the binary GMO/water system were conducted. Amar-Yuli and Garti surmised that immobilization of a TAG between GMO tails would lead to a change in the geometry of monoolein molecules from cylindrical to wedge-shaped and thereby an increase in the CPP value of the system (Amar-Yuli & Garti, 2005). This should encourage transition from lamellar or cubic phases to hexagonal structures. In addition, the immobilization of TAG in the system was expected to reduce the packing frustration, stabilizing the hexagonal LLC at room temperature. These experimental results showed that a critical and optimal chain length of the triglyceride is required to induce the formation of H_{II} at room temperature. Among the examined TAGs, tricaprylin (C_8) attained the largest region of hexagonal structure in the phase diagram (Fig. 11.10), indicating its flexible accommodation between the tails of the GMO. It was suggested that the high

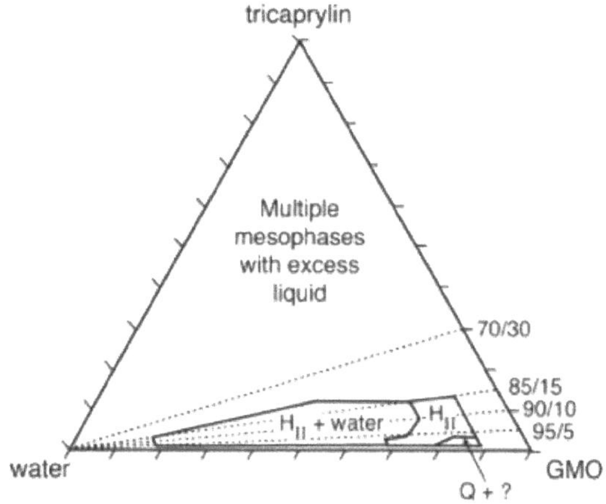

Fig. 11.10. The ternary phase diagram of GMO/tricaprylin/water at 25°C. The dilution lines represent the surfactant/oil weight ratio. *Source:* Amar-Yuli & Garti, 2005.

mobility of tricaprylin in the surfactant aggregate provided direct transformation of the lamellar mesophase to the hexagonal phase, avoiding the formation of a cubic structure.

The structural properties of ternary hexagonal mesophases composed of GMO, tricaprylin, and water were extensively and systematically studied in our lab (Amar-Yuli et al., 2009; Amar-Yuli et al., 2007a; Amar-Yuli et al., 2007b; Amar-Yuli et al., 2008a; Amar-Yuli et al., 2008b). Several additives, including dermal penetration enhancers, were solubilized to control the physical properties of these carriers (Amar-Yuli et al., 2007b; Amar-Yuli et al., 2008a; Amar-Yuli et al., 2008b), such as viscosity and thermal stability. For instance, the synergistic solubilization of two major hydrophilic (vitamin C, ascorbic acid, AA) and lipophilic (vitamin E, D-alpha-tocopherol, VE) antioxidants within H_{II} mesophases was reported by Bitan-Cherbakovsky et al. (Bitan-Cherbakovsky et al., 2009; Bitan-Cherbakovsky et al., 2010). This enabled expanding conditions to obtain stable H_{II} mesophase at room temperature. In addition, it was shown that phosphatidylcholine (PC) can be embedded into the ternary GMO/TAG/water system (Libster et al., 2007, Libster et al., 2008; Libster et al., 2009a; Ben Ishai et al., 2009). PC was incorporated into the ternary mesophases, as it is known that phospholipid-based structures possess relatively high thermal stability, and enhance transdermal drug permeation (Kurosaki et al., 1991; Touitou et al., 1994; Spernath et al., 2006) and transmembrane transport across the digestive tract (Brondsted et al., 1995; Liu et al., 1999).

Incorporation of PC to the ternary system (Fig. 11.11.A) caused competition for water binding between the hydroxyl groups of GMO and the phosphate groups of the phosphatidylcholine (PC), leading to dehydration of the GMO hydroxyls in

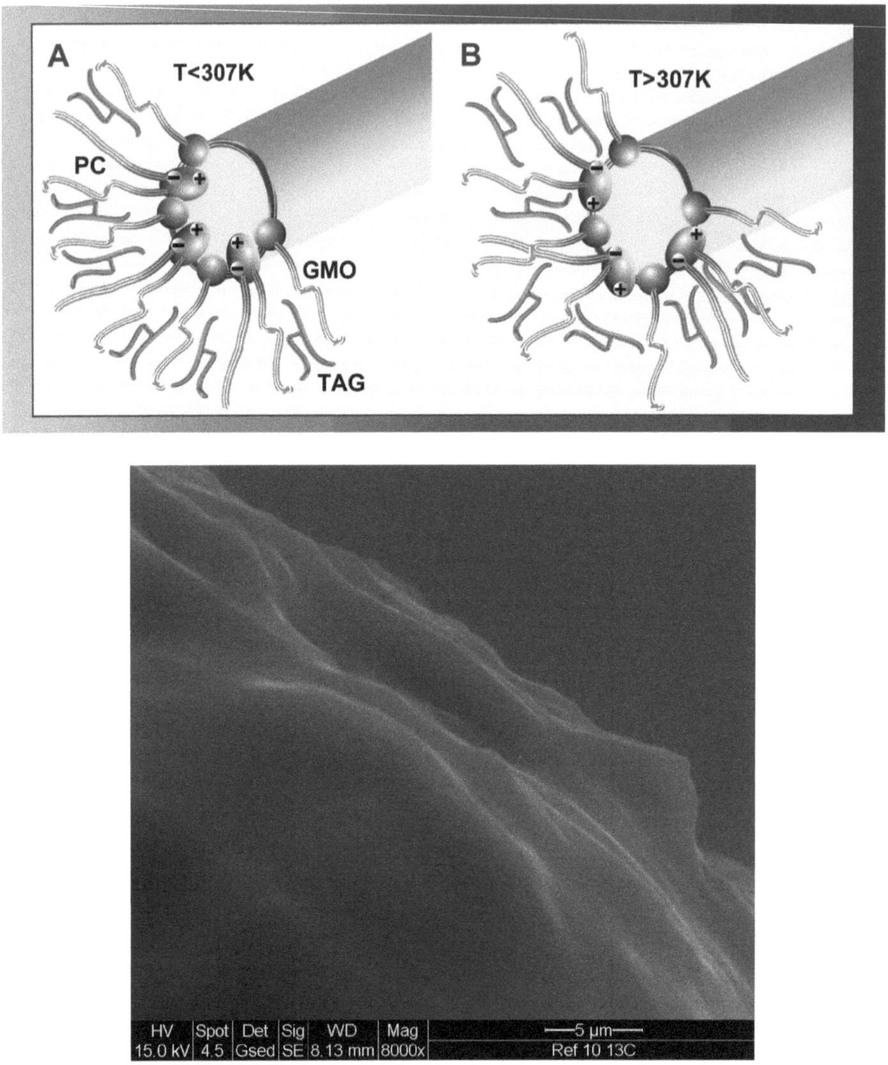

Fig. 11.11. (A) Schematic illustrations focused on one cylinder of the mixed surfactant (GMO and PC) hexagonal system organization. Note that GMO and PC polar moieties are hydrated but that TAG is located between the lipophilic chains of the surfactants. Below T = 307 K, interfacial water stabilizes the lipid layer via its interaction with the GMO head γ—OH group. Consequently, the TAG molecule remains intercalated between the tails. Dangling tails of the PC group give rise to counterion movement as well as TAG dipolar relaxation. (B) Above T = 307 K, the interface is destabilized by the breakdown of the interfacial water layer. The perturbation of the PC heads leads to greater dangling, enabling diffusive movement of the TAG molecule through the tails (Ben Ishai et al., 2009). (C) ESEM micrographs of H_{II} mesophase of GMO/tricaprylin/PC (63, 7, and 10 wt %, respectively) containing 20 wt % water at 14°C and magnification ×8000 (Libster et al., 2008).

favor of the phospholipid hydration (Libster et al., 2009a). On the macroscopic level, this was correlated with improvement of elastic properties and thermal stability of the H_{II} mesophase (Libster et al., 2007). Structural flexibility and stability is essential for a drug delivery system, especially for tuning its physical properties (such as viscosity) and composition. Dielectric spectroscopy analysis of this system showed a critical temperature, T_0 = 307 K, related to the dehydration of the GMO head groups (Ben Ishai et al., 2009). Effectively, this represented a breakdown of the interfacial layer of water. A physical picture emerged whereby at T_0 = 307 K, the "loosening " of the GMO heads accentuates the dangling motion of the phosphatidylcholine (PC) tails, as evidenced by counterion motion along the PC head (Fig. 11.11.B). Furthermore, it precipitates the percolation of the large TAG molecules that are intercalated in the GMO and PC tails (Ben Ishai et al., 2009).

Libster et al. also explored and demonstrated correlations between the microstructural and mesostructural properties of the reverse hexagonal LLC, using ESEM technique (Libster et al., 2008). It was shown that the mesoscopic organization of these systems is based on an alignment of polycrystalline domains. The topography of H_{II} mesophases, imaged directly in their hydrated state, as a function of aqueous phase concentration, was found to possess fractal characteristics, indicating a discontinuous and disordered alignment of the corresponding internal water rods on the mesoscale (Fig. 11.11.C). Fractal analysis indicated that the mesoscale topography of the H_{II} phase was likely to be influenced by the microstructural parameters and the water content of the samples (Libster et al., 2008). The fractal topology can potentially influence drug delivery through the large surface area of these carriers. Garti and coworkers also made considerable progress elucidating the solubilization of therapeutic peptides into the H_{II} mesophase (Libster et al., 2007; Libster et al., 2009a; Libster et al., 2009b; Cohen-Avrahami et al., 2010; Mishraki et al., 2005a; Mishraki et al., 2005b).

Solubilization of Lysozyme into H_{II} Mesophases

Lysozyme (LSZ) is an important enzyme that mediates protection against microbial invasion by hydrolyzing the polysaccharides found in many bacterial cell walls. LSZ is found in bacteriophages, microbes, and plants, as well as in animal secretions and tissues (Nackimbugwe et al., 2006; Matsushita & Yanase, 2008; Salehin et al., 2009; Cegielska-Radziejewska et al., 2009; Hancock & Scott, 2000). LSZ is a hydrophilic protein that has been well-characterized and widely used as a model for numerous scientific studies, such as enzymatic reactivity, protein stability, and crystallization (Mine et al., 1999; Dubin & Feher, 1986). Moreover, this enzyme was recently classified as a food additive that may be used and labeled as food ingredient (Cobro et al., 2009; Sonni et al., 2008; Conte et al., 2009), which significantly expands its application opportunities.

Hen egg white lysozyme (HEWL) is a compact globular protein (M.W.14.4 kDa) with 129 residues, cross-linked by four disulphide bridges. Its isoelectric point is ~11. HEWL, and it is composed of five α-helices, a three-stranded antiparallel β-sheet, and a large amount of random coil and β-turns (Johnson & Larson, 2004; Blake et al., 1965; Cheetham et al., 1992). The lysozyme molecule consists of two domains or lobes linked by a long α-helix, between which lies the active site of the enzyme. The lower N-terminal lobe (residues 40-88) consists of some helices and is mostly antiparallel β-sheet. The second lobe is composed of residues 1-39 and 89-129, and its secondary structure is largely α-helical. All LSZ polar groups are on its surface and the majority of nonpolar (hydrophobic) groups are buried in the interior.

Several biomacromolecules were reported to induce structural transitions from cubic to hexagonal phase. However, there were no reports in the literature on direct incorporation of a protein into monoolein-based H_{II} mesophase. Mishraki et al. addressed this issue by solubilizing LSZ protein into the ternary GMO/tricaprylin/water mesophase (Mishraki et al., 2005a; Mishraki et al., 2005b). These investigators focused on elucidating the mode of entrapment and interactions of LSZ with the mesophase components, conformation of the protein and its stability under varying pH and denaturation conditions, and structural changes in the hexagonal mesophase. Solubilization of the protein into the GMO/tricaprylin/water structures was found to be water concentration-dependent. The samples containing low water content (13 and 16 wt %), were able to accommodate only 1–2 wt % LSZ, respectively, while 3 and 3.5 wt % protein were solubilized into the water-richer structures (20 and 23 wt %, respectively). SAXS measurements of the hexagonal structures containing different water concentrations at pH 4.5 showed a decrease of up to 4 Å in the lattice parameter of the high water-content mesophases (20 and 23 wt %) in the presence of 3 wt % guest molecule. The moderate shrinking of the lattice parameter indicating partial dehydration of the hydrophilic heads of monoglyceride was assigned to competition for water binding between the LSZ and the polar moieties of the surfactant. The results showed that the diameters of the cylinders in the presence of 20 and 23 wt % water are smaller than the diameter of LSZ. Moreover, the dimensions of the water cylinders containing 13 and 16 wt % water are much smaller than the protein diameter. Hence, from a structural point of view, the protein has to be intercalated in the interface region to fit into the narrow water cylinders. In this case, the suggested embedment of LSZ into the interface region should increase the lattice parameter of the structures and this effect is supposed to be more pronounced at low water content. Nevertheless, the possibility of partial dehydration of the hydrophilic heads of GMO, which should lead to a decrease in the lattice parameter, analogous to the greater water-content mesophases, cannot be ruled out. Therefore, the net effect of these two competing processes was reflected by keeping the lattice dimensions intact (Mishraki et al., 2005a).

Based on the FTIR and SAXS experimental results, it was claimed that LSZ interacts with hydroxyls of GMO in the outer interface region and simultaneously induces dehydration of the monoglyceride carbonyls. The interactions of LSZ with hydroxyls of GMO can potentially increase the lattice parameter of the structures while the dehydration of the carbonyls is supposed to cause the opposite effect. This mechanism of molecular interactions of the two competing trends is probably responsible for the moderate decrease of the lattice parameter in the water-rich samples or no effect on the lattice dimensions in water-poor samples (Mishraki et al., 2005a).

The LSZ interactions with the polar moieties of GMO can potentially affect the conformation of the protein and its stability. Hence, the pH-dependent conformations of LSZ and urea's influence on the stability of the protein within the carrier was investigated using the ATR-FTIR technique.

In general, the α-helix to β-sheet transition in proteins is an important issue for understanding their folding and biological functions. It is recognized that the α→β transition is probably responsible for several conformational diseases, such as Alzheimer disease, where the α-helical forms are normal and the β-sheet forms are amyloidogenic. The effect of urea as a denaturating agent on the LSZ structure is demonstrated in Fig. 11.12. In water solution the α-helix and random coil content was drastically decreased from 0.73 to 0.21 and from 0.27 to 0.03, respectively (Figs. 11.12.A and 11.12.B), in favor of β-sheet formation up to 0.69 (Fig. 11.12.C), starting with a concentration of 2M urea. In contrast, within the hexagonal mesophase the α-helical conformation stayed intact up to 4M urea. The β-sheet content within the H_{II} mesophase was significantly lower and the random coil content was greater compared to the water solution (0.18 vs. 0.06). This indicates a considerably lower degree of denaturation of LSZ within the water cylinders. The hexagonal phases impeded the α→β transition in LSZ, thereby stabilizing the protein under

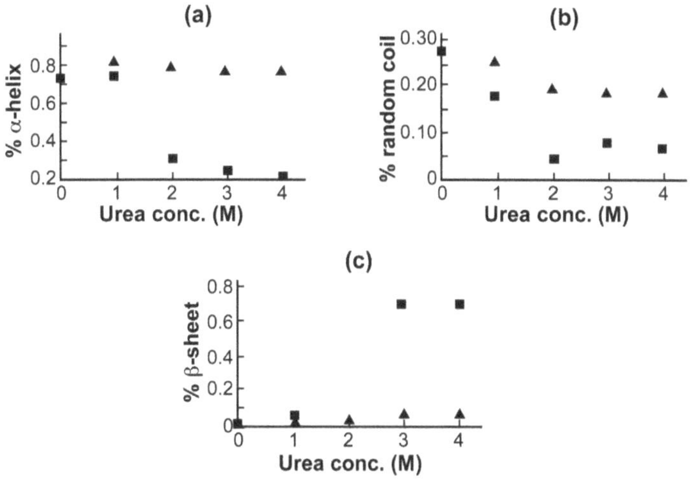

Fig. 11.12. (a) α-helix, (b) random coil, and (c) β-sheet content as a function of urea concentration (M) in the (▲) GMO/tricaprylin/LSZ/water (70.2/7.8/2.0/20.0) system and in (■) 9.1 wt % LSZ solution in D_2O. *Source:* Mishraki et al., 2005a).

denaturation conditions. It seems that the interactions between the protein and the hydroxyls of monoolein reduced the ability of urea to induce the denaturation of LSZ via direct interactions, preserving the helical structure. By analogy to the pH-dependent behavior of LSZ, the hexagonal mesophase provides a stabilizing environment for LSZ, preferring helical and random coil structures over β-sheet conformation.

Solubilization of LSZ had a significant impact on the structural properties of the mesophases. Although the same tendency was observed in the lattice parameter behavior of both empty and loaded systems as a function of temperature, different trends were detected in the domains' sizes (L_H) (Fig. 11.13). The embedment of LSZ decreased the L_H values at room temperature, suggesting smaller crystal sizes and lower order. However, while the expected decrease of domain size occurred in the empty system, the L_H values of the H_{II} mesophases containing LSZ increased at elevated temperatures. This was due to weakening of the monoolein hydroxyl hydrogen bonding with LSZ upon heating, which led to partial reconstitution of the primarily low degree of order.

Immobilization of LSZ into the proposed carriers hampered destruction of the LSZ helical structure between 65°C and 73°C, compared to the water solution, probably due to hydrogen bonding of the protein with monoolein polar heads (Mishraki et al., 2005b). Furthermore, intrinsic fluorescence measurements of the enzyme in water solution and H_{II} mesophase indicated that no structural rearrangement of the active site conformation was detected. Taken together, the results of this study suggested improved thermal stability of LSZ in hexagonal liquid crystals and show that

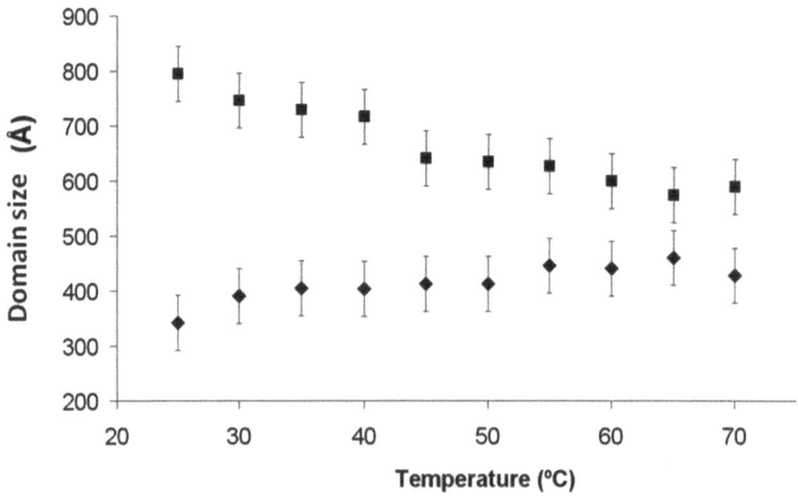

Fig. 11.13. Domain size (Å) of 17 wt % water H_{II} mesophases as a function of temperature as measured by SAXS, at (■) the empty and (◆) 2 wt % LSZ-loaded system. *Source:* Mishraki et al., 2005b.

these systems can serve as potential protein vehicles in food applications (Mishraki et al., 2005b).

Rheological measurements were carried out by Mishraki et al. to establish the correlation between the molecular level and mesoscopic organizations of these systems followed by LSZ solubilization (Mishraki et al., 2005a; Mishraki et al., 2005b). Frequency-dependent rheological measurements were performed to characterize the viscoelasticity of the H_{II} phases. This was performed by determination of the storage modulus $[G'(\omega)]$, loss modulus $[G''(\omega)]$, the complex viscosity η^*, and the relaxation time (τ_{max}) upon addition of 0–3 wt % LSZ. The storage moduli $G'(\omega)$ and the loss moduli $G''(\omega)$ were plotted against the frequency of the applied oscillations (ω). The behavior of the H_{II} phase in the absence of the additives is depicted in Fig. 11.14.A. The general trend of the $G'(\omega)$ and $G''(\omega)$ was consistent with hexagonal structure, as previously described (Mezzenga et al., 2005a). At low frequencies, the GMO/TAG/water system was found to be more viscous than elastic ($G'' > G'$). With an increase in angular frequency, both G' and G'' increased monotonically, and, finally, above

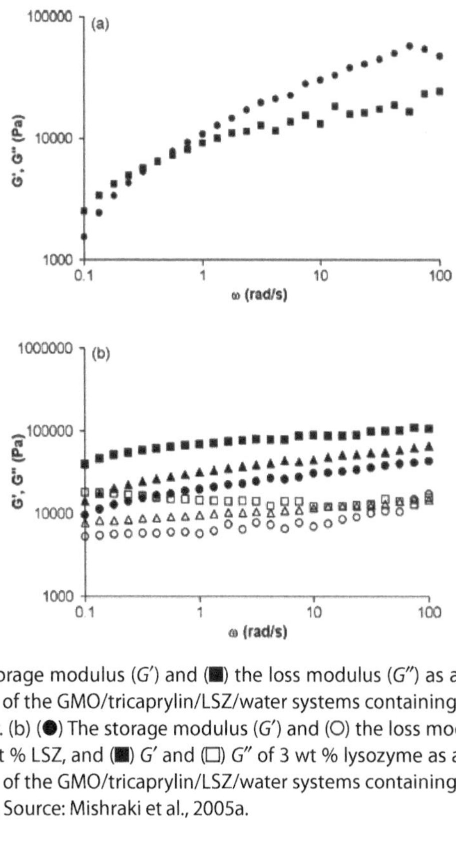

Fig. 11.14. a) (●) The storage modulus (G') and (■) the loss modulus (G'') as a function of the applied oscillation frequency (ω) of the GMO/tricaprylin/LSZ/water systems containing weight ratio GMO/tricaprylin 9:1, 20 wt % water. (b) (●) The storage modulus (G') and (O) the loss modulus (G'') of 1 wt % LSZ, (▲) G' and (△) G'' of 2 wt % LSZ, and (■) G' and (□) G'' of 3 wt % lysozyme as a function of the applied oscillation frequency (ω) of the GMO/tricaprylin/LSZ/water systems containing weight ratio GMO/tricaprylin 9:1, 20 wt % water. Source: Mishraki et al., 2005a.

the crossover point, the elastic properties of the systems dominate ($G' > G''$), indicating that the stored energy in the structure prevails over the energy that was dissipated by the viscous forces. At frequencies close to the crossover point, the H_{II} phases reveal viscoelastic behavior that can be classified as the "transition to the flow region".

The viscoelastic behavior of hexagonal structures was significantly modified in the presence of LSZ (Fig. 11.14.B). The most evident recorded changes were the disappearance of the crossover point between G' and G'', which evidently occurred below the applied frequency window ($\omega < 0.01$ rad/s), indicating greater relaxation times and increased elasticity of the samples loaded with LSZ. In addition, in the unloaded system both G' and G'' were strongly dependent on the applied frequency, suggesting a weak gel model behavior. In contrast to the empty system, the moduli of the loaded systems exhibited less pronounced dependence on the oscillation frequency, demonstrating more moderate increase of the G' and G'' as a function of frequency upon increase in LSZ concentration. This suggests a more solid-like behavior and increased strength of the intermolecular interactions as a function of increased concentration of the LSZ. Therefore, quantitative examination of the relaxation times and complex viscosities of the H_{II} mesophases should reflect the mentioned trends. The viscoelastic behavior of the H_{II} phases complies with the Maxwell model only at low frequencies, where the plot G'/G'' against ω is a straight line, resulting in τ_{max} as the slope (data not shown). At higher frequencies, the systems deviated from the Maxwell model, which should display a single relaxation time. Linear increase of τ_{max} was observed with increasing concentrations of LSZ (Table 11.A).

The embedding of even 1 wt % LSZ increased τ_{max} by approximately 100%. The maximum relaxation time was finally increased by ~180% in the presence of 3 wt % LSZ. In liquid crystals τ_{max} is the longest relaxation time and is regarded as the time-scale for relaxation to the equilibrium configuration of the water-lipid interface, following perturbation by shear deformations. Hence, the investigators suspected that linear increase in τ_{max} and therefore in the elasticity, can be attributed to stronger hydrogen bonding in the water-surfactant interface, induced only due to the solubilization of the protein. This is in agreement with ATR-FTIR results that revealed stronger hydrogen bonding between the hydroxyls of GMO and its environment on the molecular level, in the presence of LSZ. Since the major rheological properties of

Table 11.A. Effect of LSZ Concentration on τ_{max}, S, and m of the GMO/Tricaprylin/LSZ/Water Systems Containing Weight Ratio GMO/Tricaprylin 9:1 and 20 wt % Water.

LSZ (wt %)	τ_{max} (sec)	S	m
0	2.46	11303	−0.574
1	4.70	19197	−0.797
2	5.71	29959	−0.806
3	7.27	64795	−0.872

Source: Amar-Yuli et al., 2008a.

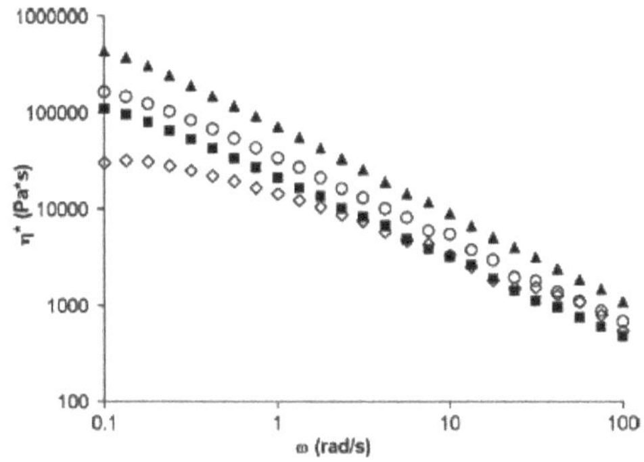

Fig. 11.15. Complex viscosity η* as a function of the applied oscillations frequency (ω), of the GMO/TAG/LSZ/water systems containing weight ratio GMO/TAG 9:1, 20 wt % water, (▲) 0, (○) 1, (■) 2, and (◇) 3 wt % LSZ. *Source:* Mishraki et al., 2005a.

LLC depend primarily on the topology of the water-surfactant interface, the complex viscosity of the mesophase was also affected by the incorporation of LSZ (Fig. 11.15). As shown in Fig. 11.15, addition of LSZ caused a considerable increase in the complex viscosity η* of the structures. The frequency dependence of complex viscosity was analyzed in terms of a power law model [eq. (1)]:

$$\eta^* = S\omega^m \quad (1)$$

where S is the gel strength parameter, which depends on the strength of intermolecular interactions, and m is the complex viscosity relaxation exponent. Values of m close to zero indicate a liquid-like behavior, while values of m approaching –1 suggest a solid-like response of the system. The obtained increase in the S values (Table 11.A) suggested enhancement of the intermolecular interactions and increased m values (from –0.57 in the empty system to –0.87 in the presence of 3 wt % LSZ), showing more pronounced solid-like response of the systems as a result of LSZ solubilization. Thus, the major effects of the entrapment of LSZ on the macroscopic rheological properties of the systems are greater elasticity and increased solid-like response of the systems. The authors assumed that that increased strength of the hydrogen bonding of the hydroxyls of GMO with the LSZ in the interface stabilized this region and therefore is responsible for the described phenomenological effects.

Low-Viscosity Reverse Hexagonal Phases Containing Transcutol or Ethanol

Amar-Yuli et al. investigated the impact of Transcutol or ethanol concentration on the viscosity of the ternary GMO/TAG/water hexagonal mesophases (Amar-Yuli et al., 2008a). Formation of low viscosity hexagonal structures would significantly enhance their application spectrum. Frequency-dependent rheological measurements were performed to determine the storage modulus [$G'(\omega)$], loss modulus [$G''(\omega)$], and the complex viscosity η^* upon addition of 0–2.75 wt % of guest molecules (Transcutol or ethanol). It should be noted that upon addition of ≥3 wt % Transcutol or ethanol, the hexagonal phase disappears and is transformed into a random distribution of micelles. Frequency-dependent rheological measurements were performed to characterize the viscoelasticity of the H_{II} phases. The storage moduli $G'(\omega)$ and the loss moduli $G''(\omega)$ were plotted against the frequency of the applied oscillations (ω). The behavior of the H_{II} phase in the absence of the additives is presented in Fig. 11.16.A, demonstrating typical behavior of the hexagonal structure already described above in the work of Mishraki et al. (Mishraki et al., 2005a).

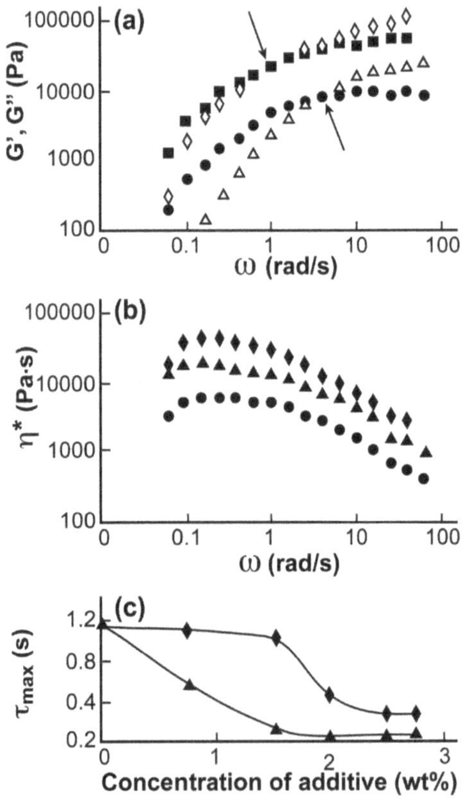

Fig. 11.16. (a) Dynamic frequency sweep test for GMO/TAG/water (G″ (■), G′ (◇) and GMO/(TAG+ethanol)/water (G″ (●), G′ (×)) systems containing weight ratio GMO/TAG or GMO/(TAG+additive) 90/10, 12.5 wt % water at 25°C. The arrow shows the crossover points. (b) Complex viscosity η^* of GMO/tricaprylin/water (◆), GMO/(TAG+Transcutol)/water(▲), and GMO/(TAG+ethanol)/water(●) mesophases, as a function of applied oscillation, all at 25°C. (c) The longest relaxation time τ_{max} as a function of Transcutol (◆) and ethanol (▲) weight concentrations, all at 25°C. *Source:* Amar-Yuli et al., 2008a.

Addition of Transcutol or ethanol resulted in a decrease in the complex viscosity η^* of the mesophases (Fig. 11.16.B). However, the general trend of the moduli was consistent with hexagonal structure. This suggested that the hexagonal mesophase was not damaged (up to 2.75 wt % additive). The complex viscosity η^* of the liquid crystals with 0 and 2.75 wt % of each additive, plotted as a function of frequency (Fig. 11.16.B), shows a decrease in the complex viscosity with addition of Transcutol or ethanol. This effect is more pronounced upon addition of ethanol. The viscosity at zero shear rate was calculated from the complex viscosity and was found to be 10^4 Pa s in the absence of guest molecules, similar to typical GMO-based H_{II} phases. The incorporation of guest molecules lowered the viscosity to 10^3 Pa s (2.75 wt % of Transcutol or ethanol). Since the major rheological properties of the different phases depend primarily on the topology of the water–lipid interface, it was suggested that the presence of the additives led to changes in the GMO–water interface during the transformation into more fluid behavior.

The "liquid-like" behavior, which was obtained upon the addition of Transcutol or ethanol, is reflected in the decrease of τ_{max} with increasing concentrations of the additives (Fig. 11.16.C). The addition of Transcutol up to 1.5 wt % (GMO/additive ratio of 52/1) did not change the viscoelastic properties of the H_{II} phase (Fig. 11.16.C). However, at 2 wt % (39/1) of added Transcutol, τ_{max} decreased by 67%, and the sample showed plastic behavior. Above 2 wt % (31–28/1), the viscoelastic properties changed only slightly. The system into which ethanol was added behaved somewhat differently (Fig. 11.16.C). The incorporation of even 0.75 wt % ethanol (104/1) decreased τ_{max} by approximately 50%. Doubling the ethanol concentration (1.5 wt % or 52/1 GMO/ethanol weight ratio) led to an additional decrease in the longest relaxation time (75%), which is attributed to more plastic behavior at the expense of elastic behavior. Further addition of ethanol (>1.5 wt %) only slightly affected the viscoelastic properties of the samples. The authors concluded that, on a weight basis, ethanol is more efficient in lowering the viscosity and is more effective in inducing the formation of a liquid-like (plastic) structure than is Transcutol.

The incorporation of either Transcutol or ethanol induced the formation of the less-ordered H_{II} structures with smaller domain size and lattice parameters at room temperature (up to 30°C), similar to the GMO/TAG/water at more elevated temperatures (35–40°C). A schematic representation of the modifications taking place in the H_{II} microstructure upon addition of each additive is presented in Fig. 11.17. This figure was drawn based on the structural understanding that was derived from the SAXS, FTIR, and NMR analysis. The typical H_{II} phase is formed in the absence of additives and exhibits free flow of water in the core of the cylindrical GMO micelles. The micelles pack in a two-dimensional hexagonal lattice (Fig. 11.17, upper part). Upon addition of Transcutol or ethanol, the hexagonal phase is liquefied with reduced domain size, yet retains its symmetry. The presence of the strong hydrogen-bonding acceptors (hydrophilic guest molecules) in the water core of the cylinders and at the interface decreased the number of water molecules available to hydrate the GMO headgroups (Fig. 11.17, lower part). This effect was more prominent in

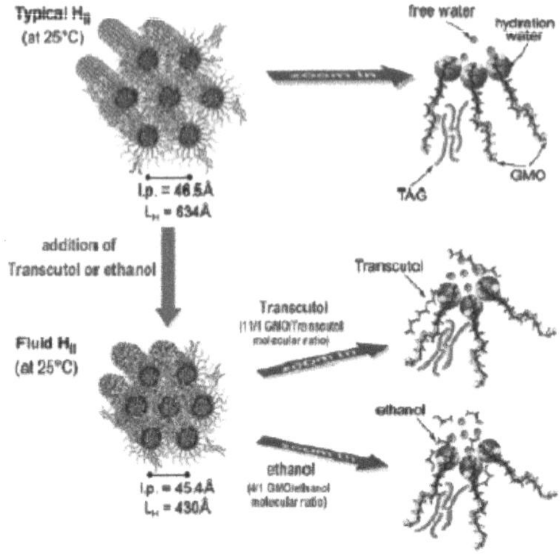

Fig. 11.17. A cartoon illustrating the room-temperature structural and physicochemical properties of the fluid mesophases containing Transcutol and ethanol as compared to those in their absence. l.p. and L_H correspond to the mean lattice parameter and the effective crystallite size in the hexagonal mesophase. It should be noted that the molecular ratio between all of the components should not be concluded from the image. The calculated ratios are GMO/TAG 14/1 in the ternary system and 12/1 in the quaternary. *Source:* Amar-Yuli et al., 2008a.

the presence of ethanol than of Transcutol. The dehydration process enhanced the mobility of the hydrocarbon chains, resulting in increased curvature and concomitant reduction of the lattice parameter.

Solubilization-dependence of the Hexagonal Structure by Introducing Hydrophilic, Amphiphilic, and Lipophilic Molecules

Amar-Yuli et al. investigated the impact of ascorbic acid (AA) or ascorbyl palmitate (AP) guest molecules on the macrostructure of the typical (Typ) and fluid hexagonal mesophases (Amar-Yuli et al., 2008b). The combined data obtained from the three methods (SAXS, thermal behavior, and FTIR) helped to clarify where these molecules are located once solubilized. The influence of the AA and AP guest molecules and their location in the three H_{II} phases is summarized by illustration in Fig. 11.18. It was found that in the H_{II} structure, the AA acts as a chaotropic guest molecule, located in the water-rich core and at the interface (Fig. 11.18). Hence, its effect was more pronounced on the water and interface behavior, including cylinder swelling, water fusion temperature, and interactions with hydroxyls and carbonyl functional groups. The AP acts as a chaotropic guest molecule as well, yet it is solubilized at the

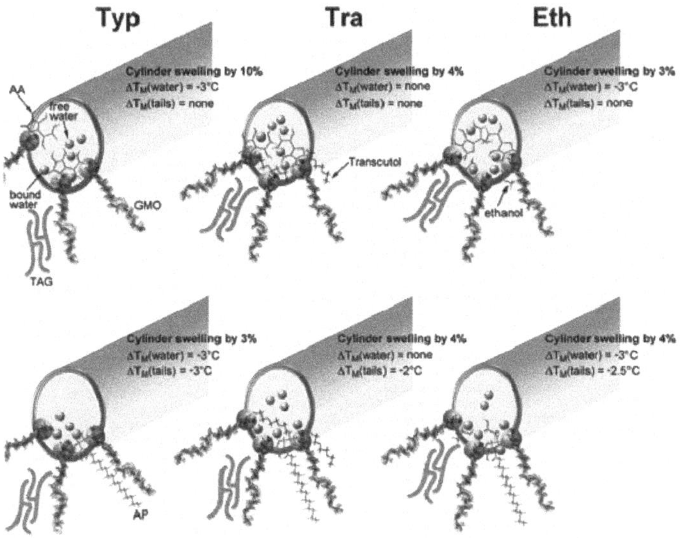

Fig. 11.18. Cartoon illustrating the solubilization location and interactions of AA (upper part) and AP (lower part) in the three H_{II} systems (from left to right) Typ, Tra, and Eth. Three major modifications in the H_{II} properties are indicated: cylinder swelling (%), based on the lattice parameter values, and $\Delta T_{M(water)}$ and $\Delta T_{M(tail)}$, the change in melting temperature of the water and GMO tail, respectively, upon the solubilization process, based on DSC thermograms. It should be noted that the molecular ratio between all the components should not be determined from the image. *Source:* Amar-Yuli et al., 2008b.

interface by its polar head while its lipophilic fatty tail is strongly solubilized and embedded within the GMO hydrophobic fatty tail (Fig. 11.18). Hence its presence affected all three regions of the H_{II} structure, the water-rich core, the interface, and the hydrophobic tail region. AP decreased water and tail fusion temperatures and interacted with carbonyl and methylene groups. It was revealed and established that in the typical system and in the presence of ethanol, both AA and AP are solubilized closer to the vicinity of the water than the interface in comparison to their solubilization location if Transcutol was part of the mesostructure. We also learned that the AA and AP induced partial migration of Transcutol from the interface to the hydrophobic region (GMO tail).

The influence of the vitamin E (VE) and vitamin E acetate (VEA) guest molecules and their location in the three H_{II} phases was also investigated (Amar-Yuli et al., 2008b) and is illustrated in Fig. 11.19. Introduction of the hydrophobic guest molecules, VE and VEA, led to their incorporation between the GMO tails, with vitamin E located closer to the interface (Fig. 11.19). Both guest molecules, especially when replacing all TAG molecules, exhibited mainly interactions with the carbonyl and methylene groups and correspondingly delayed the GMO tail fusion temperature. It was suggested that integration of the Transcutol or ethanol in the fluid H_{II}

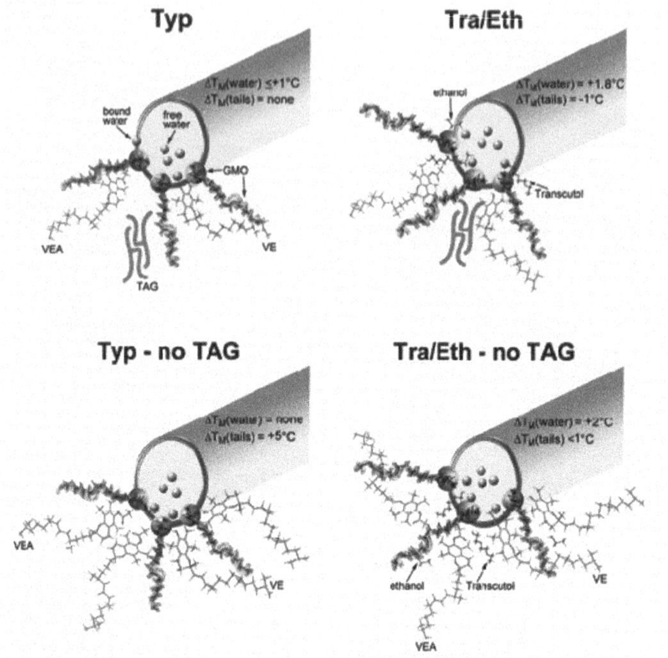

Fig. 11.19. Cartoon illustrating the solubilization location and interactions of VE and VEA in the presence of TAG (upper part) and when replacing the TAG (lower part) in the typical (left) and fluid (right) H_{II} systems. Two major modifications in the H_{II} properties are indicated: $\Delta T_{M(water)}$ and $\Delta T_{M(tail)}$, the change in melting temperature of the water and GMO tail, respectively, upon the solubilization process, based on DSC thermograms. It should be noted that the molecular ratio between all the components should not be determined from the image. *Source:* Amar-Yuli et al., 2008b.

structure led to partial migration toward the hydrophobic region, thus their impact was less pronounced.

Furthermore, the synergistic solubilization of ascorbic acid and lipophilic vitamin E antioxidants within H_{II} mesophases, composed of (GMO)/VE/AA/water, was performed (Bitan-Cherbakovsky et al., 2009; Bitan-Cherbakovsky et al., 2010). A wide range of VE concentrations was examined by the investigators (on the expense of GMO concentrations) while the AA and water concentrations remained constant (4 and 12.5 wt %, respectively) to expand the conditions for the formation of the H_{II} mesophase. SAXS and DSC combined with ATR-FTIR techniques were utilized by the investigators to study the interactions between each solubilizate and the H_{II} component that enabled the synergistic accommodation of the hydrophilic and hydrophobic molecules. It was revealed that up to 27 wt % VE solubilized within the H_{II} mesophase. This hydrophobic additive localized at the lipophilic GMO tail region solvating the surfactant tails, thereby enabling the formation of the H_{II} structure. As a result, the lattice parameter and the melting point of the hydrophobic tails decreased. Above 27 wt % VE (up to 33 wt %), once the GMO lipophilic region was homogenously solvated, additional VE molecules were located closer to the interface. At this range of concentrations, new hydrogen bonds between O–H groups of VE and O–H groups of GMO were formed. Once 35 wt % VE was introduced, the H_{II} structure transformed to face-centered reverse micellar cubic phase (*Fd3m*, Q^{227}).

Conclusions

This chapter illustrates that oleogels can be tailored for food applications, demonstrating significant advantages over other colloid systems. These universal carriers allow thermodynamic stability, high solubilization levels, protection against denaturation and severe environmental conditions.

It was shown that the composition of the oleogels can be tuned to regulate physical properties of hexagonal and micellar cubic mesophases. Series of fluid H_{II} and Q_L mesophases with complex architectures were rationally designed and prepared for the solubilization and potential delivery of food additives. It was shown that comprehensive understanding of the structural properties of these carriers is imperative for rational and successful tailoring of oleogel-based vehicles. Our principal strategy was to first characterize the different levels of organization of these special materials and then to explore the detailed relationships and possible correlations between them, their structures, and the macroscopic properties. Combination of advanced analytical techniques, including SAXS, ATR-FTIR, and rheology revealed both qualitative and quantitative insights into the structural changes induced by solubilization of guest molecules.

The incorporation of kosmotropic hydrophilic guest molecules, such as mono- and polysaccharides led to the same transformation sequence, $L_\alpha \rightarrow H_{II}$ or $Q \rightarrow H_{II}$, by decreasing the number of water molecules hydrating the surfactant polar head. As a result the CPP value increased. Moreover, the protective function of saccharides on monoolein-based cubic systems was demonstrated during dehydration by osmotic stress experiments in the presence of a series of sugars.

The entrapment of lysozyme into the H_{II} mesophases suggests that these GMO-based structures can be used for entrapment of hydrophilic proteins and can potentially be utilized for food and biotechnological purposes. Furthermore, within these carriers, biomacromolecules can improve their thermal stability and gain additional stability against denaturation and be used in an expanded pH range. H_{II} systems enabled synergistic incorporation of hydrophilic (AA) and lipophilic (VE) bioactive molecules at relatively large concentrations, exhibiting flexible behavior and promising opportunities for commercial applications. Controlling the correlation between the structures in which the guest molecules are embedded and the availability of these bioactives is important for practical applications, guiding the manufacturers to formulate the final products for best performance.

References

Alexandridis, P.; Olsson, U.; Lindman, B. A reverse micellar cubic phase. Langmuir 1996, 12, 1419–1422.

Amar-Yuli, I.; Garti, N. Transitions induced by solubilized fat into reverse hexagonal mesophases. Colloids Surf. B 2005, 43, 72–82.

Amar-Yuli, I.; Wachtel, E.; Ben Shoshan, E.; Danino, D.; Aserin, A.; Garti, N. Hexosome and hexagonal phases mediated by hydration and polymeric stabilizer. Langmuir 2007a, 23, 3637–3645.

Amar-Yuli, I.; Wachtel, E.; Shalev, D.; Moshe, H.; Aserin, A.; Garti, N. Thermally induced fluid reversed hexagonal (H-II) mesophase. J. Phys. Chem. B 2007b, 111, 13544–13553.

Amar-Yuli, I.; Wachtel, E.; Shalev, D.; Aserin, A.; Garti, N. Low viscosity reversed hexagonal mesophases induced by hydrophilic additives. J. Phys. Chem. B 2008a, 112, 3971–3982.

Amar-Yuli, I.; Aserin, A.; Garti, N. Solubilization of nutraceuticals into reverse hexagonal mesophases. J. Phys. Chem. B 2008b, 112, 10171–10180.

Amar-Yuli, I. Hexagonal liquid crystals and hexosomes structural modifications and solubilization. Ph.D. dissertation; The Hebrew University of Jerusalem, Israel, 2008.

Amar-Yuli, I.; Libster, D.; Aserin, A.; Garti, N. Solubilization of food bioactives within lyotropic liquid crystalline mesophases. Curr. Opin. Colloid Interface Sci. 2009, 14, 21–32.

Ben Ishai, P.; Libster, D.; Aserin, A.; Garti, N.; Feldman, Y. Molecular interactions in lyotropic reverse hexagonal liquid crystals: a dielectric spectroscopy study. J. Phys. Chem. B 2009, 113, 12639–12647.

Bitan-Cherbakovsky, L.; Amar-Yuli, I.; Aserin, A.; Garti, N. Structural Rearrangements and Interaction within H-II Mesophase Induced by Cosolubilization of Vitamin E and Ascorbic Acid. Langmuir 2009, 25, 13106–13113.

Bitan-Cherbakovsky, L.; Amar-Yuli, I.; Aserin, A.; Garti, N. Solubilization of Vitamin E into H-II LLC Mesophase in the Presence and in the Absence of Vitamin C. Langmuir 2010, 26, 3648–3653.

Blake, C.C.; Koenig, D.F.; Mair, G.A.; North, A.C.T.; Phillips, D.C.; Sarma, V.R. Structure of Hen Egg-White Lysozyme: A Three-dimensional Fourier Synthesis at 2 Å Resolution. Nature 1965, 206, 757–761.

Bonacucina, G.; Palmieri, G.F.; Craig, D.Q.M. Rheological and dielectric characterization of monoolein/water mesophases in the presence of a peptide drug. J. Pharm. Sci. 2005, 94, 2452–2462.

Bouchet, A.M.; Frias, M.A.; Lairion, E.; Martini, F.; Almaleck, H.; Gordillo, G.; Disalvo, E.A. Structural and dynamical surface properties of phosphatidylethanolamine containing membranes. BBA Biomembr. 2009, 1788, 918–925.

Boyd, B.J.; Whittaker, D.V.; Khoo, S.-M.; Davey, G. Lyotropic liquid crystalline phases formed from glycerate surfactants as sustained release drug delivery systems. Int. J. Pharm. 2006, 309, 218–226.

Boyd, B.J.; Khoo, S.-M.; Whittaker, D.V.; Davey, G.; Porter, C.J.H. A lipid-based liquid crystalline matrix that provides sustained release and enhanced oral bioavailability for a model poorly water soluble drug in rats. Int. J. Pharm. 2007, 340, 52–60.

Briggs, J.; Chung, H.; Caffrey, M. The temperature-composition phase diagram and mesophase structure characterization of the monoolein/water system J. Phys. II 1996, 6, 723–751.

Brondsted, H.; Nielsen, H.M.; Hovgaard, L. Drug-delivery stufies in caco-2 monolayers. 3. Intestinal transport of various vasopressin analogs in the presence of lysophosphatidylcholine. Int. J. Pharm. 1995, 114, 151–157.

Cegielska-Radziejewska, R.; Lesnierowski, G.; Kijowski, J. Antibacterial activity of hen egg white lysozyme modified by thermochemical technique. Eur. Food Res. Technol. 2009, 228, 841–845.

Cheetham, J.C.; Artymiuk, P.J.; Phillips, D.C. Refinement of an enzyme complex with inhibitor bound at partial occupancy - Hen egg-white lysozyme and tri-N-acetylchitotriose at 1-bullet-75-Angstrom resolution. J. Mol. Biol. 1992, 224, 613–628.

Cherezov, V.; Clogston, J.; Papiz, M.Z.; Caffrey, M. Room to move: Crystallizing membrane proteins in swollen lipidic mesophases. J. Mol. Biol. 2006, 357, 1605–1618.

Cohen-Avrahami, M.; Aserin, A.; Garti, N. H-II mesophase and peptide cell-penetrating enhancers for improved transdermal delivery of sodium diclofenac Colloids Surf. B 2010, 77, 131–138.

Conte, A.; Gammariello, D.; Di Giulio, S.; Attanasio, M.; Del Nobile, M.A. Active coating and modified-atmosphere packaging to extend the shelf life of Fior di Latte cheese J. Dairy Sci. 2009, 92, 887–894.

Corbo, M.R.; Bevilacqua, A.; Campaniello, D.; D'Amato, D.; Speranza, B.; Sinigaglia, M. Prolonging microbial shelf life of foods through the use of natural compounds and non-thermal approaches - a review Int. J. Food Sci. Technol. 2009, 44, 223–241.

Cullis, P.R.; De Kruijff, B. Lipid polymorphism and the functional roles of lipids in biological membranes. Biochim. Biophys. Acta 1979, 599, 399–420.

Drummond, C.J.; Fong, C. Surfactant self-assembly objects as novel drug delivery vehicles Curr. Opin. Colloid Interface Sci. 1999, 4, 449–456.

Durbin, S.D.; Feher, G. Crystal-growth studies of lysozyme as a model for protein crystallization. Biophys. J. 1986, 76, 583–592.

Efrat, R.; Aserin, A.; Kesselman, E.; Danino, D.; Wachtel, E.J.; Garti, N. Liquid micellar discontinuous cubic mesophase from ternary monoolein/ethanol/water mixtures. Colloids Surf. A 2007, 299, 133–145.

Efrat, R.; Aserin, A.; Garti, N. On structural transitions in a discontinuous micellar cubic phase loaded with sodium diclofenac. J. Colloid Interface Sci. 2008, 321, 166–176.

Efrat, R.; Kesselman, E.; Aserin, A.; Garti, N.; Danino, D. Solubilization of Hydrophobic Guest Molecules in the Monoolein Discontinuous Q(L) Cubic Mesophase and Its Soft Nanoparticles Langmuir 2009, 25, 1316–1326.

Efrat, R. Stable cubic mesophases and nanoparticles (cubosomes and micellosomes) for solubilization of active molecules. Ph.D. dissertation; The Hebrew University of Jerusalem, Israel, 2009.

Epand, R.M.; Leon, B.T. Hexagonal phase forming propensity detectedin phospholipid-bilayers with fluorescent-probes. Biochem. 1992, 31, 5, 1550–1554.

Gin, D.L.; Pecinovsky, C.S.; Bara, J.E. Kerr, L. Functional lyotropic liquid crystal materials. Struct. Bond 2008, 128, 181–222.

Gradzielski, M. Investigations of the dynamics of morphological transitions in amphiphilic systems Curr. Opin. Colloid Interface Sci. 2004, 9, 256–263.

Gruner, S.M.; Tate, M.W.; Kirk, G.L.; So, P.T.C.; Turner, D.C.; Keane, D.T.; Tilcock, C.P.S.; Cullis, P.R. X-ray-diffractionstudy of the polymorphic behavior of N-methylated dioleoylphosphatidylethanolamine. Biochem. 1988, 27, 2853–2866.

Gruner, S.M. Stability of lyotropic phases with curved interfaces. J. Phys. Chem. 1989, 93, 7562–7570.

Hancock, R.E.W.; Scott, M.G. The role of antimicrobial peptides in animal defenses. Proc. Natl. Acad. Sci. 2000, 97, 8856–8861.

Harper, P.; Mannock, D.A.; Lewis, R.N.A.H.; McElhaney, R.; Gruner, S. X-ray diffraction structures of some phosphatidylethanolamine lamellar and inverted hexagonal phases. Biophys. J. 2001, 81, 2693–2706.

Hubner, W.; Blume, A. Interactions at the lipid-water interface. Chem. Phys. Lipids 1998, 96, 99–123.

Israelachvili, J.N.; Mitchell, D.J.; Ninhan, B.W. Theory of self-assembly of hydrocarbon amphiphiles into micelles and bilayers. J. Chem. Soc. 1976, 72, 1525–1568.

Johnson, E.A.; Larson, A.E. Lysozyme. Antimicrobials in Food. Taylor and Francis: Boca Raton, FL, 2004.

Jones, J.L.; McLeish, T.C.B. Concentration fluctuations in surfactant cubic phases: Theory, rheology, and light scattering. Langmuir 1999, 15, 7495–7503.

Koynova, R.; Brankov, J.; Tenchov, B. Modulation of lipid phase behaviour by kosmotropic and chaotropic solutes. Eur. Biophys. J. 1997, 25, 261–274.

Kurosaki, Y.; Nagahara, N.; Tanizawa, T.; Nishimura, H.; Nakayama, T.; Kimura, T. Use of lipid disperse systems in transdermal delivery: comparative study of flufenamic acid permeation among rat addominal skin, silicon rubber membrane and stratum corneum sheet isolated from hamster cheek pouch. Int. J. Pharm. 1991, 67, 1–9.

Langner, M.; Kubica, K. The electrostatics of lipid surfaces. Chem. Phys. Lipids 1999, 101, 3–35.

Larsson, K. Cubic lipid-water phases: Structures and biomembrane aspects. J. Phys. Chem. 1989, 93, 7304–7314.

Lewis, R.N.A.H.; Mannock, D.; McElhaney, R.N.; Turner, D.C.; Gruner, S.M. Effect of fatty acyl chain length and structure on the lamellar gel to liquid-crystalline and lamellar to reversed hexagonal phase transitions of aqueous phosphatidylethanolamine dispersions, Biochem. 1989, 28, 541–548.

Libster, D.; Aserin, A.; Wachtel, E.; Shoham, G.; Garti, N. An H_{II} liquid crystal-based delivery system for cyclosporin A: Physical characterization. J. Colloid Interface Sci. 2007, 308, 514–524.

Libster, D.; Ben Ishai, P.; Aserin, A.; Shoham, G.; Garti, N. From the microscopic to the mesoscopic properties of lyotropic reverse hexagonal liquid crystals. Langmuir 2008, 24, 2118–2127.

Libster, D.; Ben Ishai, P.; Aserin, A.; Shoham, G.; Garti, N. Molecular interactions in reverse hexagonal mesophase in the presence of Cyclosporin A. Int. J. Pharm. 2009a, 367, 115–126.

Libster, D.; Aserin, A.; Yariv, D.; Shoham, G.; Garti, N. Concentration- and temperature-induced effects of incorporated desmopressin on the properties of reverse hexagonal mesophase. J. Phys. Chem. B 2009b, 113, 6336–6346.

Liu, D.-Z.; LeCluyse, E.L.; Thakker, D.R. Dodecylphosphocholine- mediated enhancement of paracellular permeability and cytotoxicity in Caco-2 cell monolayers. J. Pharm. Sci. 1999, 88, 1161–1168.

Lis, L.J.; McAlister, M.; Fuller, N.; Rand, R.P.; Parsegian, V.A. Interactions between neutral phospholipid bilayer membranes. Biophys. J. 1982, 37, 657–665.

Lopes, L.B.; Speretta, F.F.F.; Bentley, M.V.L.B. Enhancement of skin penetration of vitamin K using monoolein-based liquid crystalline systems. Eur. J. Pharm. Sci. 2007, 32, 209–215.

Luzzatti, V.; Vargas, R.; Gulik, A.; Mariani, P.; Seddon, J.M.; Rivas, E. Lipid polymorphism: a correction. The structure of the cubic phase of extinction coefficient Fd– consists of two types of disjointed reverse micelles embedded in a three-dimensional hydrocarbon matrix. Biochem. 1992, 31, 279–285.

Mariani, P.; Luzzati, V.; Delacroix, H. Cubic phases of lipid-containing systems. Structure analysis and biological implications. J. Mol. Biol. 1988, 204, 165–189.

Mariani, P.; Rivas, E.; Luzzati, V.; Delacroix, H. Polymorphism of a lipid extract from Pseudomonas fluorescens: structure analysis of a hexagonal phase and of a novel cubic phase of extinction symbol Fd–. Biochem. 1990, 29, 6799–6810.

Matsushita, I.; Yanase, H. A novel thermophilic lysozyme from bacteriophage phiIN93. Biochem. Biophys. Res. Comm. 2008, 377, 89–92.

Mezzenga, R.; Meyer, C.; Servais, C.; Romoscanu, A.I.; Sagalowicz, L.; Hayward, R.C. Shear rheology of lyotropic liquid crystals: A case study. Langmuir 2005a, 21, 3322–3333.

Mezzenga, R.; Grigorov, M.; Zhang, Z.; Servais, C.; Sagalowicz, L.; Romoscanu, A.I. Polysaccharide-induced order-to-order transitions in lyotropic liquid crystals. Langmuir 2005b, 21, 6165–6169.

Mine, S.; Tate, S.; Ueda, T.; Kainosho, M.; Imoto, T. Analysis of the relationship between enzyme activity and its internal motion using nuclear magnetic resonance: N-15 relaxation studies of wild-type and mutant lysozyme. J. Mol. Biol. 1999, 286, 1547–1565.

Mishraki, T.; Libster, D.; Aserin, A.; Garti, N. Lysozyme entrapped within reverse hexagonal mesophases: physical properties and structural behavior. Colloids Surf. B 2010a, 75, 47–56.

Mishraki, T.; Libster, D.; Aserin, A.; Garti, N. Temperature-dependent behavior of lysozyme within the reverse hexagonal mesophases (H(II)). Colloids Surf. B 2010b, 75, 391–397.

Montalvo, G.; Valiente, M.; Rodenas, E. Rheological properties of the L phase and the hexagonal, lamellar, and cubic liquid crystals of the CTAB/benzyl alcohol/water system. Langmuir 1996, 12, 5202–5208.

Nakimbugwe, D.; Masschalck, B.; Deckers, D.; Callewaert, L.; Aertsen, A.; Michiels, A.C.W. Cell wall substrate specificity of six different lysozymes and lysozyme inhibitory activity of bacterial extracts. FEMS Microbiol. Lett. 2006, 259, 41–46.

Pohle, W.; Selle, C.; Fritzsche, H.; Bohl, M. Comparative FTIR spectroscopic study upon the hydration of lecithins and cephalins J. Mol. Struct. 1997, 408–409, 273–277.

Pouzot, M.; Mezzenga, R.; Leser, M.; Sagalowicz, L.; Guillot, S.; Glatter, O. Structural and rheological investigation of Fd3m inverse micellar cubic phases Langmuir. 2007, 23, 9618–9628.

Rand, R.P. Interacting phospholipid-bilayers—Measured forces and induced structural-changes. Annu. Rev. Biophys. Bioeng. 1981, 10, 277–314.

Rand, R.P.; Parsegian, V.A. Hydration forces between phospholipid-bilayers. Biochim. Biophys. Acta 1989, 988, 351–376.

Rodriguez-Abreu, C.; Acharya, D.P.; Aramaki, K.; Kunieda, H. Structure and rheology of direct and reverse liquid-crystal phases in a block copolymer/water/oil system. Colloids Surf A 2005, 269, 59–66.

Sagalowicz, L.; Mezzenga, R.; Leser, M.E. Investigating reversed liquid crystalline mesophases Curr. Opin. Colloid Interface Sci. 2006a, 11, 224–229.

Sagalowicz, L.; Leser, M.E.; Watzke, H.J.; Michel, M. Monoglyceride self-assembly structures as delivery vehicles. Trends Food Sci. Technol. 2006b, 17, 204–214.

Salehin, M.; Ghosh, A.K.; Mallick, P.K.; Bhattacharya, T.K. Molecular characterization, polymorphism and association study of lysozyme gene with milk production and somatic cell trait in Bos indicus X Bos taurus cattle. Animal 2009, 3, 623–631.

Saturni, L.; Rustichelli, F.; Di Gregorio, G.M.; Cordone, L.; Mariani, P. Sugar-induced stabilization of the monoolein Pn3m bicontinuous cubic phase during dehydration. Phys. Rev. E 2001, 64, 040902/1–4.

Seddon, J.M. Structure of the inverted hexagonal (HII) phase, and non-lamellar phase-transitions of lipids. Biochim. Biophys. Acta 1990, 1031, 1–69.

Seddon, J.M.; Zeb, N.; Templer, R.H.; McElhaney, R.N.; Mannock, D.A. An Fd3m lyotropic cubic phase in a binary glycolipid/water system. Langmuir 1996, 12, 5250–5253.

Seddon, J.M.; Robins, J.; Gulik-Krzywicki, T.; Delacroix, H. Inverse micellar phases of phospholipids and glycolipids. Phys. Chem. 2000, 2, 4485–4493.

Shah, M.H.; Paradkar, A. Cubic liquid crystalline glyceryl monooleate matrices for oral delivery of enzyme Int. J. Pharm. 2005, 294, 161–171.

Shearman, G.C.; Ces, O.; Templer, R.H.; Seddon, J.M. Inverse lyotropic phases of lipids and membrane curvature. J. Phys. Condens. Matter 2006, 18, S1105–S1124.

Shui, L.L.; Wang, Z.N.; Zheng, L.Q. Rheological properties of cubic liquid crystals formed from monoglyceride/H2O systems. Chin. J. Chem. 2005a, 23, 245–250.

Shui, L.L.; Guo, P.Z.; Chen, F.; Xu, G.Y.; Zheng, L.Q. The effect of Iopamidol on rheological properties of monoglyceride/water system. Colloids Surf. A 2005b, 256, 85–90.

Siddig, M.A.; Radiman, S.; Jan, L.S.; Muniandy, S.V. Rheological behaviours of the hexagonal and lamellar phases of glucopone (APG) surfactant. Colloids Surf A 2006, 276, 15–21.

Sonni, F.; Bastante, M.J.C.; Chinnici, F.; Natali, N.; Riponi, C. Replacement of sulfur dioxide by lysozyme and oenological tannins during fermentation: influence on volatile composition of white wines. J. Sci. Food Agr. 2009, 89, 688–696.

Spernath, A.; Aserin, A.; Garti, N. Phase transition induced bywater dilution in phospholipid U-type food-grade microemulsions studied by DSC. J. Colloid Interface Sci. 2006, 299, 900–909.

Touitou, E.; Levi-Schaffer, F.; Dayan, N.; Alhaique, F.; Riccieri, F. Modulation of caffeine skin delivery by carrier design: liposomes versus permeation enhancers. Int. J. Pharm. 1994, 103, 131–136.

Tsvetkova, N.; Koynova, R.; Tsonev, L.; Quinn, P.; Tenchov, B. Influence of proline on the stability of fully hydrated and freeze-dried lipid phases. Chem. Phys. Lipids 1991, 60, 51–59.

Uddin, M.H.; Rodriguez, C.; Watanabe, K.; Lopez-Quintela, A.; Kato, T.; Furukawa, H.; Harashima, A.; Kunieda, H. Phase behavior and formation of reverse cubic phase based emulsion in water/poly(oxyethylene) poly(dimethylsiloxane) surfactants/silicone oil systems. Langmuir 2001, 17, 5169–5175.

Yaghmur, A.; De Campo, L.; Salentinig, S.; Sagalowicz, L.; Leser, M.E.; Glatter, O. Oil-loaded monolinolein-based particles with confined inverse discontinuous cubic structure (Fd3m). Langmuir 2006, 22, 517–521.

Yang, L.; Ding, L.; Huang, H.W. New phases of phospholipids and implications to the membrane fusion problem. Biochem. 2003, 42, 6631–6635.

Zasadzinski, J.A.N.; Helm, C.A.; Longo, M.L.; Weisenhornt, A.L.; Gould, S.A.C.; Hansmat, P.K. Atomic force microscopy of hydrated phosphatidylethanolamine bilayers. Biophys. J. 1991, 59, 755–760.

12

Protein-templated Oil Gels and Powders

Raffaele Mezzenga
ETH Zurich, Food & Soft Materials Science,
Institute of Food, Nutrition & Health, 8092 Zürich, Switzerland

Introduction

The use of solid hydrophobic matrices in functional materials, pharmaceutical and food formulations, cosmetics, lubrication, delivery applications, and several other branches of soft condensed matter is required to both increase viscoelasticity properties and slow diffusion processes of active molecules (Turchiuli et al., 2005; Weiss et al., 2005). An ideal material for these applications should, in principle, accomplish the demanding task of simultaneously exhibiting good elastic properties and low viscosity behavior. Emulsion and dispersion technology have both been widely employed to design functional materials of this type. High internal phase oil-in-water emulsions (HIPEs) have been designed, where the balance of elasticity versus viscous properties can be tuned to some extent by controlling the volume fraction of the continuous phase, that is, the amount of total interfaces. However, the increase in elastic modulus for these blends generally remains limited even when the volume fraction of the continuous phase is reduced below 10% (Mason et al., 1995). Further reduction of the continuous phase volume fraction generally results in coalescence of the emulsions under shear (Dimitrova & Leal-Calderon, 2004). A sharp increase in elasticity of the oil can also be reached by hydrogenation of the liquid oil. However, in addition to irreversibly modifying the oil phase, this also corresponds to a strong increase of viscosity of the oil phase with a subsequent reduction of processability.

Oil-rich solid-like powders share many similarities with solid oils, except that the procedures and the systems by which these are obtained differ substantially from those used on bulky, solid oil materials. The most widely used method is spray drying an oil-in-water emulsion precursor to rapidly evaporate the water continuous phase (Faldt et al., 1996; Vega et al., 2005; Fuchs et al., 2006; Bruckner et al., 2007; Baranauskiene et al., 2007; Jayasundera et al., 2009). This method is typically used to encapsulate hydrophobic compounds and aroma into the oil droplets or to increase stability of oil against oxidation (Klinkesorn et al., 2005; Gu et al., 2004; Shaw et al., 2007).

To prevent the oil droplets from collapsing into a macro-phase separated liquid state during either the spray drying processing (early stage instabilities) or during the shelf-life of the powder (late stage instabilities), a solid hydrophilic carrier is normally added to the water solution. After water has evaporated, this hydrophilic carrier, together with the surfactant used to stabilize the liquid emulsion, constitutes the continuous phase of the powder, also referred to as "solid dry base". The minimum amount of solid hydrophilic carrier needed varies from study to study but it is in general comprised between 30 and 80 wt % of the total final powder (Klinkesorn et al., 2005; Gu et al., 2004; Jost et al., 1989). Typical examples of the constituents of the solid dry base are lactose, glucose, maltodextrin, starch, and cellulose. The presence of this hydrophilic carrier in the formulation alters the composition of the powder with respect to the original emulsion, concentrating the solid carrier into the final powder, and thus reducing the amount of liquid dry base (the oil droplets). Without the presence of this solid carrier, however, the emulsion droplets coalesce either during the spray drying processing or afterwards, leading to the collapse of the powder and the leaking of the oil. For example, 35–60% of maltodextrin has been used to spray dry fruit juices such as apricot, raspberry, and orange juice (Bandhari et al., 1993; Adhikari et al., 2009).

A possible alternative way to the introduction of this solid carrier is the stabilization of the oil droplets interfaces in a physico-chemical way, providing enough elasticity to the interfaces to survive the spray drying process. A typical example is the multi-step deposition of positively charged polyelectrolytes capable of cross-linking by ionic complexation onto the droplet surfaces, thus providing the interfaces with the needed elasticity (Klinkesorn et al., 2005; Gu et al., 2004; Moreau et al., 2003). A typical example would be to stabilize an oil-in water emulsion by a protein and to adjust the solution pH to sufficiently acidic values to positively charge the protein. Then the emulsion is dialyzed or diluted with another water solution containing a negatively charged polyelectrolyte capable of forming an ionic complex with the positively charged oil droplet interfaces (made positive by the protein layer). Another variant is to stabilize the droplets by a low molecular weight anionic surfactant such as lecithin and then expose the emulsion to a polycationic polysaccharide, such as chitosan (Mun et al., 2005). Although this is generally considered a reliable method, the main drawback is that multiple steps are needed to stabilize the droplet surfaces, which makes this technology rather expensive and unsuitable for large scale processing. Thermal or enzymatic cross-linking of the protein-stabilized interface is another efficient method to provide elasticity to the interface (Romoscanu et al., 2005; Kellerby et al., 2006; Cho et al., 2003). Thermal cross-linking of whey protein stabilized by oil-in-water emulsion has been employed using a large excess of protein, in combination with spray drying, leads to a dry oil powder. The calculated amount of protein in the final powder was, however, still of the order of 28–30 wt % of the total powder (Jost et al., 1989).

To summarize, the dried solid oil powders available today are formulations composed of oily compounds, which are liquid at room temperature, that are stabilized by a relatively high weight percentage of non-oily compounds. If a dry solid oil powder containing a high percentage of oily compounds at room temperature were available, it would have several significant advantages, such as simpler formulations, absence of water-transport layers, increased hydrophobic encapsulation efficiency, etc.

In the present chapter, we review the recent progresses made by the authors to convert low-viscous apolar liquids (oil) into either an elastic solid or an oil powder without the need to add additional compounds than those already present in the initial protein-stabilized oil-in-water emulsion template. The method involves no chemical modification of the oil, which remains in its native liquid state, with unchanged saturation level of the glyceride molecules. When applied to food products as an alternative to oil hydrogenation, this method avoids potential adverse effects of hydrogenated fats in relation to cardiovascular diseases (Willett et al., 1993; Han et al., 2002; Lichtenstein et al., 2003; Tsai et al., 2005). In principle, it can be applied to any apolar fluid, which yields a stable protein-stabilized emulsion.

The hydrophobic liquid solidification method presented in this chapter consists of self-assembling a monodisperse oil-in-water emulsion, where the oil droplets are stabilized by a cross-linked protein monolayer absorbed at their interface. A similar process to design percolating structures for organic semiconductors has been employed by Mezzenga et al. by evaporating solvent from a dispersion of polymeric colloidal particles and block copolymers and annealing the final blend (Mezzenga et al. 2003a; 2003b; 2003c). In the present case, however, the intermediate step of annealing the blend is not necessary since all components are already below their glass transition temperature, and equilibrium is achieved directly upon solvent evaporation. Depending on whether a slow evaporation is used (under controlled atmosphere) or a fast evaporation process is preferred (spray drying), the method is equally suited to design, in a single easy step, both oil gels and powders, respectively.

Oil Gels

Experimental Procedures

The main constituents are the oil phase, the water buffer continuous phase, and a protein, which is absorbed at the oil-water interface during the emulsification process. In a typical experiment, all solutions are prepared in a 20 mM imidazole buffer, adjusted to pH=7.0 with 1 M NaOH. The pH of 7.0 ensures good solubility of the protein as well as the absence of flocculation and inter-particle cross-linking in emulsions.

β-Lactoglobulin (β-Lg), the principal whey protein for nutritional and functional properties, has been used as emulsifying protein. β-Lg monomers have a molecular weight ca. $18 \cdot 10^3$ g/mole and are highly soluble in water, except around pH=5.1 at the protein's isoelectric-point (IEP). The cross-linking behavior of β-Lg in solution or at interfaces has been the subject of numerous studies. Interfacially adsorbed β-Lg cross-link via intermolecular disulfide bonds at room temperature with relatively slow kinetics (Dickinson & Matsumura, 1991). Cross-linking kinetics are considerably accelerated when the interfacially adsorbed protein layer is heated at 80°C (Rodríguez Patino et al., 2001). β-Lg monomers can also be chemically cross-linked with glutaraldehyde, which reacts with amino acid sidechains, particularly with the lysine group. Protein cross-linking via glutaraldehyde has also been investigated in detail (Monsan et al., 1975; Marquié et al., 1998; Gerrard et al., 2002; Gerrard et al., 2003).

Two different oils have been used to demonstrate the feasibility of the concept: paraffin oil and a common olive oil. In contrast to highly refined triglycerides, olive oil is not chemically pure and contains aroma compounds, which account for as much as 1% of the overall oil composition. These compounds, which consist mainly of aldehydes, ketones, esters, and organic acids, may possibly influence the cross-linking behavior of proteins adsorbed at the oil-water interface. For comparison purposes, highly refined olive oil was also used in interface characterization experiments. Glutaraldehyde, 50% solution in water solution, was used as 1 wt % buffered solution.

A typical procedure followed to convert the liquid oil into a gel is described by Romoscanu & Mezzenga in *Langmuir* 2006 and can be summarized in the following main steps.

Monodisperse oil droplets, first obtained using the technique described in Umbanhowar et al. (2000), were dispersed in a 1 wt % buffered protein water solution. For scale-up purposes, emulsions of identical composition were also realized by homogenization of the oil phase in the buffered protein solution using a Polytron mixer, or for the finest droplet sizes, a homogenizer.

The emulsion is left for about one hour to allow complete protein adsorption onto the oil-water interfaces. To remove un-adsorbed protein the following procedure was followed: the emulsion was washed with water using a 10:1 dilution factor, which allowed decreasing the concentration of the un-adsorbed protein concentration by one order of magnitude. A dense oil emulsion was then collected and separated by the rest of the water by creaming or centrifugation processes and diluted again by repeating the washing procedure. The washing/separation process was reiterated until the concentration of un-adsorbed protein in water was negligible. Because of the irreversibility of protein adsorption on the time-scale of the experiment, typically three iterations were sufficient to decrease the un-adsorbed protein concentration to vanishing amounts. The washing steps were performed with pH=7.0 buffer to avoid emulsion flocculation.

Cross-linking of adsorbed protein was performed, either thermally by heating the concentrated, washed emulsion at 80°C for 10 minutes or chemically with glutaraldehyde. In this case the emulsion was poured in the same volume of 1 wt % glutaraldehyde pH=7.0 buffered solution to ensure the cross-linking of adsorbed protein molecules while avoiding inter-particle cross-linking. The dilute emulsion is left for 10 minutes under gentle stirring and washed 3 times to separate non-reacted glutaraldehyde as described above.

To prevent the protein layers from collapsing in the dried state, a low amount (0.5 wt %) of polar, low-molecular non-volatile compound (glycerol) was added to water to increase its chemical potential, slow down evaporation process, and reduce internal stresses caused by water evaporation.

The concentrated emulsion was allowed to dry for a period of 72 h under ventilation at room temperature to yield a fully transparent lipidic gel.

Theoretical Aspects

The mechanical properties of the lipidic gels can be interpreted directly using the analogy of dry foams, i.e., the structure obtained upon dispersion of gas bubbles in a vanishing small volume of fluid. The fact that the cell-filling fluid is a gas in the case of dry foams, and a liquid in the present case, does not alter the qualitative similarity of the structures, nor their common mathematical approach, since the magnitude of the Laplace pressure $\Delta P=\sigma/R$ in a typical dry foam is smaller than the atmospheric pressure by some orders of magnitude. In both cases, the resulting structure consists of a space-filled stacking of polyhedral cells. The films, which delimitate the individual cells, obey a given number of rules, called *Plateau laws*, which originate in the requirement for pressure and surface forces to be balanced at the film level. These rules, together with the energy-minimization driven reduction of the internal surface, confer a solid, elastic behavior to the bulk material at low strains.

Available models of dry foams based on the storage of energy in area changes all predict an elastic shear modulus $G'(f)$, following, in the zero frequency limit:

$$G' = \xi\sigma R^{-n} \tag{1}$$

where $n = 1$, $0.50 < \xi < 0.54$, and σ is the interfacial tension (Princen, 1983; Kraynik & Reinelt, 1996; Weaire & Hutzler, 1999). With respect to the applicability of these models to lipidic gels, the following comments can be made:

i. The proportionality constant between G' and R^{-1} is approximately equal to $\sigma/2$. Provided that lipidic gels follow the same scaling law $G' \sim R^{-1}$, the interfacial tension σ can be derived from the proportionality constant between G and R^{-1}.
ii. All models are based on the assumption of a monodisperse cell volume distribution. Their application to polydispersed systems can only be made considering the relevant average radius needed to minimize the induced error. The surface-volume mean radius $R_{32} = \sum_i n_i R_i^3 / \sum_i n_i R_i^2$ usually used as relevant average radius (Princen & Kiss, 1986).
iii. The above model (Equation 1) only takes into account the energy stored during deformation in changes in film area and is based on variation of the sum of individual film area changes, which collectively scales linearly with the bulk shear strain γ. In contrast to wet foams, energy can also be stored in the variation of film thickness in dry foams since the dry foam morphology allows work to be done against the disjoining pressure Π between two adjacent protein layers belonging to two different cells. Buzza et al. (1995) estimate the modulus arising from film compression to be of order σ/R, thus comparable with that arising from changes in area (Equation 1). Because both contributions scale as R^{-1}, evaluation of the

interfacial tension from data best-fitting with Equation 1 will eventually lead to higher values than those measured on single droplets, if work done against the disjoining pressure between two adjacent cells is not accounted for in the fitted interfacial tension parameter. Narsimhan (1992) predicts an order of magnitude of 10^{-1} μm for the distance between two protein layers belonging to two different droplets/bubbles for disjoining pressure phenomena to become relevant.

iv. The increase of the interfacial tension with increasing film area cannot be taken into account within a linear viscoelastic formalism (Equation 1), since this would imply a strain dependent value of the shear modulus. A discussion of the relevance of the increase of the interfacial tension with increasing film area in the present context is given below. For now we anticipate and stress the fact that local interfacial tension variations will increase with increasing frequency when the material is subjected to oscillatory strain, so that an the storage modulus G' is expected to increase with increasing frequency f.

Interfacial Properties

In the lipidic gels, the interface consists of a cross-linked protein bilayer with an aqueous core. Qualitative differences between proteins and low molecular weight surfactants may affect the relevancy of the dry foam model for the present structures. These are outlined below.

Single Droplet Experiment: Dilatational Elasticity of Adsorbed Protein Layer

Protein adsorption is a highly irreversible process on the time-scale of our experiments (Svitova et al., 2003; Dimitrova et al., 2004), due to protein unfolding (denaturation) upon adsorption, the existence of many adsorption sites per molecule, and strong protein–protein interactions. This irreversibility allows systematic dilution of the un-adsorbed protein concentration in water by washing the emulsions prior to cross-linking (Romoscanu & Mezzenga, 2005). Upon cross-linking, individual protein molecules are covalently bound to their adsorbed neighboring proteins, preventing any further adsorption/desorption phenomena between the interface and the liquid phase.

Under these conditions, the protein interfacial concentration is determined by the total amount of interfacial area. Changes of the total interfacial area occur at two stages: 1) during the droplet shape transition from spheres to polyhedrons (i.e., during evaporation of the aqueous matrix) and 2) during the shear deformation of the solid material. In a monodisperse system the droplet surface increases by a factor 1.1 upon transition from a sphere to a space-filling Kelvin cell (Weaire & Phelan, 1994). This area increase is significantly larger than the interfacial area increase occurring during bulk shear deformations of the solid material, since the latter goes as $(1+\gamma^2/3)$ for small strains, where γ, the shear deformation of the bulk material, is of $O(10^{-2})$. The increase of interfacial tension upon increase of the total available interfacial area (at constant protein amount) for one single protein layer can be measured by dynamic tensiometry (Romoscanu & Mezzenga, 2005). The interfacial elasticity, defined in Equation 2 is a usual measure of the increase of interfacial tension with increasing interfacial area:

$$E_D = \frac{\mathrm{d}\sigma}{\mathrm{d}\ln A}. \tag{2}$$

The interfacial elasticity E_D cannot be used as an explicit parameter in the derivation of the elastic modulus of the material (Equation 1), since this would automatically imply a strain-dependent elastic modulus, i.e., a non-linear elastic bulk behavior. However, since the interfacial area increase during the transition from a sphere to a space-filling polyhedron is much larger than that arising upon shearing of the bulk material, a sensible way to consider the influence of E_D in the final expression of G is to take into account the increase of interfacial tension during the transition from a sphere to a polyhedron for the determination of the relevant interfacial tension, following:

$$\sigma_A = \sigma_0 + \int_{A_0}^{A} E_D \mathrm{d}\ln(A) = \sigma_0 + E_D \ln\left(\frac{A}{A_0}\right) \cong \sigma_0 + 0.1 \cdot E_D, \tag{3}$$

where σ_A is the interfacial tension of the strained interface of area A, and σ_0 the interfacial tension corresponding to the initial adsorption interface of area A_0. The last result in Equation 3 is obtained under the assumption of A=1.1A_0.

Interfacial Properties in the Gel: Protein Bilayer

The percolating internal interface of the lipidic gels consists of a film, which is a bilayer of cross-linked proteins belonging to the faces of two adjacent polyhedrons with an aqueous polar core. As mentioned above, the most relevant parameter of these films in the context of the dry foam analogy is the overall interfacial tension of the film, since the amount of interfacial energy that can be stored upon deformation of the structure is directly proportional to this quantity.

Because of the complex interactions, which are expected to take place between two polyelectrolytic layers within sub-micrometric distance (Narsimhan, 1992), the assumption of the overall film tension as the double of the monolayer (adsorbed protein) value does not necessarily apply (Soos et al., 1994; Xu et al., 2003). Moreover, these films are characterized by a disjoining pressure whose magnitude controls the equilibrium film thickness, the overall film tension, and the amount of work stored in the strained structure. Narsimhan (1992) proposed a model for the nature and extent of the disjoining pressure in emulsions concentrated by centrifugation. Dimitrova et al. (2001; 2004) used Scheludko and Mysels-type cells, as well as the magnetic chain technique involving apolar ferrofluid emulsions, to investigate disjoining pressures. A thorough thermodynamical investigation of surfactant film structure and internal interactions can be found in Eriksson and Toshev (1982).

Unfortunately, little information on the mechanical behavior of protein-covered thin films is available, and quantitative values for the overall tension of such films (i.e., protein bilayers with a polar core), to our knowledge, have yet to be published.

In the absence of reference values for the overall film tension, we estimate in section 4.2 the interfacial tension of the protein bilayer by best-fitting the experimental data of the elastic shear modulus *G* versus the radius of the cell following Equation 1.

Structure-Properties Relationship in Oil Gels

Structure

The three dimensional nature of a lipidic gel templated by a monodisperse emulsion of 80 μm droplet diameter is shown in Fig. 12.1. The internal structure is revealed by confocal optical microscopy, where the protein has been labeled with rhodamine. Morphologically, the final structure of the material resembles a dry foam, where the protein bilayer interfaces constitute the walls of the foam, and air has been replaced by the oil phase. In this configuration, the liquid, chemically unmodified oil is restricted within closed polyhedral cells with sizes equivalent to the droplet size of the original emulsion template, thus conferring to the lipidic gel a solid viscoelastic behavior.

An optical micrograph of a monodisperse emulsion template with droplet diameter 24 μm, together with the structure of the lipidic gel resulting from drying

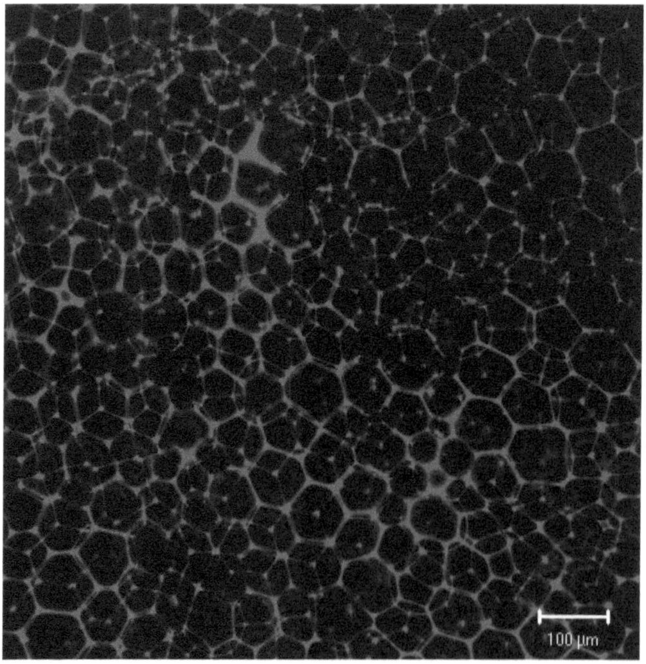

Fig. 12.1. Internal structure of a gel resulting from a monodisperse emulsion template with droplet diameter of 80 μm, as revealed by confocal microscopy. To image the protein phase, rhodamine, in a concentration of 10^{-10} M is added to the pH=7.0 buffered water phase used in the final washing step. (Reproduced from Romoscanu & Mezzenga, 2006).

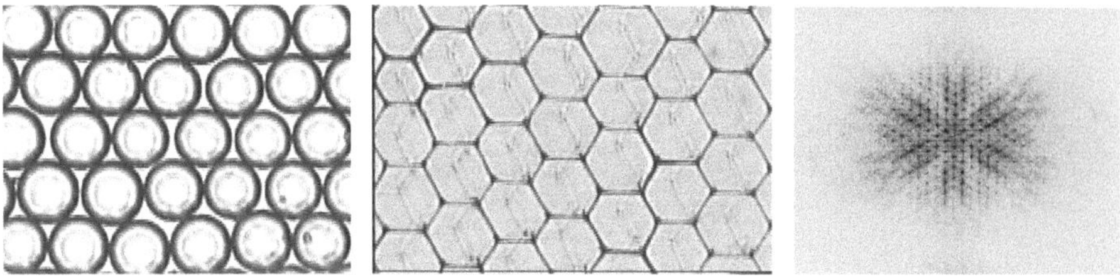

(a) (b) (c)

Fig.12.2. (a) Emulsion template (droplet radius 24 μm) (b) Twofold-layered thin film of polyhedron gel obtained upon water evaporation on a glass substrate (c) Diffraction pattern simulated by fast Fourier Transformation of Figure 12.2.B. (Reproduced from Romoscanu & Mezzenga, 2006).

two single layers of the emulsion template on a glass substrate, are shown in Fig. 12.2.A and 12.2.B, respectively. Clearly, twofold polyhedron layers can be created by this process, which can then be also viewed as a technique to hydrophobically modify substrates or lubricate interfaces. Single-layered films were also attempted, without success. This is probably due to the minimization of total interfacial area, which for a bilayer of polyhedrons is more efficient than for a single layer. The light diffraction pattern of the gel, simulated by a Fast Fourier Transform (FFT) of the two-dimensional gel micrograph is also displayed in Fig. 12.2.C.

The water content of the lipidic gel samples determined by Karl-Fischer lies below 0.25%. This value increases with decreasing droplet size and is somewhat lower (max 0.18%) for paraffin oil than for olive oil samples. Considering the low solubility of water in oils (typically in the 30–80 ppm range), a simple calculation shows that the thickness of the water layer is of $O(10^{-1}$ μm) or less, which is consistent with the transparent nature of the lipidic gels.

Bulk Rheology

Figs. 12.3 and 12.4 show the storage shear modulus G' at low frequency (f=1 Hz) as a function of the surface-averaged mean cell radius R_{32} for four different gels realized with different oils or cross-linked by different processes (paraffin and olive oils; thermal and glutaraldehyde cross-linking). Scaling exponents n as well as interfacial tensions σ obtained by fitting the experimental data with Equation 1 are shown in Table 12.A. The value of σ is obtained by fitting the shear modulus vs. cell averaged mean cell radius R_{32} with a scaling exponent n=1, to ensure unit consistency. The agreement between the least square root-determined exponent and the theoretical value of –1 is excellent in all cases. We therefore conclude that at small deformations the rheological behavior of the lipidic gels is ruled by the same laws that are valid for dry foams; elasticity is provided by the increase of the specific interfacial area together with work done against the film disjoining pressure Π. Because of the dependence of the gel elasticity on both the interfacial tension of the bilayer of proteins, as well as on the average size of the polyhedral cell, the mechanical properties of the lipidic gels can be tuned by either the cross-linking process or the average diameter size of the emulsion template.

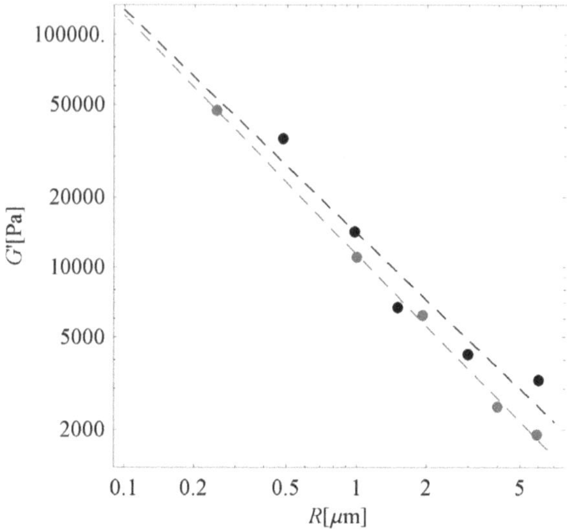

Fig. 12.3. Shear elastic modulus *G'* (*f*=1 Hz) as a function of the surface-averaged mean cell radius R_{32} for gels based on thermally cross-linked β-Lg. The dashed lines represent least-squares-fitting of experimental data with Equation 1. Black dots: paraffin oil, *n*=0.96. Gray dots: olive oil, *n*=1.02. (Reproduced from Romoscanu & Mezzenga, 2006).

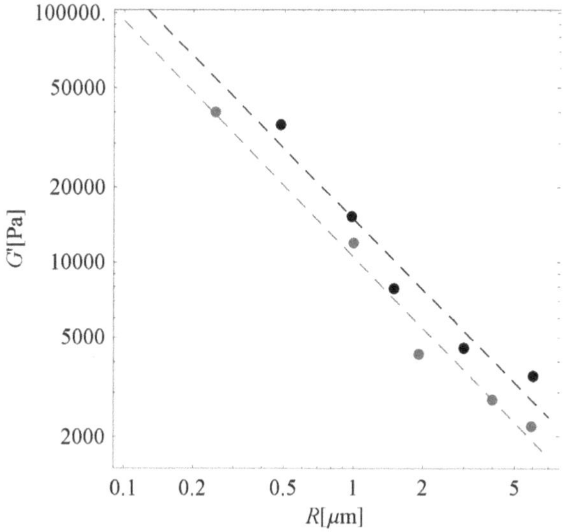

Fig. 12.4. Shear elastic modulus *G'* (*f*=1 Hz) as a function of the surface-averaged mean cell radius R_{32} for gels based on glutaraldehyde cross-linked β-Lg. The dashed lines represent least-squares-fitting of experimental data with Equation 1. Black dots: paraffin oil, *n*=0.94. Gray dots: olive oil, *n*=0.95. (Reproduced from Romoscanu & Mezzenga, 2006).

Table 12.A. Experimentally Determined Scaling Exponents and Interfacial Tensions.

Substrate	Cross-linking Method	n[–]	σ[mN/m]
olive oil	thermal	1.02	23
olive oil	glutaraldehyde	0.95	21
paraffin oil	thermal	0.96	32
paraffin oil	glutaraldehyde	0.94	33

The real and imaginary parts of the complex shear modulus G^* as well as the loss angle δ=ArcTan(G″/G′) for two paraffin oil-based gel samples with 6 μm original droplet radius in the 10^{-1} to 10^2 Hz frequency range are displayed in Fig. 12.5.

At low frequencies, the observed behavior is typical for a dry foam, with G' being a weak function of the frequency and G'' lower than G' by one order of magnitude over a wide frequency band. As pointed out by Buzza et al. (1995), the weak but measurable frequency dependence of G' in dry foams can be explained by a finite interfacial elasticity as well as local variations in interfacial tension during shear. In

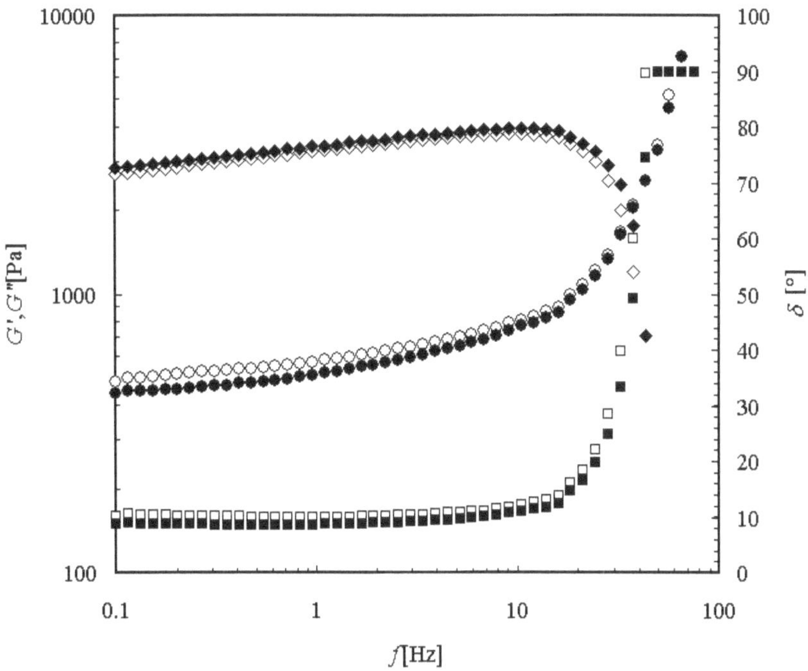

Fig. 12.5. Frequency dependence of complex elastic modulus for two thermally cross-linked paraffin oil gels with R_{32}=6.1 μm. ◆, ↓: G′; ˜, ™: G″; ′, ≤: loss angle δ. Measurement performed at 1% strain and 23°C. (Reproduced from Romoscanu & Mezzenga, 2006).

contrast to compressed (but non-dry) emulsions, which show an almost frequency-independent storage modulus (Mason et al., 1995), relaxation of interfacial tension during shear is hindered in dry systems, leading to a more pronounced frequency dependence. While the increase of G'' with increasing frequency in the higher frequency range can be explained based on the theoretical behavior of a simple Maxwell solid, the sudden decrease of G' in the higher frequency range is difficult to explain in the same context. In the present case, we attribute the observed decrease of G' at higher frequencies to inertia phenomena. From a rheometric point of view, the 180° phase shift between acceleration (resulting from geometry inertia) and elastic forces is expected to result in an artifact consisting of an apparent decrease of the sample elasticity.

Interfacial Rheology

Interfacial tensions listed in Table 12.A, obtained by fitting elasticity moduli data displayed in Fig. 12.3 and 12.4 with Equation 1 are higher (ca two-fold) than interfacial tensions that typically characterize β-Lg-stabilized oil-in-water systems (ca. 10 mN/m; Romoscanu & Mezzenga, 2005).

In an attempt to explain this discrepancy, dilatational interfacial rheological experiments were performed on the systems used here. In particular, the interfacial tension increase of ca. 10% implied by the interfacial area increase incurred by the droplets during their transition from a spherical to a polyhedral shape was quantified. Fig. 12.6 illustrates the increase of the interfacial tension as a function of interfacial area for a β-Lg layer adsorbed at the oil/water interface of an oil droplet suspended in a protein-free buffered aqueous matrix. Three different cases are considered: adsorbed and uncross-linked β-Lg, chemically (glutaraldehyde) cross-linked β-Lg, and thermally (80°C) cross-linked β-Lg.

In absence of a particular interfacial protein cross-linking treatment (no glutaraldehyde nor thermal cross-linking), the weak increase in interfacial tension reflects the reduction of protein interfacial concentration induced by the increase of interfacial area, as well as residual cross-linking via disulfide bonds following the denaturation of the protein upon adsorption (Dickinson & Matsumura, 1991). As shown in Fig. 12.6, the increase in interfacial tension with interfacial dilatational strain is more pronounced if the adsorbed protein layer is processed either thermally or chemically, as a consequence of extensive cross-links between individual protein molecules.

The interfacial tension was measured on the unstrained interface (i.e., on the area where adsorption is initially performed) at 23°C (σ_0), and Table 12.B shows the low strain interfacial elasticity (E_D) as well as the interfacial tension values at an interfacial strain of 10% (σ_A) for the various systems investigated. Generally, elasticity values for thermally-processed protein films obtained in the present work compare well with those reported by Rodríguez Platino et al. (2001). Interfacial elasticity values appear to depend on the nature of the oil phase. In particular, extensional elasticity values suggest that proteins adsorbed at the (non-refined) olive oil-water interface may undergo a cross-linking process even in the absence of external cross-linking step

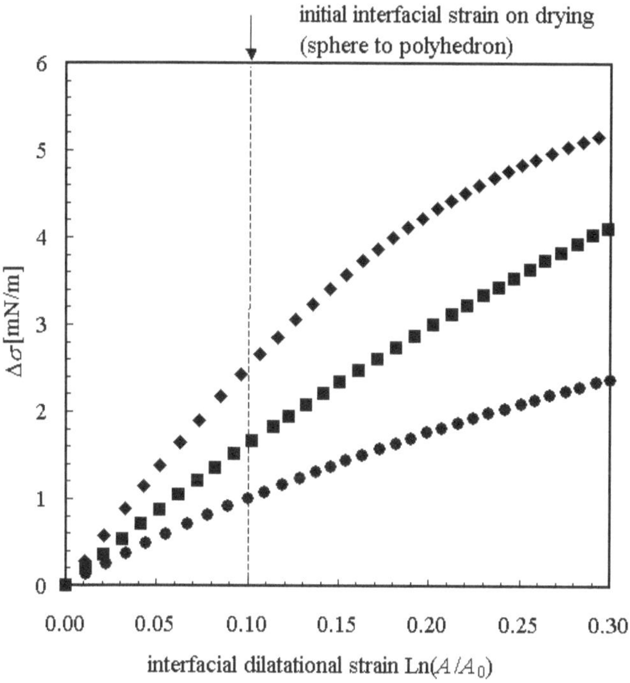

Fig. 12.6. Increase of the interfacial tension of a β-Lg covered paraffin oil droplet in a protein-free matrix. (●: no treatment after adsorption; ■: thermally cross-linked protein (80°C, 600 s); ◆: chemically (glutaraldehyde) cross-linked protein). Initial protein adsorption was performed from 0.1 wt % protein solution in 20 mM imidazole, pH =7.0 buffer. (Reproduced from Romoscanu & Mezzenga, 2006).

Table 12.B. Interfacial Tension of Unstrained β-Lg-Covered Oil/Water Interfaces (σ_0), Dilatational Elasticity Values (E_D), and Interfacial Tension of β-Lg-Covered Oil/Water Interfaces at 10% Dilatational Strain (σ_A) for Different Oil Substrates (Virgin Olive Oil, Paraffin Oil) and Interfacial Treatments (No Cross-linking Treatment, Thermal Cross-linking, and Glutaraldehyde Cross-linking). The Refined Olive Oil Values Are Provided for Comparison Purposes.

Substrate	Cross-linking Method	σ_0[mN/m]	E_D[mN/m]	σ_A[mN/m]
virgin olive oil	no treatment	7.5	23	9.6
virgin olive oil	thermal	7.2	26	9.6
virgin olive oil	glutaraldehyde	6.6	33	9.8
refined olive oil	no treatment	7.7	15	9.1
paraffin oil	no treatment	10.2	8	11.3
paraffin oil	thermal	9.8	21	11.2
paraffin oil	glutaraldehyde	7.1	27	9.6

(thermal or chemical). Indeed, elasticity values that are measured at olive oil-water interface in the absence of external cross-linking are typical of cross-linked interfaces on other oils. In contrast, proteins adsorbed at the refined olive oil-water interface do not show elasticity values typical of cross-linked interfaces. This supports the assumption that the aldehydes and ketones contained in virgin olive oil (but absent from refined oil) have a cross-linking effect on interfacially adsorbed proteins.

A comparison between the interfacial tension values measured on strained interfaces (A=1.1A_0) in single-drop experiments (Table 12.B) and interfacial tension values determined in situ via the rheological characterization of the bulk materials (Table 12.A) reveals a subsisting discrepancy of a factor of ca. 2.4 (for olive oil-based samples) to 2.8 (for paraffin oil samples). It is highly likely that the additional work performed against the disjoining pressure of neighboring droplets is at the origin of this discrepancy. As mentioned above, Buzza et al. (1995) estimate this additional work to result in an additional modulus of the same order of magnitude as the modulus resulting from internal interfacial area increase in the case of dry foams. The low water content (ca. 0.2%) of the systems investigated here supports this explanation, since precisely the double of the interfacial tension is determined from the bulk rheological experiments. The thickness of the interstitial layer of O(10^{-1} μm) determined from the water and glycerol contents and the specific interface also supports this assumption. As mentioned above, Narsimhan (1992) predicted an order of magnitude of 10^{-1} μm for the distance between two protein layers belonging to two different droplets/bubbles for disjoining pressure phenomena to become relevant.

Finally, we observed that, as can be expected by analysis of Equation 1, the cross-linking process constitutes an efficient way for tuning the interfacial tension, and thus the elastic modulus of the resulting material.

Re-hydration

To examine the reconstitution of the original emulsion from the lipidic gel, the re-hydration process was investigated for the various experimental conditions used to design the gels. Re-emulsification upon re-hydration is successful for any model oil, such as paraffin oil, highly refined olive oil, and medium chain tryglycerides (MCT). The droplet size distribution of re-emulsified gels is close to the original emulsion template, as shown in Fig. 12.7 for a 0.5 μm droplet radius paraffin oil-based gel. Gels generated from less pure oils, such as virgin olive oil-based gels, however, do not re-emulsify as successfully as standard model oil-based gels. Again, this is likely due to oils containing aldehydes and ketones, such as olive oils, triggering β-Lg cross-linking at the oil-water interface even in absence of external cross-linking agents (see Table 12.B). Thus, in contrast to model oil-based gels, inter-droplet cross-linking may occur during storage time of the gel in common oil-based gels, resulting in a clear swelling, but only for a partial re-emulsification of the gel. Nevertheless, the present technique generally allows design of oil-based protein gels, which exhibit tuneable elastic properties, and which can re-emulsified back to the original emulsion template upon re-hydrating the material.

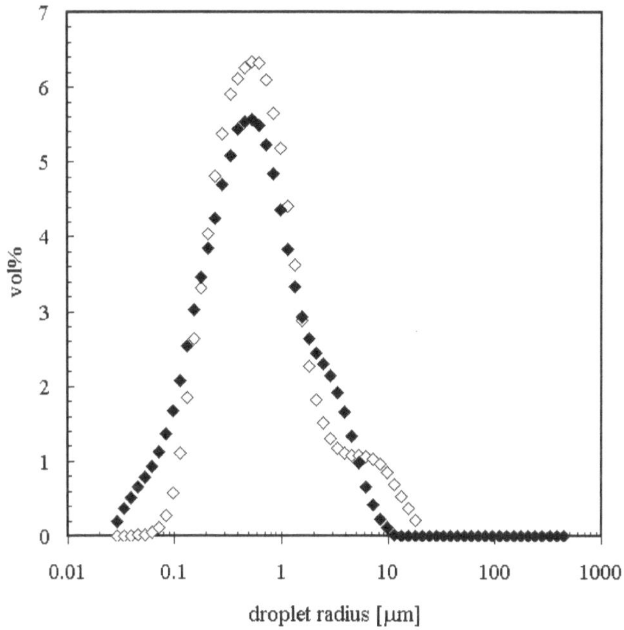

Fig. 12.7. Re-hydration of paraffin oil-based, thermally cross-linked gel with R_{32}=0.5 μm. ◇: Droplet radius distribution of emulsion template (after cross-linking); ◆: Droplet radius distribution of emulsion obtained after re-hydration of the dried gel with 20 mM imidazole, pH =7.0 buffer. (Reproduced from Romoscanu & Mezzenga, 2006).

Oil Powders

Experimental Section

When a liquid oil has to be converted into a powder rather than a gel, the choice of oil, protein, buffer solution, emulsification process, and cross-linking procedure are essentially identical to those for the production of a gel. The evaporation step, however, is different, since a fast water-evaporation process is followed by means of spray drying. Spray drying of the 10% cross-linked emulsion was carried out on a Büchi 190 mini spray dryer, using 125°C for the temperature inlet and 84°C for the temperature outlet, respectively.

The dry oil powder obtained as described above was analyzed by examining the visual powder aspect and by studying the structure of the powder by UV light microscopy and Brightfield light microscopy. A detailed experimental procedure and characterization of oil powder is given in Mezzenga & Ulrich, 2010.

Structural Properties

Fig. 12.8 and Table 12.C give the particle size distributions and main characteristics for the 10% oil-in-water emulsion prior to and after the cross-linking procedure. As it can be observed, the cross-linking treatment at 80°C has a little impact on both

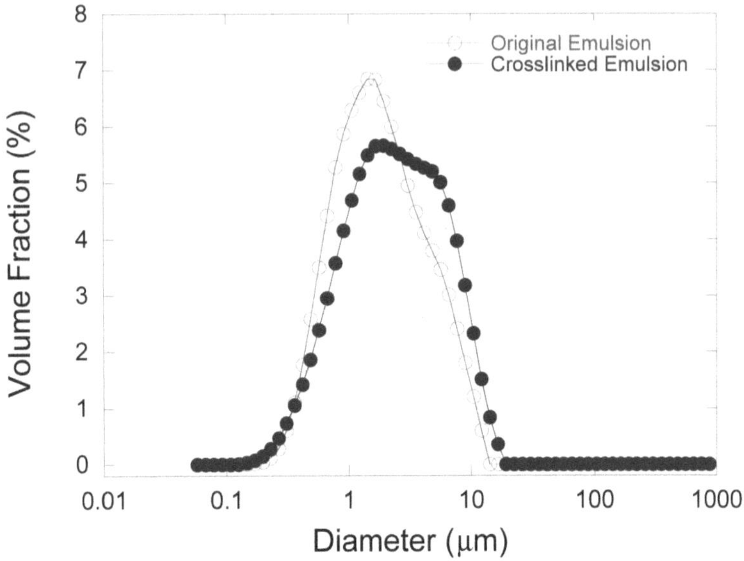

Fig. 12.8. Particle size distributions of the oil-in-water emulsion prior and after the β-lactoglobulin thermal cross-linking process at 80°C. (Reproduced from Mezzenga & Ulrich, 2010).

Table 12.C. Particle Size Distribution and Main Characteristics for the 10% Oil-in-water Emulsion Used for the Spray Drying Process.

10% oil-in-water original emulsion						
D[4,3]/um	D[3,2]/um	D(v,0.5)/um	D(v,0.1)/um	D(v,0.9)/um	Span	Residual/%
2.37	1.21	1.60	0.57	5.44	3.05	0.54
10% oil-in-water original emulsion after cross-linking						
D[4,3]/um	D[3,2]/um	D(v,0.5)/um	D(v,0.1)/um	D(v,0.9)/um	Span	Residual/%
3.17	1.40	2.19	0.63	7.19	3.00	0.33

the size distribution and on the average size diameter, leading to an increase of $\overline{D_{4,3}}$ from 2.37 to 3.17 μm. After the cross-linking procedure, the emulsion is found to be stable, maintaining the same particle size distribution over a period of several weeks to months.

The cross-linked emulsion can be efficiently converted into a solid oil powder upon spray drying as such, without the addition of hydrocolloids in the water phase,

Fig. 12.9. Visual aspect of the spray dried oil powder from the emulsion described in Fig. 12.8. (Reproduced from Mezzenga & Ulrich, 2010).

nor of any other hydrophilic solid dry basis. Fig. 12.9 shows the visual aspect of the resulting powder after spray drying. We further note that spray drying of emulsions with identical particle size distribution, which did not undergo the thermal cross-linking process of the protein shell, resulted in a liquid oil phase at the outlet of the spray drier. This confirms that cross-linking the protein shell is the main step needed for stabilizing the solid oil powder.

The powder appears as a light and dry granular media, and on touch, it does not present any stickiness. No evidence of moderate or partial oil leakage of any kind is found even after several months of storage. Only upon shearing among fingers is a lubricating oil film released, which reflects the nature of the oil-rich phase in the powder and the efficiency of the encapsulation.

To better assess the microstructure of the resulting powder, various optical microscopy procedures were performed. Figs. 12.10.A and 12.10.B show the aspect of the powder as observed by bright field microscopy and UV-light microscopy, respectively. The powder observed is the pristine one as obtained by the spray drying process.

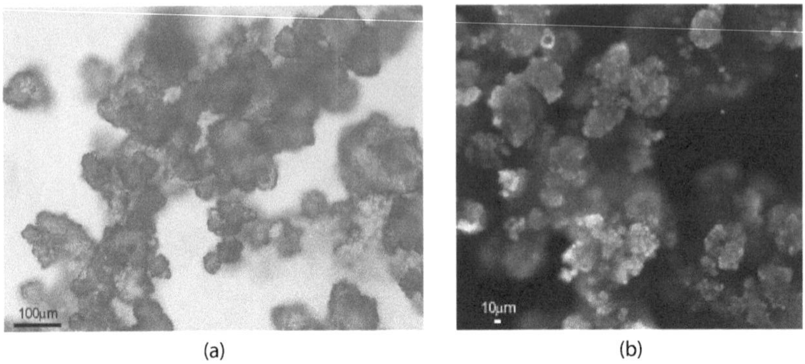

Fig. 12.10. (**A**) The oil-powder shown in Fig. 12.9, as observed by Bright field light microscopy. (**B**) The oil-powder shown in Fig. 12.9, as observed by UV Light Microscopy (excitation: BP450-490, emission LP 520, Nile Red staining, Mounting agent: glycerol 40%). (Reproduced from Mezzenga & Ulrich, 2010).

In Figs. 12.10.A and 12.10.B, the powder granules appear as irregular fractal-like aggregates, which infer that each granule is constituted by several oil droplets. This is also consistent with the size of the granules, which is of the order of 10^2 μm, and thus, 20 to 50 times larger than the individual emulsion oil droplets. Exact identification of the protein and the oil phases, however, is difficult in Fig. 12.10, owing to the depth of the images and their three-dimensional nature.

To more precisely identify the protein constituent in the spray dried powders, bright field microscopy was performed on thin sections of the powders, whose grains were first chemically fixed, then embedded in an epoxy resin as continuous phase, and finally cut in the form of 1 μm-thick sections by a microtome. The presence of the proteins is then highlighted by selectively staining them with toluidine blue. Fig. 12.11.A shows a representative micrograph, which indicates the blue-labeled proteins forming a homogeneous shell around each powder granule; these images confirm that the protein-rich protective shell alone is sufficient to stabilize the powder granules.

Furthermore, because the section thickness is one order of magnitude thinner than the powder granule size, the apparent volume fraction of the protein observed in the two-dimensional transmission projection of the micrograph directly reflects the three-dimensional, bulk volume fraction of the protein in the globule (Mezzenga et al., 2001; Plummer et al., 1995). The amount of visible, labeled protein in Fig. 12.11.A is consistent with the expected total weight fraction of solid dry basis of ca 8%. Again, the low amount of protein with respect to globules supports the findings that a low content of protein is sufficient to stabilize the powder, provided that the protein monolayer is cross-linked prior to spray drying so as to impede phase inversion (the minority phase separating within the majority phase).

Fig. 12.11.B, shows a higher magnification image of the granules. It can be observed that within the rather thick blue protein shell enveloping each powder

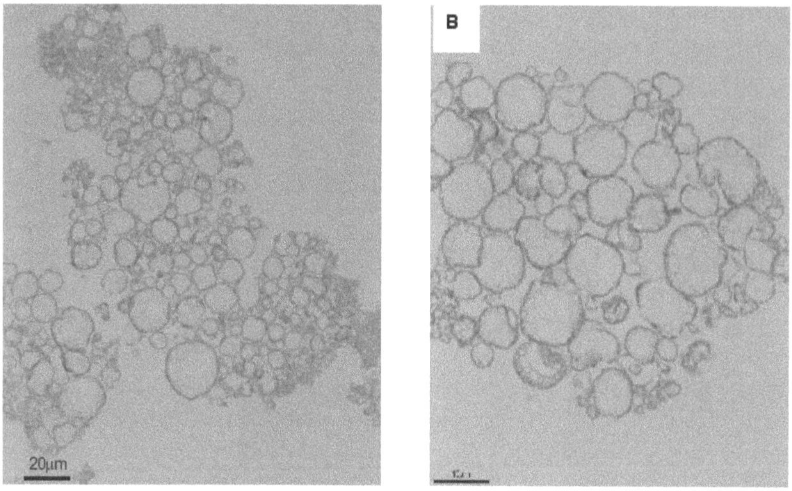

(a) (b)

Fig. 12.11. (**A**) The same powder granules of Fig. 12.2, as observed by Bright Field Light Microscopy (Protein Staining). Cryofixation by anhydrous Glutaraldehyde/ Osmium tetraoxide in methanol. Thin sections 1 micron. Staining of protein by Toluidine blue. (**B**) Larger magnification image showing the internal structure of the powder granules. (Reproduced from Mezzenga & Ulrich, 2010).

granule, the internal structure of neighboring pristine oil droplets emerges, which appear to be compressed together and separated by a thin layer of proteins. Preservation of individual oil droplets is indeed expected due to the cross-linked viscoelastic protein monolayers stabilizing their surfaces and impeding coalescence of individual oil droplets. Furthermore, the compression of the individual droplets to polyhedral-like cellular structures is also an expected feature in colloidal dispersions in which plastic rearrangement is allowed, as by this mechanism, particles can minimize the high interfacial energy associated with their interfaces/surfaces (Romoscanu et al., 2005; Mezzenga et al., 2003a; 2003b; 2003c; Keddie, 1997). More informative is the difference in thicknesses between the protein layers protecting individual oil droplets within a powder granule and the protein shell of the granule itself. This difference suggests that during the spray drying process, individual oil droplets with a cross-linked protein monolayer on their surfaces aggregate first to form granules, and that granules are finally further stabilized by the excess protein remaining in solution, which concentrate at their surfaces during the final stages of water evaporation in the spray drying process. As a consequence of this agglomeration process, only partial re-dispersibility of the powder in water was observed. It needs to be further stressed that powders generated from an emulsion stabilized by either a buffer solution or with a pH adjusted by HCl or NaOH produce indistinguishable results, and thus the total nominal amount of solid dry basis in the final oil powder can be maintained below

10% by weight. It has to be envisaged that this process can be further improved so that the final content of protein in the powder can be decreased further. For example, one could easily dialyze the unabsorbed protein molecules from the cross-linked liquid emulsions, thus further improving the oil content in the final powder.

Outlook and Conclusions

Protein-stabilized oil-in water emulsions can be converted into protein-in-oil gels or powders by cross-linking the interfacial protein layer and subsequently removing the water. Depending on the specific process followed to remove water, either compact gels or oil powders are obtained. In the case of the gel, the final morphology can be viewed as a oil-in-protein high internal phase emulsion (HIPE), where the oil phase can be as high as 99.9%. Yet, the resulting gel exhibits elastic properties similar to those of a rubbery material, and these can be tuned by either controlling the average diameter size of the emulsion template or the cross-linking process. Following the HIPE/dry foam analogy, rheological properties of the gels can be studied using models previously developed for heterogeneous phases where energy is stored essentially in percolating interfaces and bilayer disjoining pressure, such as in the case of dry foams. The present procedure changes the physical properties of the emulsion while preserving the chemical nature of the oil phase. In particular, the saturation level of glyceride molecules remains unchanged, in contrast to hydrogenation. Therefore this procedure appears to have a high potential for purposes of encapsulation of hydrophobic components. As far as the protein film is concerned, two alternative processes were presented in the present work as viable routes to cross-link the protein film, that is, chemical and thermal cross-linking, although other routes, such as enzymatic cross-linking, may also be pursued. This demonstrates that the protein film percolating through the oil phase can be altered to an acceptable extent for pharmaceutical and food applications.

Finally, under well-defined conditions, the gels obtained as described in the present work can be re-hydrated leading back to emulsions that are practically undistinguishable from the original emulsion template used to design the gel. The reversibility of the process thus allows, in principle, designing elastic solid oil phases without jeopardizing the processability of the original emulsion, which makes them attractive for many possible applications in the fields of encapsulation, formulation, aromas, foods, and pharmaceutics.

By repeating the same steps used to design the gels, but performing a fast water evaporation, using for example spray-drying, oil powders without the usual addition of any hydrophilic molecules, can be obtained. This offers all the typical advantages that powders have over bulk products, while maintaining an affordable and cheap processing pathway. Due to the simplicity of the approach, the spray-drying of oil emulsion into oil-powder with ultra high oil content can open up new possibilities for pharmaceutical, cosmetic, and food technologies.

References

Adhikari, B.; Howes, T.; Wood, B.J.; Bhandari, B.R. Effect of addition of proteins on the production of amorphous sucrose powder through spray drying. J. Food Eng. 2009, 94, 135.

Baranauskiene, R.; Bylaite, E.; Zukauskaite, J.; Venskutonis, R.P. Flavor retention of peppermint Mentha piperita L. essential oil spray-dried in modified starches during encapsulation and storage. J. Agric. Food Chem. 2007, 55, 3027.

Bhandari, B.R.; Senoussi, A.; Dumouilin, E.D.; Lebert, A. Spray-drying of concentrated fruit juices. Drying Technol. 1993, 11, 1081.

Bruckner, M.; Bade, M.; Kunz, B. Investigations into the stabilization of a volatile aroma compound using a combined emulsification and spray drying process. Eur. Food Res. Techn. 2007, 226, 137.

Buzza, D.M.A.; Lu, C.-Y. D.; Cates, M.E. Linear shear rheology of incompressible foams. J. Phys. II 1995, 5, 37–52.

Cho, Y.H.; Shim, H.K.; Park, J. Encapsulation of fish oil by an enzymatic gelation process using transglutaminase cross-linked proteins. J. Food Sci. 2003, 68, 2717.

Derjaguin, B. The elastical characteristics of foam. Kolloid Zeitschrift 1933, 64, 1–6.

Dickinson, E. Adsorbed protein layers at fluid interfaces: interactions, structure and surface rheology. Colloids Surf. B: Biointerfaces 1999, 15, 161–176.

Dimitrova, T.D.; Leal-Calderon, F.; Gurkov, T.D.; Campbell, B. Disjoining pressure vs thickness isotherms of thin emulsion films stabilized by proteins. Langmuir 2001, 17, 8069–8077.

Dimitrova, T.D.; Leal-Calderon, F.; Gurkov, T.D.; Campbell, B. Surface forces in model oil-in-water emulsions stabilized by proteins. Adv. Colloid Interface Sci. 2004, 108–109, 73–86.

Dimitrova, T.D.; Leal-Calderon, F. Rheological properties of highly concentrated protein-stabilized emulsions. F. Adv. Colloid Interface Sci. 2004, 108–109, 49–61.

Exerowa, D.; Kashchiev, D.; Platikanov, D. Stability and permeability of amphiphile bilayers. Adv. Colloid Interface Sci. 1992, 40, 201–256.

Exerowa, D.; Todorov, R.; Nikolov, L. Amphiphile bilayer films from DPPC: bilayer lipid membranes and Newton black films. Colloids Surf. A: Physicochem. Eng. Aspects 2004, 250, 195–201.

Faldt, P.; Bergenstahl, B. Spray-dried whey protein/lactose/soybean oil emulsions. 1. Surface composition and particle structure. Food Hydrocolloids, 1996, 10, 421.

Fuchs, M.; Turchiuli, C.; Bohin, M.; Cuvelier, M.E.; Ordonnaud, C.; Peyrat-Maillard, M.N.; Dumoulin, E. Encapsulation of oil in powder using spray drying and fluidised bed agglomeration. J. Food Eng. 2006, 75, 27.

Gerrard J.A. Protein-protein crosslinking in food: methods, consequences, applications. Trends Food Sci. Technol. 2002, 13, 391–399.

Gerrard, J.A.; Brown, P.K.; Fayle, S.E. Maillard crosslinking of food proteins I: the reaction of glutaraldehyde, formaldehyde and glyceraldehyde with ribonuclease. Food Chem. 2002, 79, 343–349.

Gerrard, J.A.; Brown, P.K.; Fayle, S.E. Maillard crosslinking of food proteins II: the reactions of glutaraldehyde, formaldehyde and glyceraldehyde with wheat proteins in vitro and in situ. Food Chem. 2003, 80, 35–43.

Gu, Y.S.; Decker, E.A.; McClements, D.J. Influence of pH and iota-carrageenan concentration on physicochemical properties and stability of beta-lactoglobulin-stabilized oil-in-water emulsions. J. Agric. Food Chem. 2004, 52, 3626.

Ivanova, R.; Balinov, B.; Sedev, R.; Exerowa, D. Formation of a stable, highly concentrated O/W emulsion modeled by means of foam films. Colloids Surf. A: Physicochemical and Engineering Aspects 1999, 149, 23–28.

Jayasundera, M.; Adhikari, B.; Aldred, P.; Ghandi, A. Surface modification of spray dried food and emulsion powders with surface-active proteins: A review. J. Food Eng. 2009, 93, 266.

Jost, R.; Dannenberg, F.; Rosset. J. Heat-set gels based on oil-water emulsions: an application of whey protein functionality. Food Microstructure 1989, 8, 23.

Keddie, J.L. Film formation of latex. Mater. Sci. Eng. Rep. 1997, 21, 101.

Kellerby, S.S.; Gu, Y.S.; McClements, D.J.; Decker, E.A. Lipid oxidation in a menhaden oil-in-water emulsion stabilized by sodium caseinate cross-linked with transglutaminase. J. Agric. Food Chem. 2006, 54, 10222.

Kelvin, L.; Thompson, W. On the division of space with minimal partitional area. Philos. Mag. 1887, 24, 503–514.

Klinkesorn, U.; Sophanodora, P.; Chinachoti, P.; McClements, D.J.; Decker, E.A. Stability of spray-dried tuna oil emulsions encapsulated with two-layered interfacial membranes. J. Agric. Food Chem. 2005, 53, 8365.

Kraynik, A.M.; Reinelt, D.A. Linear elastic behavior of dry soap foams. J. Colloid Inter. Sci. 1996, 181, 511–520.

Marquié, C.; Tessier, A.M.; Aymard. C.; Guilbert, S. How to monitor the protein cross-linking by formaldehyde, glutaraldehyde or glyoxal in cotton-seed protein-based films. Short communication, Nahrung / Food 1998, 42, 264–265.

Mason ,T.G.; Bibette, J.; Weitz, D.A. Elasticity of compressed emulsions. Phys. Rev. Lett. 1995, 75, 2051–2054.

Mezzenga, R.; Plummer, C.J.G.; Boogh, L.; Manson, J.A.E. Morphology build-up in dendritic hyperbranched polymer modified epoxy resins: modelling and characterization. Polymer 2001, 42, 305.

Mezzenga, R.; Ruokolainen, J.; Fredrickson, G. H.; Kramer, E.J. High internal phase polymeric emulsions by self-assembly of colloidal systems. Macromolecules, 2003c, 36, 4466–4471.

Mezzenga, R.; Ruokolainen, J.; Fredrickson, G.H.; Kramer, E.J.; Moses, D.; Heeger, A.J.; Ikkala, O. Templating organic semiconductors via self-assembly of polymer colloids. Science 2003b, 299, 1872–1874.

Mezzenga, R.; Ruokolainen, J.; Hexemer, A. On the role of block copolymers in self-assembly of dense colloidal polymeric systems. Langmuir 2003a, 19, 8144–8147.

Mezzenga, R.; Ulrich, S. Spray-dried oil powder with ultrahigh oil content. Langmuir 2010, 26, 16658.

Monsan, P.; Puzo, G.; Mazarguil, H. Mechanism of formation of glutaraldehyde-protein bonds. Biochimie 1975, 57, 1281–1292.

Moreau, L.; Kim, H.J.; Decker, E.A.; McClements, D.J. Production and characterization of oil-in-water emulsions containing droplets stabilized by beta-lactoglobulin-pectin membranes. J. Agric. Food Chem. 2003, 51, 6612.

Mun, S.; Decker, E.A.; McClements, D.J. Influence of droplet characteristics on the formation of oil-in-water emulsions stabilized by surfactant-chitosan layers. Langmuir 2005, 21, 6228.

Narsimhan, G. Maximum disjoining pressure in protein stabilized concentrated oil in water emulsions. Colloids Surf. 1992, 62, 31–39.

Plummer, C.J.G.; Hilborn J.G.; Hedrick, J.L. Transmission electron microscopy methods for the determination of void content in polyimide thin-film nanofoams. Polymer 1995, 36, 2485.

Princen, H.M. Rheology of foams and highly concentrated emulsions 1. Elastic properties and yield stress of a cylindrical model system. J. Colloid Interface Sci. 1983, 91, 160–175.

Princen, H.M.; Kiss, A.D. Rheology of foams and highly concentrated emulsions 3. Static shear modulus. J. Colloid Interface Sci. 1986, 112, 427–432.

Reinelt, D.A.; Kraynik, A.M. Large elastic deformations of 3-dimensional foams and highly concentrated emulsions. J. Colloid Interface Sci. 1993, 159, 460–470.

Rodríguez Patino, J.M.; Rodríguez Niño, M.R.; Carrera Sánchez, C.; Navarro García, J.M.; Rodríguez Mateo, M.; Cejudo Fernández, M. Dynamic properties of beta-casein-monoglyceride mixed films at the air-water interface. Long-term relaxation phenomena. Colloids Surf. B: Biointerfaces 2001 21, 87–99.

Romoscanu, A.; Mezzenga, R. Cross linking and rheological characterization of adsorbed protein layers at the oil-water interface. Langmuir 2005, 21, 9689–9697.

Romoscanu, A.I.; Mezzenga, R. Emulsion-templated fully reversible protein-in-oil gels. Langmuir 2006, 22, 7812.

Shaw, L.A.; McClements, D.J.; Decker, E.A. Spray-dried multilayered emulsions as a delivery method for omega-3 fatty acids into food systems. J. Agric. Food Chem. 2007, 55, 3112.

Soos, J.M.; Koczo, K.; Erdos, W; Wasan, D.T. An automatic apparatus for measuring interfacial and film tension under static and dynamic conditions. Rev. Sci. Instr. 1994, 65, 3555–3562.

Stamenovic, D.; Wilson, T.A. The shear modulus of liquid foam. J. Appl. Mech. 1984, 51, 229–231.

Svitova, T.F.; Wetherbee, M.J.; Radke C.J. Dynamics of surfactant sorption at the air/water interface: continuous-flow tensiometry. J. Coll. Interf. Sci. 2003, 261, 170–179.

Turchiuli, C.; Fuchs, M.; Bohin, M.; Cuvelier, M.E.; Ordonnaud, C.; Peyrat-Maillard, M.N.; Dumoulin, E. Oil encapsulation by spray drying and fluidised bed agglomeration. Innovat. Food Sci. Emerg. Technol. 2005, 6, 29–35.

Umbanhowar, P.B.; Prasad, V.; Weitz, D.A. Monodisperse emulsion generation via drop break off in a coflowing stream. Langmuir 2000, 16, 347–351.

Vega, C.; Dalgleish, D.G.; Goff, H.D. Effect of kappa-carrageenan addition to dairy emulsions containing sodium caseinate and locust bean gum. Food Hydrocolloids 2005, 19, 187.

Weaire, D.; Hutzler, S. Physics of Foams. Oxford University Press: New York, 1999.

Weaire, D. Phelan, A counterexample to Kelvin conjecture on minimal-surfaces. R. Philos. Mag. Lett. 1994, 69, 107–110.

Weiss, J.; Scherze, I.; Muschiolik, G. Polysaccharide gel with multiple emulsion. Food Hydrocolloids 2005, 19, 605–615.

Xu, W.; Nikolov, A.; Wasan, D.T.; Gonsalves, A.; Borwankar, R.P. Foam film rheology and thickness stability of foam-based food products. Colloids Surf., A, 2003, 214, 13–21.

13

Ethylcellulose Oleogels

Tania Dey[1], Dennis A. Kim[2], and Alejandro G. Marangoni[1]
[1]University of Guelph, Department of Food Science, Guelph, Ontario, N1G 2W1, Canada, [2] Kraft Foods R&D Center, Glenview, IL

Structured oils offer an alternative to improving the fatty acid profile of foods by decreasing saturated and eliminating *trans* fatty acids, while increasing mono and polyunsaturated fatty acid content. This will improve the nutritional properties of the foods we consume. This can be achieved by gelling oils into "oleogels" or "organogels". The building blocks of these organogels can vary widely, mainly falling under three categories:

(1) Crystalline particles:

Gelling can be obtained by means of network formation of colloidal crystalline triacylglycerol (TAG) particles, trapping the liquid TAG phase within. The size and shape of the crystals and the interactions between them determine the mechanical properties of the network (Bot et al., 2007; Marangoni, 2005). It is also possible to structure edible oils with diacylglycerols (DAG), monoacylglycerols (MAG), and fatty acids (Pernetti et al., 2007a; Ojijo et al., 2004; Wright & Marangoni, 2006). Structuring can also be achieved with wax esters/sorbitan monostearate, ceramides (Daniel & Rajasekharan, 2003; Rogers, 2009), and lecithin/sorbitan tristearate (Pernetti et al., 2007b).

(2) Crystalline Fibers:

Self-assembled fibrillar network (SAFIN) of phytosterols + oryzanol (Bot & Agterof, 2006; Bot et al., 2008), 12-hydroxystearic acid and ricinoleidic acid have been reported in literature (Rogers et al., 2008; Rogers et al., 2009), where the structuring agent forms helical and twisted crystalline ribbons, hundreds of micrometers long.

(3) Polymeric Strands:

Polymers can be added to promote gelation of a medium. These gels can be of two types: gels formed by covalent bonding (chemical gel) or gels formed by self-assembly, e.g., van der Waal's forces or hydrogen bonding (physical gel). Formulation of fatty acid-based co-monomers (Dey, 2007) and vegetable oil-based polymers (Lu

& Larock, 2009; Yilmaz & Kusefoglu, 2005; Sharma & Kundu, 2006) have already been reported in literature due to their importance as renewable resources, but these systems lack the potential for food applications. An abundance of literature can be found on gelation of biopolymers such as cellulose, starch, gelatin, etc. in aqueous solutions as well as in dispersions (Ruan et al., 2008; Ostrovskii et al., 1999; Frohoff-Hülsmann et al., 1999; Brouillet-Fourmann et al., 2003; Lee et al., 2008; Gilsenan & Ross-Murphy, 2000; Borchard & Burg, 1990), but similar studies in edible oil medium are scarce (Aiache et al., 1992; Ruiz Martinez et al., 2003).

Aiache et al. reported oleogel preparation from ethylcellulose (EC) in an ethoxylated glyceride medium, indicating that the color, smell, and transparency of EC remains intact upon gelation, along with some structural characterization by X-ray diffractometry (Aiache et al., 1992). Although the report by Martinez et al. where they described oleogel formation from EC in olive oil in presence of "olivem" branded surfactants and elucidated the effect of polymer concentration and temperature on rheology and creep-recovery (Ruiz Martinez et al., 2003) is a little more elaborate, the study requires more physical insight into properties and interactions before oleogels can move toward general application.

Work in our group has touched upon all these areas of oil structuring, however, lately we have been systematically exploring the third approach, i.e., oil structuring via polymer gelation. This is an emerging strategy and holds promise in the area of food and nutrition. We aim to identify the most promising polymer to structure edible oils; to understand the structural, compositional, and interaction parameters associated with the gelation process; and to characterize a novel class of soft materials with enhanced oil binding capacity and thermal stability.

The non-aqueous system reported here comprises three food-grade components—oil, polymer, and surfactant—that each have a specific role. While polymers form the backbone of networks in gels, surfactants act as plasticizers (Ghebremeskel et al., 2007) and oils behave like solvents. This gelation process is shown in Fig. 13.1.

Although surfactants have been reported to form oleogels in the absence of polymer (Almeida & Bahia, 2006), this kind of gel suffers from the problem of syneresis even upon slight temperature change. Moreover, the relatively high concentration of surfactant required to gel the oil almost precludes the consideration of these types of materials as "gels", where the concentration of structurant is usually low. In this study, we found that the best polymer system for oil gelation was ethylcellulose, inferred from our wide range of scope-trial experiments, in accordance with previous literature reports. Cellulosic gels are known to contain ether-linked polymer strands laterally cross-linked by hydrogen bonding (Physical Gel). Structure-wise, the surfactants were systematically varied by choosing different chain-lengths, head groups, unsaturation, sidechains, and hydrophilic lipophilic balance (HLB) (Fig. 13.2). The edible oils used in our study had various types of fatty acid content and were screened to determine any impact from the intrinsic composition (Table 13.A).

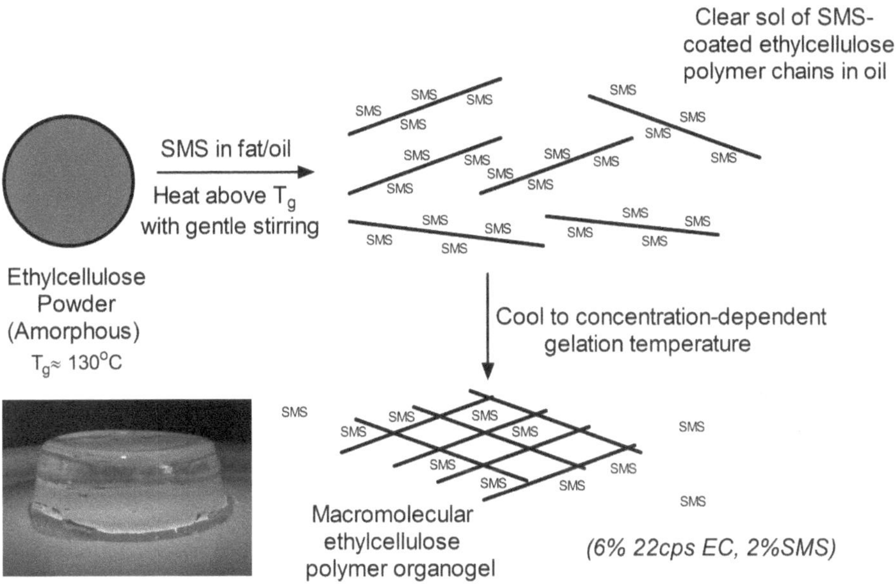

Fig. 13.1. Gelation scheme for ethylcellulose gelation in edible oil, plasticized by a surfactant.

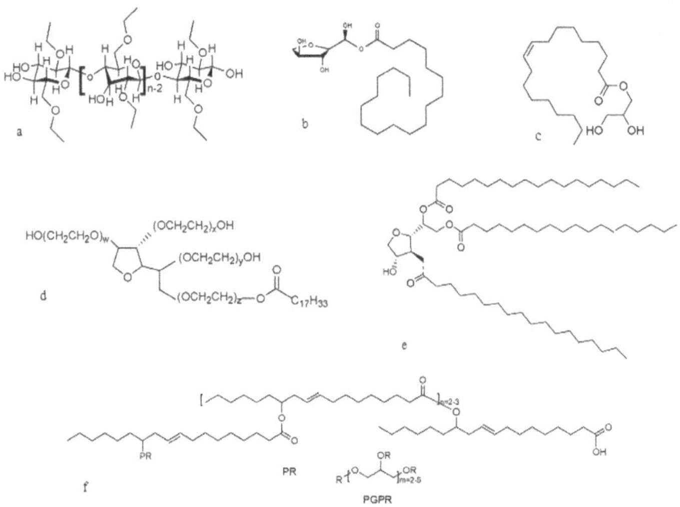

Fig. 13.2. Chemical Structure of (a) Ethyl Cellulose (b) Sorbitan monostearate (SMS)/ SPAN 60, HLB 4.7 (c) Gycerol Monooleate, HLB 3.8 (d) Tw 80/Polyoxyethylene(20) sorbitan monooleate, HLB = 15.0 (x+y+z+w = 20) (e) Sorbitan Tristearate/SPAN 65, HLB = 2.1 (f) Polyglycerol Polyricinoleate (PGPR) (in PGPR, at least one of the groups marked R is PR, while the rest are either hydrogen, fatty acid residue or PR), HLB = 0.4.

Table 13.A. Fatty Acid Composition of the Edible Oils Used in this Study.

		Saturated					*Monounsaturated*	*Polyunsaturated*	
Oils	Unsat./Sat. ratio	Capric Acid C10:0	Lauric Acid C12:0	Myristic Acid C14:0	Palmitic Acid C16:0	Stearic Acid C18:0	Oleic Acid C18:1	Linoleic Acid (ω6) C18:2	Alpha Linolenic Acid (ω3) C18:3
Canola Oil	15.7	—	—	—	4	2	62	22	10
Corn Oil (Maize)	6.7	—	—	—	11	2	28	58	1
Fish Oil (Cod Liver)	2.9	—	—	8	17	—	22	5	—
Flaxseed Oil	9.0	—	—	—	3	7	21	16	53
Peanut Oil	4.0	—	—	—	11	2	63	32	—
Soybean Oil	5.7	—	—	—	11	4	24	54	7

The fish oil under study was Menhaden oil with similar composition, but with a high content of Eicosapentaenoic Acid EPA, C20:5 (15.01%) and Docosahexaenoic Acid DHA, C22:6(10.07%) as mentioned in the product bulletin. Source: Scientific Psychic, 2010.

It is rather surprising to think a high molecular weight polymer would form a macromolecular gel, similar in nature to a hydrogel, except for the type of solvent present—an organic liquid rather than water. Fig. 13.1 illustrates the gelation process in the form of a schematic diagram—a sequential process of mixing, heating, and cooling of the components. Interestingly, the clearing temperature was always at or above the Glass Transition Temperature of the pure polymer. The clearing temperature was in the range of 125°C to 155°C, and was observed to be composition-dependent, i.e., the type of oil, surfactant, and polymer influenced this clearing temperature.

Fig. 13.3 and Table 13.B report the viscoelastic measurements performed on the different gel samples. For simplicity, only the storage modulus, G′, stress sweeps

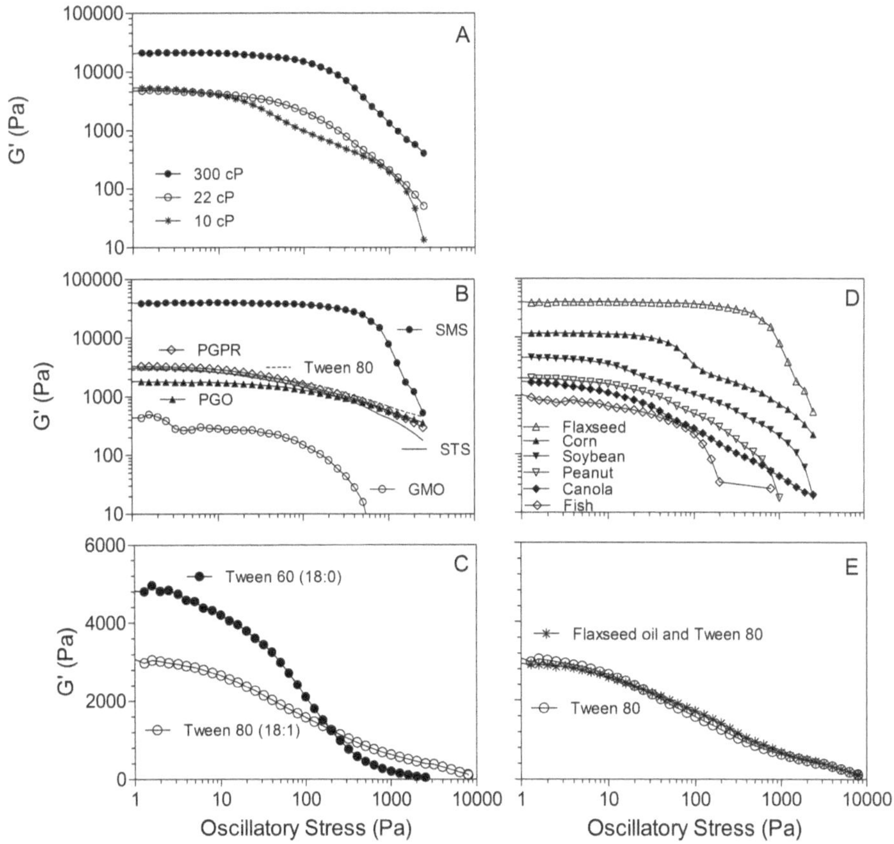

Fig. 13.3. Viscoelastic properties (G′ values shown) of gels comprising (a) Ethyl Cellulose of varying viscosities + 95% Tw 60 (effect of polymer MW (b) 5% Ethyl Cellulose 22cp + 5 % Surfactant + 90% Flaxseed Oil (effect of surfactant variation) (c) 5% Ethyl Cellulose 22cp + 95 % Surfactant (effect of unsaturation in surfactant) (d) 5% Ethyl Cellulose 22cp + 95% Tw80 vs. 5% Ethyl Cellulose 22cp + 5% Tw80 + 90% Flaxseed Oil (e) 5% Ethyl Cellulose 22cp + 5 % Sorbitan MonoStearate + 90% Edible Oil (effect of oil variation).

Table 13.B. Viscoelastic Parameters of Oleogel Samples of Different Compositions.

Gel Sample		G′ at LVE region (Pa)	G″ at LVE region (Pa)	tan δ = G″/ G′ (calculated)	yield stress (Pa)
Polymer variation	5% ETHOCEL 300 + 95% Tw60	21095 ± 4829.54	13990 ± 2927.42	0.6632	112.94 ± 18.34
	5% EC 22cp + 95% Tw60	4321.8 ± 4856.69	3862.65 ± 4231.82	0.8938	15.07 ± 14.28
	5% EC 10cp + 95% Tw60	4239 ± 3635.94	2948 ± 2266.98	0.6954	7.56 ± 7.17
Surfactant variation	5% EC 22cp + 5% Tw80 + 90% Flaxseed oil	2829.5 ± 386.79	461.3 ± 101.12	0.1630	142.2 ± 23.05
	5% EC 22cp + 5% GMO + 90% Flaxseed oil	264.7 ± 48.41	80.71 ± 51.04	0.3049	44.06 ± 8.58
	5% EC 22cp + 5% PGE + 90% Flaxseed oil	1765 ± 626.50	389.25 ± 122.54	0.2205	39.79 ± 0
	5% EC 22cp + 5% PGPR + 90% Flaxseed oil	3180.5 ± 1300.37	672.05 ± 274.85	0.2113	112.94 ± 18.34
	5% EC 22cp + 5% STS + 90% Flaxseed oil	2931.5 ± 437.70	663.5 ± 92.21	0.2263	50.13 ± 0
Oil variation	5% EC 22cp + 5% SMS + 90% Flaxseed oil	39570 ± 11257.14	3440.5 ± 2281.83	0.0869	56.6 ± 9.15
	5% EC 22cp + 5% SMS + 90% Peanut oil	1827.5 ± 515.48	528.2 ± 230.94	0.2890	15.82 ± 0
	5% EC 22cp + 5% SMS + 90% Corn oil	11560 ± 198.00	1812.5 ± 2.12	0.1568	50.13 ± 0
	5% EC 22cp + 5% SMS + 90% Soybean oil	4412.5 ± 2269.11	815.0 ± 193.75	0.1847	5.62 ± 0.91
	5% EC 22cp + 5% SMS + 90% Fish oil	749.4 ± 528.35	282.35 ± 223.09	0.3768	7.11 ± 1.20
	5% EC 22cp + 5% SMS + 90% Canola oil	1588.5 ± 562.15	454.4 ± 100.55	0.2861	7.11 ± 1.20

Values represent means and standard deviations.

are shown. Our results show that the gel stiffness, G′, decreases with decreasing viscosity of the pure polymer, i.e., with decreasing molecular weight of the polymer (Fig. 13.3.A). This is easily understandable given that in our gel samples the polymer molecules provide the backbone of the gel. The viscosity and molecular weight (MW) of a pure polymer can be considered as synonymous according to Mark-Houwink's equation. The Mark–Houwink equation gives a relationship between Intrinsic Viscosity, (η) and Molecular Weight, M, of a polymer solution expressed as (η) = KM^a, where the constants, a and K, depend on the particular polymer-solvent system. A value of $a = 0.5$ is indicative of "theta solvent". A value of $a = 0.8$ is typical for "good solvents". For most flexible polymers, $0.5 \leq a \leq 0.8$. For semi-flexible polymers, $a \geq 0.8$. For polymers with an absolute rigid rod, such as Tobacco Mosaic Virus, $a = 2.0$. Fig. 13.3A shows that the G′ value remains almost invariant between 10 and 22cP ethylcellulose, but increases significantly for 300cP EC. In this work, we also determined the stress at the limit of linearity of the G′ vs. stress sweeps. This critical stress is referred to here as the "yield stress". An exponential increase in yield stress as a function of increasing MW of the pure polymer was also observed and reported in Fig. 13.4.A. The yield stress trends mirrored the G′ vs. polymer MW trends.

Surfactants of different molecular structures were used as plasticizers, as these can play a significant role in this gelation process. For a given polymer and oil, the SMS-containing gel was found to be the strongest (G′ = 31610 Pa) whereas the GMO-containing one was the weakest (G′ = 319.8 Pa). The rest of the surfactants contributed somewhere in between, as shown in Fig. 13.3.B. These trends were, however, not mirrored in the yield stress trends (Fig. 13.4.B). An interesting finding, however, was that surfactants with a smaller head group had a stronger plasticizing effect than ones with larger head groups (Fig. 13.5), and oleic acid-containing surfactants offered a greater plasticizing effect, i.e., a lower G′ than its stearic acid counterparts (Fig. 13.3.C).

In a physical gel, gel junctions are formed by weak Van der Waal's interactions or hydrogen bonding between polymer strands. According to plasticization theory (Wypych, 2004), a plasticizer molecule interacts with the gel junctions by causing the polymer chains to fall apart at certain places so that they can glide over each other and hence offers a plasticizing effect. This causes an increase in free volume and a reduction in the glass transition temperature, T_g of the pure polymer. Therefore a more branched surfactant is predicted to have a slower diffusion rate and hence better retention inside a polymer network, thus causing a greater reduction in the elastic modulus. For example, Tween 80 and STS have more similar head groups than SMS, but their lower G′ values, as seen in Fig. 13.3.B, can be attributed to the dangling chains attached to the sorbitan moiety. It is also highly recommended to consider the molar content of surfactants in the gel composition, rather than the mass content, for more meaningful comparison of plasticization effects. Moreover, a hydrophilic head group may cause more hydrogen-bonding interactions with the polymer strands, leading to an increased plasticization. Although we attempted to correlate the G′ trends to surfactant molecular weight, three-dimensional bulkiness, and

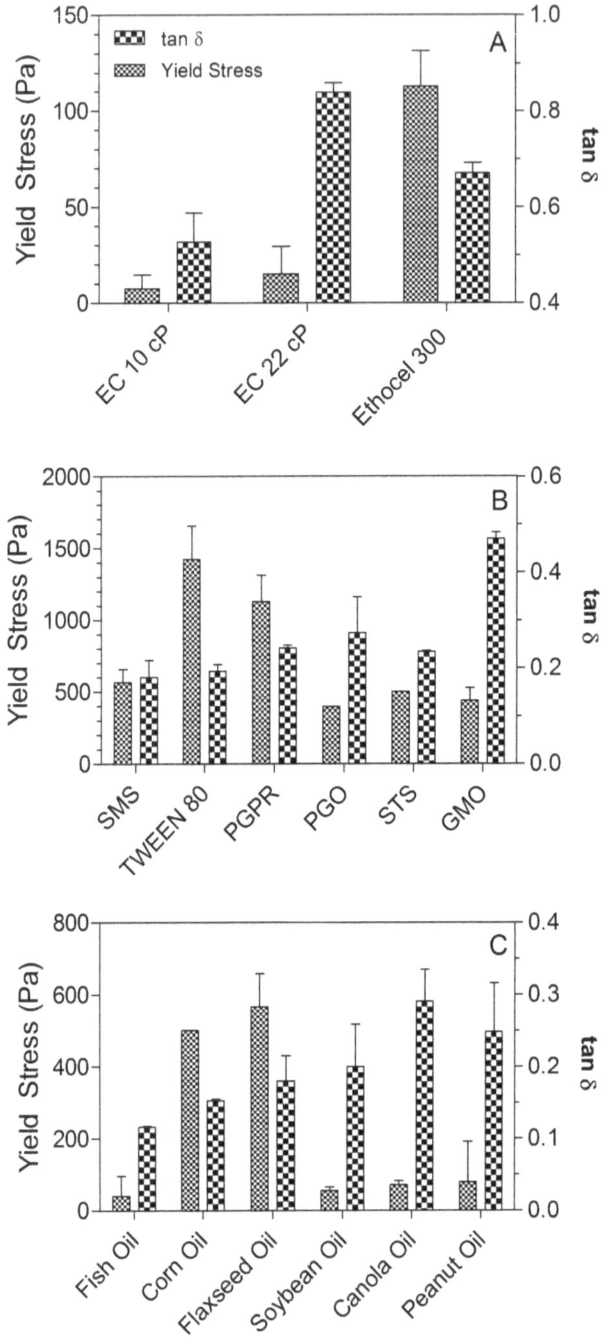

Fig. 13.4. Bar diagrams showing viscoelastic properties (G′, tan δ, yield stress/cross-over point) as a function of (a) different polymer MWs, (b) different surfactant types, and (c) different oil types.

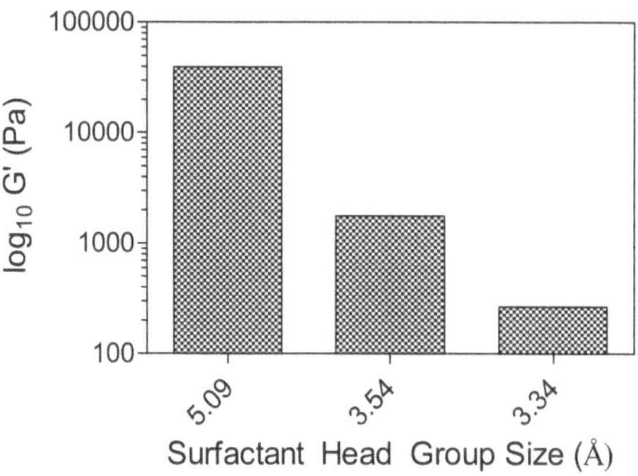

Fig. 13.5. Correlation between size of surfactant head group and G′ values.

hydrophilicity, no clear results were obtained. A more systematic approach is necessary where one factor is varied at a time.

Common edible oils contain a wide variety of saturated and unsaturated fatty acids, as shown in Table 13.A. The G′ values varied over a wide range with oil variation (starting from 1123 Pa for unrefined fish oil to 31610 Pa for flaxseed oil) for a given polymer and surfactant. It was found that the degree of unsaturation of the oils had a significant effect on gel-strength, G′ (Figs. 13.3.E and 13.4.C). This result was also correlated to the IR –C–H stretching frequencies of alkane (2923 and 2853 cm^{-1}) and alkene (3007–3011 cm^{-1}) peaks obtained from the gel samples (Fig. 13.6.A).

The area under the latter peak is a measure of degree of unsaturation of oil (Leskovian et al., 2010; El-Bahy, 2005). This parameter for various oil-containing ternary gels was plotted against the corresponding gel strength G′, as shown in Fig. 13.7. The trend was in accordance with Fig. 13.3.E. An exponential relationship was observed between degree of oil unsaturation and gel strength, particularly for multiple double bonds. This can be attributed to the fatty acids with multiple double bonds having multiple bends, whereas fatty acids with no double bond are somewhat straight and stiff. Hence, as the amount of multiple bend-containing unsaturation increases, the amount of nodular structures in oil increases as well, which allows for a greater polymer chain-chain interaction and thus a stronger network (Laredo et al., 2011).

Moreover, positions and shifts in alkene –C–H frequency were identical in oil-containing ternary gels and in the corresponding pure oils, indicating that the oil unsaturation remained intact upon gelation. This was further proved by rheological study on a pair of gels varying only in presence/absence of oil. The gel strength G′ remained unaffected by oil (Fig. 13.3.D), indicating the non-interactive nature of the oil.

A broad O–H stretching frequency in the range of 3200–3550 cm^{-1} (commonly attributed to hydrogen bonding) was observed in pure polymer (EC 22) as well as in pure surfactant (SMS), but not in Flaxseed oil and the corresponding ternary gel (Fig. 13.6.B), indicating a disruption of hydrogen bonding taking place during oleogel

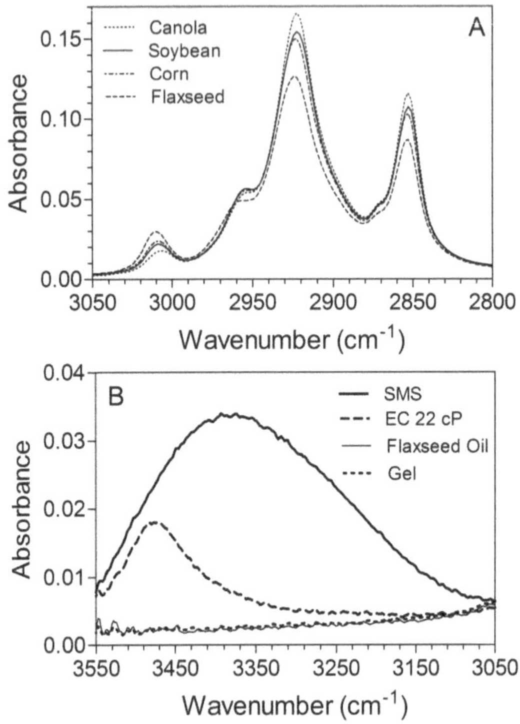

Fig. 13.6. IR spectra: (a) –C-H stretching vibrations associated with alkane and alkene, as obserbed in gels having 5% Ethyl Cellulose 22cp + 5 % Sorbitan MonoStearate + 90% Edible Oil (effect of oil variation) (b) broad -OH stretching frequency attributed to Hydrogen bonding as observed in pure polymer, pure surfactant, pure oil, and their ternary gel.

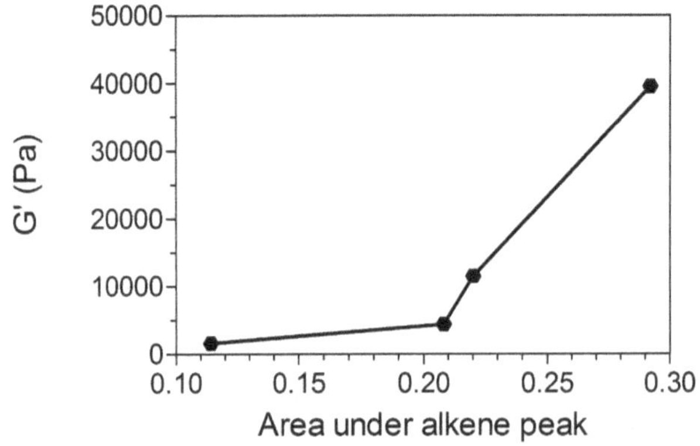

Fig. 13.7. Correlation between viscoelastic and IR results: effect of oil unsaturation on G′ values.

formation. This kind of compromise in hydrogen bonding may result from freely rotating single bonds in solvent (oil), proving the molecular conformation of solvent to be the determining factor for polymer-solvent interaction strength (Heng et al., 2005). The C=O ester stretching frequencies for pure surfactant and pure oil appeared at 1735.93 cm^{-1} and 1743.65 cm^{-1} respectively, which is in fair agreement with literature (Freeman, 1963). The difference in peak position can be attributed to the electron-releasing inductive effect offered by one vs. three alkyl chains attached to carbonyl carbon, respectively. The ternary gel showed an ester peak in the same position as that of the pure oil, again supporting the non-interactive role of oil (data not shown).

As far as elasticity of the gel samples was concerned (smaller tan δ values), the surfactant contribution followed a trend of SMS:Tw80>PGPR>PGE>STS>GMO with Tw80 and PGPR-containing gels having significantly high cross-over values (5012 Pa), as shown in Fig. 13.4.B. Elasticity of gel samples with oil variation (Fig. 13.4.C) followed a trend of fish oil:corn oil:flaxseed oil:soybean oil:canola oil:peanut oil with corn and flaxseed oils showing comparatively high cross-over point value (1585 Pa). Interestingly the trend observed for elasticity in terms of surfactant variation was the same as that of the G′ values, but no such correlation was observed with oil variation. This is understandable given that for a given oil medium (i.e., fluidic character, G″ being constant), the strongest gel will show the lowest tan δ value, i.e., the maximum elasticity, as tan δ is a ratio of G″ to G′. This again supports the extent of plasticizing role being played by different surfactants. However, as oils affect the fluidity of the medium (G″), a similar correlation does not exist for oil variation.

Our DSC results are unique and this section will serve as a prelude. The pure polymers showed a Glass Transition Temperature (T_g) in the range of 125–131°C (not shown) in accordance with literature (Morflex, 1995). An endothermic peak typically associated with gel melting was expected to occur but in certain cases was not observed during gel heating. Fig. 13.8 shows a specimen DSC curve for pure polymer, pure oil, polymer + surfactant solidified mixture, polymer + oil gel, and polymer + surfactant + oil gel. The exothermic peak around 130–135°C corresponds to a degradation event of the vegetable oil, while the peak at ~165°C corresponds to the thermal degradation of the SMS. As stated above, no thermal gel-melting event could be detected by DSC. However, a large melting event of SMS could be observed at ~55°C in the EC-SMS sample. Interestingly, for the oil–EC–SMS sample, one could observe two melting transitions, one at 55°C and one at 70°C. The 55°C peak could correspond to free SMS, while the one at 70°C could correspond to "bound" SMS. The ratio of free to bound SMS was determined from the ratio of the areas under the curve, which was 3.0. Since EC concentration in DSC experiment was 10 wt % while SMS concentration was 5 wt%, considering that only 1/3 of the SMS was bound, this would suggest that the maximum binding ratio of polymer to surfactant was 6:1 (w/w).

Cryo-SEM images were collected to correlate the mechanical strength of gels with the extent of network formation. Some of the gel samples indeed had a dense network corresponding to high G′ values and all of the samples showed oil globules encaged by the polymer strands (Fig. 13.9).

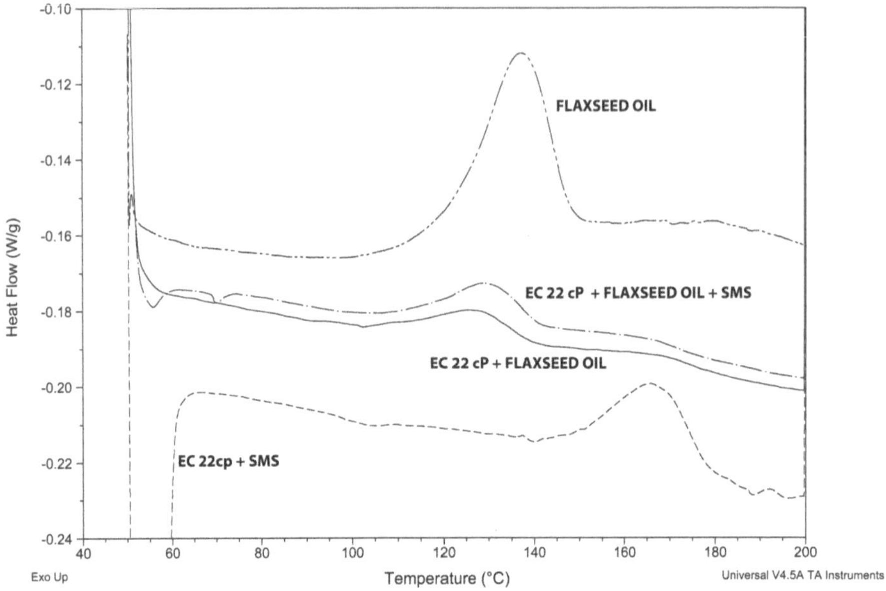

Fig. 13.8. Differential Scanning Calorimetry (DSC) curves of pure polymer EC 22cp, 10% Ethyl Cellulose 22cp + 90% Sorbitan MonoStearate and 10% Ethyl Cellulose 22cp + 5% Sorbitan MonoStearate + 85% Flaxseed Oil (overlaid graphs).

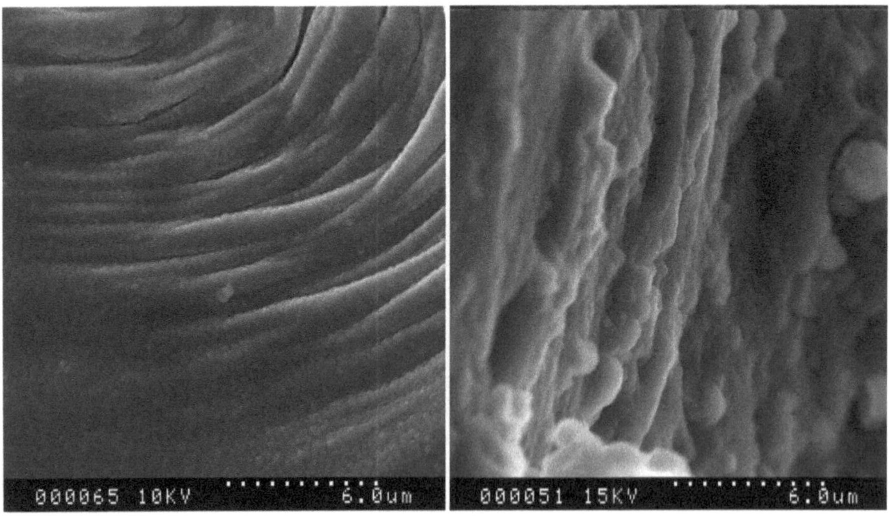

Fig. 13.9. Cryo-SEM images of gels obtained from 5% Ethyl Cellulose 22cp + 5 % Sorbitan MonoStearate + 90% Edible Oil: (a) Peanut Oil (b) Flaxseed Oil.

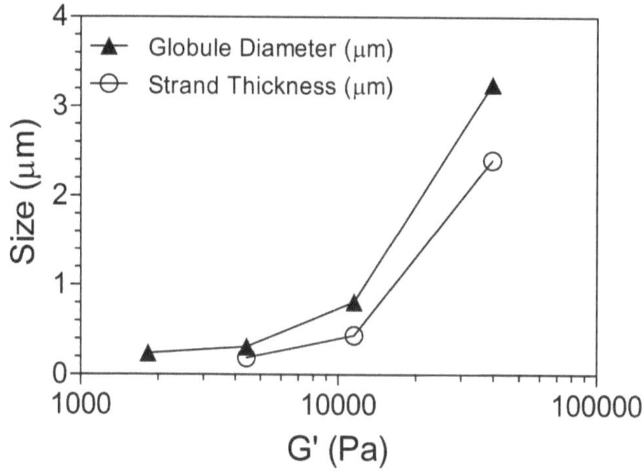

Fig. 13.10. Polymer strand thickness and oil globule size as a function of gel strength.

Peanut oil-containing gels showed a tendency for several polymer strands to be bundled up together, forming a secondary polymer strand (Fig. 13.9.A). Of flaxseed, peanut, corn, and soybean oils, the former showed the largest polymer strand thickness and oil globule diameter (2.4 μm and 3.25 μm, respectively), whereas the latter one was the smallest (0.19 μm and 0.31 μm, respectively). In general, the thickness of polymer strands as well as the oil globule size were a strong function of the type of oil used. Increases in both oil globule size and strand thickness were positively correlated to increase in the storage modulus (G′) of the gels, as shown in Fig. 13.10.

Conclusion

This chapter reports the formulation of an edible polymeric gel, plasticized by surfactants and containing entrapped oil globules, that can be used as a structured oil with fat-like functionalities. The properties of these oil-containing polymer gels was influenced by the structure and the formulation of the components. Viscoelastic parameters of gels were correlated to polymer size, surfactant structure, and oil composition. The storage modulus and the yield stress of the gels increased with increasing molecular weight of the pure polymer. Oleic acid-containing surfactants offered more plasticizing effect than their stearic acid counterparts, and surfactants with smaller (less bulky) head groups were more plasticizing than ones with larger head groups. Rheological as well as IR studies showed that the degree of unsaturation of the oil was directly proportional to gel-strength, G′, especially for multiple double bonds. Cryo-SEM images suggested a qualitative relation between gel structure and gel strength, G′. Polymer strand thickness and oil globule size were both found to affect the G′ values, although the effect was more pronounced for the latter.

Acknowledgments

The authors would like to thank Kraft Foods Inc. Glenview, IL, USA for their generous funding. A special thanks to Dr. Alexandra K. Smith for assisting us during the cryo-SEM experiments.

References

Aiache, J.-M.; Gauthier, P.; Aiache, S. New gelification method for vegetable oils I: cosmetic application. Int. J. Cosmetic Sci. 1992, 14, 228–234.

Almeida, I.F.; Bahia, M.F. Evaluation of the physical stability of two oleogels. Int. J. Pharm. 2006, 327, 73–77.

Borchard, W.; Burg, B. Progr. Molecular mechanisms during the thermoreversible gelation of gelatin-water-systems.Colloid Polym. Sci. 1990, 83, 200–210.

Bot, A.; Agterof, W.G.M. Structuring of edible oils by mixtures of γ-oryzanol with β-sitosterol or related phytosterols. J. Am. Oil Chem. Soc. 2006, 83, 513–521.

Bot, A.; Flöter, E.; Lammers, J.G.; Pelan, E.G. Understanding and controlling the microstructure of complex foods. McClements, D.J., Ed.; Woodhead Publishing: Cambridge, 2007; 575–599.

Bot, A.; den Adel, R.; Roijers, E.C. Fibrils of γ-Oryzanol + β-Sitosterol in Edible Oil Organogels. J. Am. Oil. Chem. Soc. 2008, 85, 1127–1134.

Brouillet-Fourmann, S.; Carrot, C.; Mignard, N. Gelatinization and gelation of corn starch followed by dynamic mechanical spectroscopy analysis. Rheol Acta 2003, 42, 110–117.

Daniel, J.; Rajasekharan, R. Organogelation of plant oils and hydrocarbons by long-chain saturated FA, fatty alcohols, wax esters, and dicarboxylic acids. J. Am. Oil Chem. Soc. 2003, 80, 417–421.

Dey, T. Properties of vinyl ester resins containing methacrylated fatty acid comonomer: the effect of fatty acid chain length. Polymer Int. 2007, 56, 853–859.

El-Bahy, G.M.S. FTIR and Raman spectroscopic study of Fenugreek (Trigonella foenum graecum L.) seeds. J. Appl. Spectrosc. 2005, 72, 111–116.

Freeman, N.K. Simultaneous determination of triglycerides and cholesterol esters in serum by infrared spectrophotometry. J. Lipid Res. 1963, 5, 236–241.

Frohoff-Hülsmann, M.A.; Lippold, B.C.; McGinity, J.W. Aqueous ethyl cellulose dispersion containing plasticizers of different water solubility and hydroxypropyl methyl cellulose as coating for diffusion pellets. Part 2. Properties of sprayed films. Eur. J. Pharm. Biopharm. 1999, 48, 67–75.

Ghebremeskel, A.N.; Vemavarapu, C.; Lodaya, M. Use of surfactants as plasticizers in preparing solid dispersions of poorly soluble API: Selection of polymer-surfactant combinations using solubility parameters and testing the processability. Int. J. Pharm. 2007, 328, 119–129.

Gilsenan, P.M.; Ross-Murphy, S.B. Viscoelasticity of thermoreversible gelatin gels from mammalian and piscine collagens. J. Rheol. 2000, 44, 871–883.

Heng, P.W.S.; Chan, L.W.; Chow, K.T. Development of Novel Nonaqueous Ethylcellulose Gel Matrices: Rheological and Mechanical Characterization. Pharm. Res. 2005, 22, 676–684.

Kaufman, M.H. Nitroglycerine Migration of Ethylcellulose. J. Chem. Eng. Data 1962, 7, 426–428.

Laredo, T.; Barbut, S.; Marangoni, A.G. Molecular interactions of polymer oleogelation. Soft Matter 2011, DOI: 10.1039/c0sm00885k.

Lee, K.Y.; Kim, Y-R.; Park, K.H.; Lee, H.G. Rheological and gelation properties of rice starch modified with 4-α-glucanotransferase. Int. J. Biol. Macromol. 2008, 42, 298–304.

Leskovjan, A.C.; Kretlow, A.; Miller, L.M. Fourier Transform Infrared Imaging Showing Reduced Unsaturated Lipid Content in the Hippocampus of a Mouse Model of Alzheimer's disease. Anal. Chem. 2010, 82, 2711–2716.

Lu, Y.; Larock, R.C. Novel Polymeric Materials from Vegetable Oils and Vinyl Monomers: Preparation, Properties, and Applications. ChemSusChem 2009, 2, 136–147.

Marangoni, A.G. Fat Crystal Networks. Marcel Dekker: New York, 2005.

Morflex, Inc. Influence of Plasticizers on the Dissolution and Physical Properties of Ethyl Cellulose Films and Coated Beads. Pharmaceutical Coatings Bulletin 102–3 1995.

Ojijo, N.K.O.; Neeman, I.; Eger, S.; Shimoni, E. Effects of monoglyceride content, cooling rate and shear on the rheological properties of olive oil/monoglyceride gel networks. J. Sci. Food Agric. 2004, 84, 1585–1593.

Ostrovskii, D.; KjØniksen, A.-L.; Nyström, B.; Torell, L.M. Association and Thermal Gelation in Aqueous Mixtures of Ethyl (hydroxyethyl) cellulose and Ionic Surfactant: FTIR and Raman Study. Macromolecules 1999, 32, 1534–1540.

Pernetti, M.; van Malssen, K.F.; Flöter, E.; Bot, A. Structuring of edible oils by alternatives to crystalline fat. Curr. Opin. Colloid and Interface Sci. 2007a, 12, 221–231.

Pernetti, M.; van Malssen, K.; Kalnin, D.; Flöter, E. Structuring edible oil with lecithin and sorbitan tri-stearate. Food Hydrocolloids 2007b, 21, 855–861.

Rogers, M.A.; Wright, A.J.; Marangoni, A.G. Engineering the oil binding capacity and crystallinity of self-assembled fibrillar networks of 12-hydroxystearic acid in edible oils. Soft Matter 2008, 4, 1483–1490.

Rogers, M.A. Novel structuring strategies for unsaturated fats—Meeting the zero-trans, zero-saturated fat challenge: A review. Food Res. Int. 2009, 42, 747–753.

Rogers, M.A.; Wright, A.J.; Marangoni, A.G. Nanostructuring fiber morphology and solvent inclusions in 12-hydroxystearic acid / canola oil organogels. Curr. Opin. Colloid Interface Sci. 2009, 14, 33–42.

Ruan, D.; Lue, A.; Zhang, L. Gelation behaviors of cellulose solution dissolved in aqueous NaOH/thiourea at low temperature. Polymer 2008, 49, 1027–1036.

Ruíz Martínez, M.A.; Muñoz de Benavides, M.; Morales Hernández, M.E.; Gallardo Lara, V. Influence of the concentration of a gelling agent and the type of surfactant on the rheological characteristics of oleogels. Il Farmaco 2003, 58, 1289–1294.

Scientific Psychic. www.scientificpsychic.com/fitness/fattyacids1.html. May 24,2010.

Seal Oil. www.omega3sealoil.com/Chapter2.html. May 24,2010.

Sharma, V.; Kundu, P.P. Prog. Addition polymers from natural oils—A review. Polym. Sci. 2006, 31, 983–1008.

Wypych, G., Ed. Handbook of Plasticizers. William Andrew Publishing: Norwich, N.Y., 2004.

Wright, A.J.; Marangoni, A.G. Formation, structure, and rheological properties of ricinelaidic acid-vegetable oil organogels. J. Am. Oil Chem. Soc. 2006, 83, 497–503.

Yilmaz, M.; Kusefoglu, S.H. Gelation of soybean oil with polybutadiene. J. Appl. Polymer Sci. 2005, 96, 2240–2246.

Appendix: Experimental Procedure

Materials

Amorphous Ethyl Cellulose (EC) of varying viscosities (10, 22, and 300 cp) were obtained from Aldrich (EC 10 and EC 22) and Dow Chemical (Ethocel 300) with ethoxyl content of 48%, a product commonly derived from cellulose by etheration [48.5% corresponding to maximum disorder and least crystallinity (Kaufman, 1962), due to reduction in hydrogen bonding and introduction of bulky ethoxy groups]. The oils under study were 100% pure vegetable oils purchased from the local grocery store, with an exception of fish oil (Menhaden) being obtained from Swimco-Canada Inc., Ontario, CA. Polyoxyethylene(20) sorbitan monooleate or Tween 80 (Glycoperse O-20 K FG) and polyoxyethylene(20) sorbitan monostearate or Tween 60 (Glycoperse S-20 K FG), both Kosher Food Grade, were purchased from Lonza Inc., Allendale, NJ, USA. Polyglycerol Ester (PGE), Glycerol Mono Oleate (GMO), Sorbitan Mono Stearate (SMS, or SPAN 60) and PolyGlycerol Polyricinoleate (PGPR) were obtained from Danisco USA Inc. New Century, KS, USA under the trade names of GRINSTED PGE 20 VEG, DIMODAN MO 90/DK, GRINSTED SMS-M K and DIMODAN PGPR 90, respectively. Sorbitan TriStearate (STS, or SPAN 65) was obtained from Sigma. The chemical structures of polymer and surfactants used in this study are shown in Fig. 13.1.

Gel Sample Preparation

All of the ingredients (5 wt % polymer + 5 wt % surfactant + 90 wt % oil, unless otherwise mentioned) were taken in a beaker and heated with continuous stirring. For this purpose, the Fisher Scientific hotplate-cum-stirrer was used. A thermometer was constantly dipped inside the sample to note the onset of clearing of the solution. Both magnetic stir-bar (125 RPM) as well as glass rod (manual stirring) were employed for homogenous mixing. Once the clearing point was reached completely (referred as "clearing temperature" onwards, in the range of 125–155°C depending on the specific sample), the heat source was removed and the gel was allowed to set at ambient condition during cooling (referred as "gelation temperature" onwards).

Instrumentation

Viscoelastic measurements were performed at 20°C and 1Hz frequency using a controlled stress Rheometer (Model AR-2000, TA Instruments, New Castle, DE, USA) fitted with a cone-and-plate geometry (8 mm steel plate, 0° angle). A special scooper was used to cut the gel samples into cylindrical shapes of exactly 8mm diameter. Gel samples were placed over a piece of sandpaper (aluminium oxide sandpaper, 150 fine, from MasterCraft) presoaked by the respective solvent oil. This sandpaper was tightly affixed over the peltier plate of the rheometer prior to the experimental runs to avoid sample slippage problem. A portion of this sandpaper was also cut into a circular shape with an 8mm diameter and was glued on the bottom side of the geometry.

During experimental setup, this part of the geometry was allowed to lower and touch the sample surface without squishing. Various important parameters such as Storage (G′) and Loss (G″) Modulus values at LVE (linear viscoelastic) region (which are the real and imaginary parts of complex modulus and represent gel and sol characters respectively), tan δ (=G″/G′, lower value indicating more elasticity), yield stress (the stress value at which LVE region starts losing its horizontal linearity, limiting stress before permanent deformation) and cross-over point (where G′ = G″, another means of stating elasticity retention limit) were noted down from the Oscillatory Stress experiments, run in triplicate for each sample (Table 13.A).

Thermal analysis of gel samples were conducted in a Differential Scanning Calorimeter (Model DSC Q-1000, TA Instruments, New Castle, DE, USA). Approximately 9–13 mg of gel sample was placed inside a properly sealed Hermetic Alodined Pan. A linear heating rate of 5°C/min over a temperature range of 50–200°C was used for both heating and cooling cycles.

Infra Red bands of samples were recorded by Shimadzu IR Prestige-21 FTIR spectro-photometer (purchased from Mandel Scientific Co. Ltd., Guelph, ON, Canada). Small and equal amounts of sample were placed over the single-crystal diamond ATR (attenuated total reflection) device and the press was brought down to ensure contact. This device has the advantage of applying the same technique to different kinds of samples and thus did not require KBr (potassium bromide) pellet preparation for pure polymer.

Hitachi Scanning Electron Microscope (SEM) S-570 equipped with EMITECH K1250X cryo preparation unit was used to collect cryo-SEM images of our samples. Sample preparation technique involved a solvent-wash process prior to cryo-SEM experiments. For this purpose, ~0.2 g of gel sample was immersed in 40 mL of 90:10 (v/v) 2-butanol/methanol mixture and was allowed to stand (without any agitation) for 2 hours. This process was necessary to expose the stranded polymer structure for optimal viewing under microscope.

Photoshop 5.5 (Adobe Systems Inc., San Jose, CA, USA) was used to calculate the polymer strand thicknesses and oil globule sizes from cryo-SEM images and an average over 25 measurements were reported for each sample. ChemDraw Ultra Version 5.0 from CambridgeSoft Corp. (Cambridge, MA, USA) was employed to represent the chemical structures of the molecules under this study, as shown in Fig. 13.1.

14

Clinical Study on 12-hydroxystearic Acid Organogel Ingestion

Naomi Hughes[1], James W. Rush[2] and Alejandro G. Marangoni[3]
[1]CSM Bakery Products, Tucker, Georgia; [2]Department of Kinesiology, University of Waterloo, Canada; and [3]Dept. of Food Science, University of Guelph

Introduction

The unique combination of physical, functional, and nutritional properties of edible oil organogels has lead to the recognition of their applicability in both the food and pharmaceutical industries. The ability of organogelators to create a structural network and immobilize liquid triacylglycerols, or edible oils, at low concentrations could be exploited for a variety of different products and purposes within these two industries. The intention of this chapter is to outline the potential applications of edible oil organogels in the food and pharmaceutical industries and to summarize the work that has been done to evaluate their functionality within these areas. More specifically, this chapter will focus on the work that was done at the Universities of Guelph and Waterloo to assess the ability of organogelator networks to replace the saturated and *trans* fatty acids in conventional spreads and to control the rate of release of lipophilic bioactive ingredients.

Controlled Delivery of Pharma- and Nutraceuticals from Organogels

The characteristic pattern of immediate release features high plasma concentrations of a bioactive compound almost immediately following its ingestion, followed by sharp decrease to a low point, as it is quickly absorbed into the bloodstream and cleared from the system (see Fig. 14.1) (Turner et al., 2002).

The presence of a bioactive compound at concentrations either above or below its therapeutic range can lead to negative side effects and the re-emergence of the original symptoms (Turner et al., 2002). However, if the bioactive compound is released into the bloodstream at a steady or controlled rate, its concentration will not deviate significantly above or below the therapeutic range throughout the absorption process, thus preventing the re-emergence of negative symptoms and side effects. Furthermore, controlled release technology in the pharmaceuticals industry allows for less-frequent dosing, which has been associated with increased patient compliance and therefore improved drug safety and efficiency.

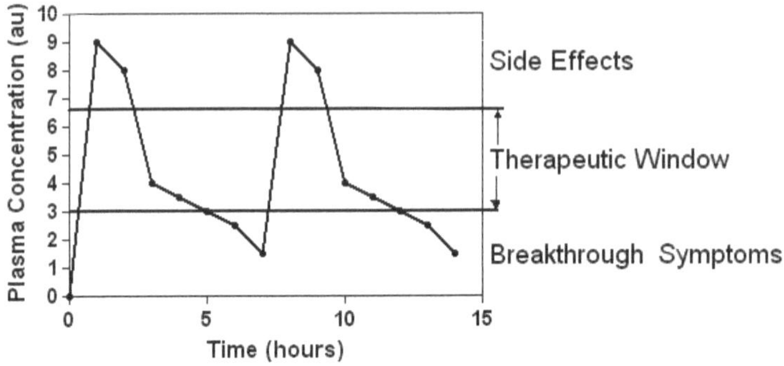

Fig. 14.1. Typical Plasma Concentration Profile of Immediate Release Absorption (Adapted from Turner et al., 2002).

Unfortunately, due to its dependence on a number of highly variable factors including the composition of digestate and gastric-emptying rate, controlling the rate of release of orally-administered bioactive compounds is not always simple or straightforward. Thanks to decades of research, however, controlled release technology is now widely used in the pharmaceutical industry, where drugs are either encapsulated by materials or incorporated into tablets that are designed specifically to delay or control the rate of drug release into the bloodstream after oral administration (Felt, 1998). Hydrogels formed with either natural or synthetic biodegradable polymers are commonly used for this purpose (Ghosh, 2004). Hydrogels, which are cross-linked polymeric three-dimensional (3D) networks with the ability to entrap water, can be used to protect a drug in the acidic environment of the stomach and carry it through to the intestine (Vogelson, 2001). The ability of hydrogels to protect bioactives through the early stages of digestion is a particularly valuable property as it prevents the loss of the compound by partial degradation prior to reaching its intended target (Vogelson, 2001).

In addition, depending upon their chemical composition, these hydrogels can be tailored to decompose or to swell and release their entrapped compounds at a predetermined point in response to an internal or external stimulus like pH, magnetic or electric field, temperature, or ultrasonic irradiation (Vogelson, 2001). For example, a glucose-sensitive hydrogel based on poly [methacrylic acid-g-poly (ethylene glycol)] has been developed to deliver insulin to diabetic patients in response to an internal pH trigger (Peppas, 2004). The development of smart or stimuli-responsive gels is currently one of the most exciting areas of soft materials science (Ishi & Shinkai, 2005).

Although hydrogels have been used in oral delivery for decades, the application of their non-polar counterparts is a relatively new field of research. Since many important bioactive compounds are hydrophobic, it has been proposed that organogels be used both to increase the solubility and control the release of non-polar pharma- and nutraceuticals (Turner et al., 2002; Bot et al., 2007; Rogers et al., 2008). The reversibility and sensitivity of organogel networks makes them promising materials for the controlled delivery or release of lipophilic compounds (van Esche & Feringa, 2000). The incorporation of a fat-soluble bioactive compound into an organogel could be achieved either by its dispersion within the sol phase prior to gelation or, alternatively, its inclusion within an organogelled capsule. Since organogel networks are formed and held together via non-covalent and physical bonding interactions rather than covalent chemical bonds, they are considered to be reversible. Depending upon the physico-chemical properties of the network, some external chemical or physical stimulus can trigger the sol-gel phase transition, causing these secondary interactions to break down. Network interactions may be weakened or broken by changes in temperature, pH, light, chemical composition/concentration, or mechanical stimuli (Tsutomu & Shinkai, 2005). In the theoretical case of delivering bioactive compounds from organogels, the breakdown or dissolution of the network would lead to the release of solubilized nutrients. Furthermore, an organogel that breaks down under the specific conditions of a particular phase of the human digestive process might be designed so that the bioactive components are released at a pre-determined point to maximize their health benefits. For the food industry, the rate of release of nutrients from food materials and their bioavailability are both important considerations in the development of value-added food products and nutraceuticals. Rather than the bioactive compound being completely solubilized and available immediately following oral ingestion, it is often desirable to control the rate at which it is released into the bloodstream (Turner al., 2002; Felt, 1998).

For non-polar nutraceuticals, the delivery and bioavailability of carotenoids, an important class of lipid soluble phytochemicals, might benefit significantly from this type of delivery system. These lipid-soluble plant-derived pigment molecules with anti-oxidant properties have been linked to the reduced risk of several chronic diseases (Cooper, 2004; Rao & Rao, 2007). However, the functionality and potency of these compounds is complicated by limitations on their absorption, such that their bioavailability is often restricted in the range of 10–30% (Rao & Rao, 2007). One factor that affects the bioavailability of carotenoids is the presence, amount, and type of lipid in the digestive tract at the time of absorption (Ryan et al., 2008; Wright et al., 2008). In the digestive tract, lipids dissolve hydrophobic compounds and also increase bile secretion from the gallbladder. Furthermore, the products of lipid digestion increase the size of bile salt micelles, which aids in the solubilization of carotenoids (Furr, 1997). As a result, the consumption of carotenoids with lipids increases their absorption and bioavailability. Theoretically, the incorporation of carotenoids into vegetable oil organogels should raise their bioavailability and should also slow or control their rate of release and absorption since they are physically entrapped within a gel network. Increasing the bioavailability of carotenoids would maximize their positive effects on human health, and the development of materials with the ability to delay the rate of release of lipophilic bioactive compounds would contribute to the development of novel pharmaceutical products and soft materials.

To explore the rate and extent of release of carotenoids, it is important to first understand their absorption process. After the ingestion of fruits and vegetables (the major dietary sources of carotenoids) and their gastric digestion, carotenoids are transferred to the micelles that are formed by dietary fat and bile acids and are subsequently absorbed by the intestine via passive diffusion. The carotenoids are then integrated into the chylomicrons and released into the lymphatic system, where they are incorporated into lipoprotein in the liver and released into the blood-stream, then becoming bioavailable (Ryan et al., 2008). The efficiency and extent of beta carotene micellarization during the intestinal phase of digestion has been shown to be an accurate representation of carotenoid bioavailability (Ryan et al., 2008; Wright et al., 2008).

Delivery of Beta Carotene from 12-HSA/Canola Oil Organogel

The ability of an organogel network to control or delay the release of carotenoids was evaluated at the University of Guelph where the rate of micellarization of beta carotene (βC) from the oily to the aqueous phase during the intestinal phase of digestion and its dependence on lipid structure were studied using an in vitro model of the human digestive system. Results from this experiment might be extrapolated to represent the micellarization/release of other similar bioactive compounds from organogels. Based on its ability to restrict the mobility of liquid triacylglycerols, 12-hydroxystearic acid (12-HSA), a derivative of castor seed oil, was selected as a model organogelator molecule for our investigation of the potential application of edible oil organogels as delivery systems in the food and pharmaceutical industries. Although 12-HSA has not yet been approved as a food-grade additive, it is derived from a natural product (castor seed oil) and has been approved by Health Canada (2001) for use in cosmetics and personal care products. It has often been used as a prototype for the study of the network formation process as well as the material properties of the resulting organogels (Rogers et al., 2008; Rogers et al., 2009; Terech et al., 2000; Terech & Weiss, 1997). In addition, it is an ideal organogelator for the structuring of edible oils and consistently produces strong, thermoversible gel networks at low concentrations (0.5–2.0 wt %). For these reasons, 12-HSA was used as a model low molecular weight organogelator molecule for our study of the ability of organogels to control or delay the release of carotenoids.

The experimental methods used in this study were adapted from an in vitro model of the upper intestinal stage of digestion (Wright et al., 2008; Garrett et al., 1999; Hedren et al., 2002). An important modification to the procedure was the re-introduction of the gastric step of the digestion process. This phase was not included in Wright's model because it was previously shown to have no effect on the proportion of βC transferred to the aqueous phase (Garrett et al., 1999). However, the effects of the gastric phase (composition, pH, time, temperature, and mechanical stress) on the 12-HSA gel network were unknown, so this step was included in our modified experimental procedure. For more experimental details, please refer to Hughes et al., 2009 and Hughes, 2009.

The concentration of beta carotene micellarized within the aqueous samples was determined by absorption spectroscopy following its solvent extraction. Beer's law [with a molar extinction coefficient of 137, 400 mL/mol·cm (Rich et al., 1998)] was used to calculate the concentration of βC in each sample, and the proportion of βC transferred to the aqueous phase was then calculated based on the initial amount of βC present in each sample. Two sets of duplicate measurements were taken for each of six replicates (24 absorbance readings per treatment). One way ANOVA analyses were performed to identify any statistically significant differences between the samples.

The transfer of beta carotene from the oily to the aqueous fraction of digestate is highly dependent upon the conditions of the gastrointestinal tract, including the presence and type of food, pH, pancreatin, and bile salt concentration (Wright et al., 2008). Based on the conditions and methods used in this experiment, the small proportion of beta carotene (<10 wt %) that was micellarized into the aqueous phase was expected. Under similar conditions, Wright (2008) reported the transfer of ~40 wt % of beta carotene from the oily to the aqueous fraction of digestate. However, Wright's experimental procedures featured the homogenization of the lipid sample and the aqueous phase prior to the remainder of the in vitro digestion process. As expected, the exclusion of this step from our experimental procedure resulted in a significantly lower transfer of beta carotene (Wright et al., 2008). The fact that (even with homogenization) the majority of beta carotene that is ingested is not actually bioavailable further demonstrates the importance of increasing bioavailability of carotenoids.

Despite the low transfer from all samples, results from this experiment confirmed the original hypothesis: the gelation of canola oil with 12-hydroxystearic acid affected the rate of βC micellarization from the oil to the aqueous phase during intestinal digestion. The observed difference between the gelled canola oil and the liquid canola oil control was in the amount of time it took for the maximum amount of βC to be micellarized from the sample into the aqueous phase of the digestate.

As shown in Fig. 14.2, there was no significant increase (P value > 0.05) in the amount of βC micellarized from the liquid canola oil samples beyond 30 minutes

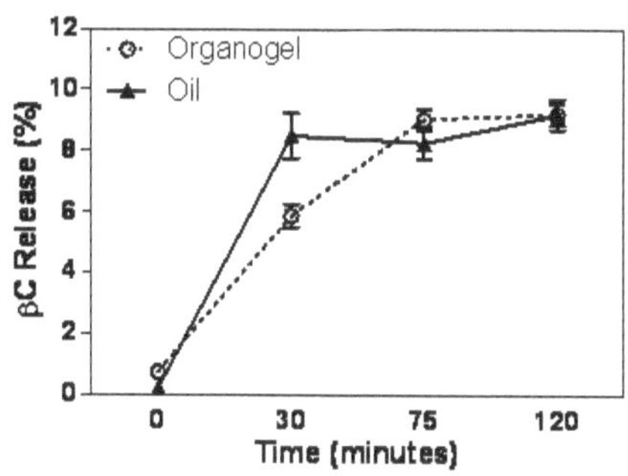

Fig. 14.2. Extent of Beta Carotene Micellarization over Time throughout a Two-Hour In Vitro Intestinal Digestion Period. Note: The two-hour intestinal digestion period followed one hour of gastric digestion after which no significant differences were observed between the samples (see Time 0). Values represent the mean and standard error of six replicates (Adapted from Hughes et al., 2009).

of intestinal digestion. In contrast, there was a statistically significant increase in the amount of βC micellarized from the gelled canola oil between the times of 30 and 75 minutes (*P* value < 0.05) with no further increases beyond this time point. This indicates that the kinetics of the release and subsequent micellarization of βC during the intestinal phase of lipid digestion is dependent upon its initial structure. In the gelled sample, the liquid oil (containing βC) is entrapped within the 12-HSA network, and therefore it is less available for micellarization. Although it is an amphiphilic molecule, 12-HSA is highly insoluble in water and does not aggregate at the oil-water interface of water-in-oil emulsions (Hughes et al., 2009). Based on these observations, it can be assumed that 12-HSA does not have any significant chemical effects on the enzymatic hydrolysis that occurs at the micellar interface during digestion and that the delay in beta carotene micellarization is solely the result of the physical entrapment of oil within the 12-HSA gel network.

The gel did, however, release a significant amount (*P* value < 0.05) of βC during the first thirty minutes of the experiment, which indicates that the oil is not completely protected or entrapped within the gel matrix. Over time (0–75 minutes), under the conditions of intestinal digestion, the oil is gradually released from the network, making the βC available for micellarization. The slow release of oil is also demonstrated by the fact that after 75 minutes of intestinal digestion, there is no significant difference (*P* value > 0.05) in the amount of βC micellarized from the gel and the canola oil control. This is similar to the ability of some hydrogels to protect their contents through the gastric phase of the digestive process and to release them only after reaching the intestinal phase of the digestive process (Vogelson, 2001). In contrast to these hydrogels, however, there was no significant difference (*P* value > 0.05) in the amount of βC micellarized after the hour-long gastric period between the organogel and the oil control (see Time 0 in Fig. 14.2), which indicates that the difference between the samples was initiated during the intestinal phase. This supports the finding that a negligible amount of βC is micellarized from oil during the gastric phase of digestion (Garrett et al., 1999) and also shows that the gastric phase has little or no effect on the 12-HSA network. This is supported by the comparison of images of the 12-HSA network acquired via polarized light microscopy before and after gastric digestion (images not shown) where no significant differences in the birefringence patterns of the gel network were observed after this step, which indicates that the gel network remained intact (Hughes et al., 2009).

The ability of the 12-HSA network to slow and control the release of carotenoids demonstrates the potential application of organogels in the delivery of lipophillic pharma and nutraceuticals. While the gradual release of vegetable oil from a 12-HSA organogel network is a fairly simple controlled release system, there is significant potential to build on this type of system in the engineering and development of smart or responsive organogel networks with the ability to delivery bioactive compounds at pre-defined rates or in response to pre-determined conditions (Hughes et al., 2009).

Nutritional Health Benefits of Edible Oil Organogels

When it first became apparent that the over-consumption of saturated fats leads to increased low density lipoprotein (LDL) cholesterol levels and risk of coronary heart disease (CHD), the food industry responded by replacing SAFAs with hydrogenated vegetable oils (Decker, 2006). At the time, hydrogenated oils seemed like the ideal alternative to SAFAs because they perform all of the important structural functions of saturated fats and, due to their derivation from highly unsaturated vegetable oils, were originally considered to be healthier than saturated fats. It was soon discovered that the physico-chemical and material properties of hydrogenated oils could be controlled by modifying the hydrogenation reaction conditions of time, temperature, and concentration of catalyst (Decker, 2006). This discovery led to the mass production of *trans*-fat-containing partially hydrogenated oils, which have unrivaled structural versatility and functionality in food products. It was not until the mid 1990s that *trans* fats were confirmed to be worse than SAFAs for CHD risk due to their simultaneous elevation of LDL and lowering of HDL cholesterol. The consumer's response to this discovery has put enormous pressure on the food industry to remove all *trans* fats from their products. Many food manufacturers have responded by simply re-introducing saturated fats back into their products (Decker, 2006; Sundram & Basiron, 2007). However, with public health in mind, Health Canada has recently implemented strict labeling regulations whereby, to claim that a food product is "Trans Fat Free", one serving must not only contain <0.2g *trans*-fats per serving, but also <2g total of saturated and *trans*-fatty acids combined per serving (Health Canada, 2006). Similar regulations are being implemented in other countries and have forced the global food industry to seek functional and nutritionally acceptable alternatives to replace both *trans*- and saturated fatty acids. Unfortunately, due to their excellent functionality (including texture, stability, mouth feel, flavor, plasticity, shortening power, creaming ability, and air holding capacity) the simultaneous removal or replacement of both SAFAs and TFAs in complex lipid-containing food products has proven to be a very difficult task."

A comprehensive up-do-date summary of oil-structuring alternatives to the conventional fat crystal network was recently published in a recent issue of "Current Opinions of Colloids and Interface Science" (Bot et al., 2007). According to this review, structuring techniques for organic solvents (including edible oils) can be sub-divided into two categories: suspensions and networks (Bot et al., 2007). Suspensions are formed via the dispersion of a large amount of filler, in the form of solid particles, crystallized solids, or aqueous droplets, throughout the continuous organic phase. Depending upon their volume fraction, these filler particles either form a network via mechanical contact, or aggregate via weak physical interactions, or join by partial coalescence. Two food products that are structured by this method are peanut butter, a suspension of 50 wt % solid non-fat particles in peanut oil, and mayonnaise, a space-filling emulsion of water droplets in canola oil. In contrast, structuring networks formed by specific molecular mechanisms are held together via covalent, electrostatic, steric, van der Waals, or hydrogen-bonding interactions. The solid fat crystal network of chocolate is an example of network structuring in a food product (Bot et al., 2007). In addition to the conventional solid-fat crystal network, there are a number of other compounds that are capable of forming networks to structure organic solvents including macromolecules, modified bio-polymers, and amphiphillic molecules that undergo self-assembly. The self-assembled low molecular weight organogelators are currently considered to be the most exciting option for application in food systems (Bot et al., 2007). The current food-grade alternatives to SAFAs and TFAs are mono- and diglycerides, fatty acids, fatty alcohols, waxes, wax esters, sorbitan esters, and phytosterols (Bot et al., 2007). Combinations of these non-TAG systems often exhibit synergism and are more effective when paired together than in their pure forms. However, due to the complexity of both edible oils and the process of gel network formation, it is difficult to predict or design structuring systems (pure or mixed) that would be applicable in composite food systems. Some mixed systems that are currently being studied are fatty acids and fatty alcohols (Schaink et al., 2007), lecithin and sorbitan tri-stearate (Pernetti et al., 2007), and phyto-sterols and oryzanol (Bot & Agterof, 2006).

Developed as a healthier alternative to conventional spreads that are structured with SAFAs and TFAs, a product that employs a mixture of sitosterol and oryzanol (<2 wt %) to solidify edible oil was patented in 2005 (Ritter et al., 2005). Furthermore, this new product is considered to be an appetite suppressant due to its extended survival in the human digestive tract; it would reach the ileum unabsorbed, which triggers the ileal brake and leads to feelings of satiety (Maljaars et al., 2007). Although this type of product has already been designed, more work will be required to exploit the full potential of structuring edible oils via organogelation. Extensive research needs to be done to determine nutritional benefits, processing considerations, as well as the potential for application in more complex composite food products, like emulsions.

Since the 12-HSA canola-oil organogel described in the previous section is composed primarily of canola oil (~98 wt %), its lipid profile is highly unsaturated, made up of approximately 59% mono- and 30% poly-unsaturated fatty acids (American Dietetic Association, 2007). The health benefits of both mono- and poly-unsaturated fatty acids have been well documented in the literature, and it has been recommended that they should represent the main source of fat in the human diet (American Dietetic Association, 2007). In canola oil, even at low concentrations of 0.5–2.0 wt %, 12-HSA forms a continuous 3D network that traps the liquid triacylglycerols and forms a gel. The solid-like physical properties of the organogel allow it to be used like a spread, while the gel network prevents the canola oil from leaking into the food product onto which it is applied. In addition, based on the study conducted at the University of Guelph described in the previous section, it was hypothesized that the gel network might also control the release of the lipids during digestion and therefore cause a more gradual or steady nutritional response.

Postprandial Nutritional Response to 12-HSA/Canola Oil Organogel

A recent study conducted by the Universities of Guelph and Waterloo compared the nutritional responses to the acute ingestion of the organogel (canola oil gelled with 2 wt % 12-HSA) to two commercially available spreads (butter and non-hydrogenated margarine), as well as a liquid canola oil-control treatment. The ingredients of the commercial butter, margarine, and canola oil used in this experiment are as follows (see Table 14.A for a detailed nutritional breakdown):

- **Neilson Salted Butter:** Cream, salt (may contain color)
- **Becel Margarine (Original):** Canola and sunflower oils 74%, water, modified palm and palm kernel oils 6%, salt 1.8%, whey protein concentrate 1.4%, soy lecithin 0.2%, vegetable monoglycerides, potassium sorbate, vegetable colour, artificial flavour, citric acid, vitamin A palmitate, vitamin D3, alpha-tocopherol acetate (vitamin E)
- **No Name Canola Oil:** Pure Canola Oil

Table 14.A. Nutritional Breakdown of 10 g Servings of Neilson Salted Butter, Becel Original Margarine, and No Name Canola Oil (Hughes et al, 2009 & Hughes, 2009).

Per 10g serving	Butter	Margarine	Canola Oil
Calories (kcal)	70.0	70.0	87.4
Total Fat (g)	8.0	8.0	9.8
Saturated (g)	5.0	1.0	0.7
Trans(g)	0.2	0.0	0.0
Polyunsaturated (g)	0.0	2.5	3.3
n-6 (g)	0.0	2.0	3.3
n-3 (g)	0.0	0.6	1.0
Monounsaturated (g)	0.0	4.0	6.6
Cholesterol (mg)	25.0	0.0	0.0
Sodium (mg)	80.0	70.0	0.0
Total Carbohydrate (g)	0.0	0.0	0.0
Dietary Fiber (g)	0.0	0.0	0.0
Sugar (g)	0.0	0.0	0.0
Protein (g)	0.1	0.0	0.0
Vitamin A (%)	8.0	10.0	21.9
Vitamin C (%)	0.0	0.0	0.0
Calcium (%)	0.0	0.0	0.0
Iron (%)	0.0	0.0	0.0
Vitamin D (%)	0.0	30.0	0.0
Vitamin E (%)	0.0	15.0	0.0

A clinical trial was conducted to compare the postprandial response to each of the above treatments in terms of some of the common indicators of a number of chronic diseases.

Six male and six female, young, healthy individuals volunteered to participate in a clinical trial, approved by the Office of Research Ethics at the University of Waterloo. The group (mean ± SEM) age, height, weight and BMI were 25.67 ± 0.69 years, 174.4 ± 1.82 cm, 74.95 ± 3.19 kg, and 24.6 ± 0.89 kg/m^2, respectively. Variability in the fasting values of the metabolic markers was controlled by feeding the subjects a ~780 calorie pre-meal 12 hours prior to the beginning of the trial comprised of approximately 29 g fat (8 g saturated fat), 98 g carbohydrates, and 34 g of protein. After consuming the pre-meal, subjects fasted overnight and abstained from caffeine, alcohol, and strenuous exercise for 12 hours prior to blood sampling. Before test meal consumption on trial days, a catheter was placed in the antecubital vein of the subjects preferred arm and fasting blood samples were obtained (t=0). In addition to one of the four sample treatments, subjects consumed a test meal consisting of two slices of Country Harvest 100% Whole Wheat Bread, 200 g of Danone Silhouette 0^+ Fat Free yogurt, and 30 g of President's Choice Raspberry Jam. The bread was topped with a spread of either butter (60 g), margarine (60 g), canola oil (48 g), or canola oil gelled with 12-HSA (48 g) to standardize for a fat content of 48 g across all conditions. The relatively high fat content of the test meal is consistent with many studies of postprandial lipid metabolism (Lairon et al., 2007) and was selected to provide ample metabolic stimulus to challenge the normal capacity of postprandial lipid handling. The macronutrient content of the test meal (excluding the sample fat spread) consisted of 400 kcal from 2 g of fat, 0 g of saturated fat, 82.4 g of carbohydrates, and 14.6 g of protein. The macronutrient content of the test meals including each of the four sample treatments is detailed in Table 14.B. Note that the saturated fat content of the canola oil and organogel samples are significantly lower than that of butter, and that the mono- and poly-unsaturated fat content of both of the canola oil-based samples is higher than both butter and margarine.

The four treatments were assigned to each participant in random order once a week for four weeks in order to prevent bias and minimize the influence of confounding factors.

After the consumption of one of the four treatments with the test meal, venous blood samples were collected every hour for six hours and were analyzed for

Table 14.B. Macronutrient Content of the Test Meal for Each of the Sample Treatments (Hughes et al., 2009 and Hughes, 2009).

Sample	Calories (kcal)	Fat (g)	Saturated Fat (g)	Carbohydrates (g)	Protein (g)
Butter	820.0	50.0	30.0	82.4	15.2
Margarine	820.0	50.0	6.0	82.4	14.6
Canola Oil	826.0	50.0	2.7	82.4	14.6
Organogel	826.0	50.0	2.7	82.4	14.6

triacylglycerols, free fatty acids, glucose, and insulin. These CHD markers were plotted against time (0–360 minutes) and the Area under the Curve (AUC) for the six hour time period was calculated using GraphPad Prism 4.0. Both the response curves and the AUC data were compared to identify any significant differences between the treatments. Statistical significance at the 95% confidence interval was evaluated using one way ANOVA followed by a two-tailed paired t-test. The fasting values, measured before the consumption of the test meal (Time 0) for each individual participant, were subtracted from all subsequently collected data points to standardize for variability in fasting blood samples. As a result of this correction, the serum concentrations of biomarkers values that are presented in the following response curves represent their average change from baseline.

While no significant differences (*P* value > 0.05) were detected in terms of insulin or glucose response levels, triacylglycerol and free fatty acid response curves corresponding to the organogel and canola oil treatment were significantly lower than those of both butter and margarine.

In general, after the consumption of any of the treatments with the test meal, serum triacylglycerol concentrations increased for approximately two hours, gradually decreased for one to two hours, then increased again through the last two hours of testing.

In general, the serum triacylglycerol response curves corresponding to the canola oil and organogel treatments were lower than those corresponding to butter and margarine (see Fig. 14.3). Although there were no significant differences between the treatments in terms of the maximum serum TAG concentration reached at the end of the experiment, the organogel and the canola oil response curves were lower throughout the majority of the experiment (particularly between t = 120 and 240

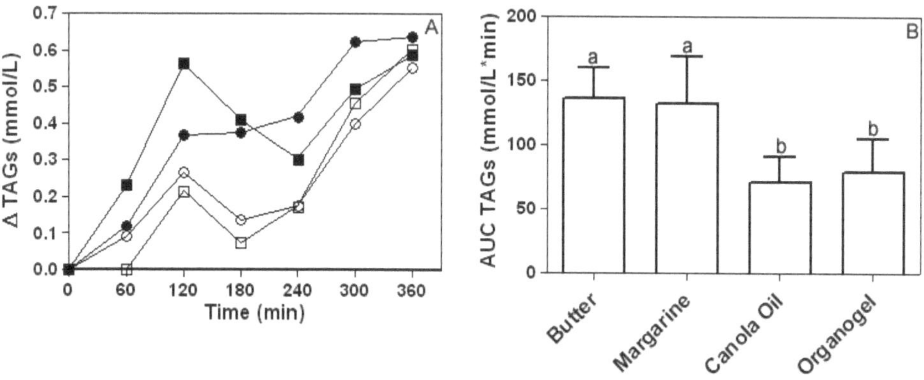

Fig. 14.3. A: Baseline Corrected Postprandial Serum Triacylglycerol Response Curves for Butter (■), Margarine (●), Canola Oil (□) and Organogel (○); Canola Oil gelled with 2 wt % 12-HSA). Note: values represent the mean of twelve subjects and error bars have been excluded for clarity; B: A Comparison of the Area under the Curve Values. Values represent the mean and standard error of twelve subjects (Adapted from Hughes et al., 2009 and Hughes, 2009).

minutes). As a result, the area under the postprandial serum triacylglycerol response curves corresponding to the organogel (79.25 ± 25.58 mmol/L*min) was significantly lower than that of butter (135.7 ± 24.27 mmol/L*min) and margarine (132.5 ± 36.63 mmol/L*min) but not significantly different from canola oil (71.10 ± 20.31 mmol/L*min). Despite the fact that atherosclerosis is more closely and directly linked to high serum levels of low-density-lipoprotein (LDL) cholesterol, high serum triacylglycerol levels are also considered to be an indicator of this dangerous condition, which often leads to coronary heart disease and stroke (Miller, 2005). Furthermore, high postprandial serum triglycerides are also considered to indicate the risk of metabolic syndrome and pancreatitis (Miller, 2005). For these reasons, lowering serum triacylglycerols is considered to be a strong defense tactic in the prevention of a number of chronic diseases. It has been well-documented that the type of fat in the diet has a direct effect on both acute postprandial- and longer term serum triglyceride levels (Mozaffarian et al., 2006). The consumption of a diet that is high in saturated and *trans* fatty acids leads to increased acute postprandial serum triglycerides as well as a chronic elevation of serum TAG levels. It follows that the replacement of saturated and *trans* fatty acids with mono- and poly-unsaturated fatty acids decreases both acute postprandial and long-term serum triacylglycerol levels (Jackson et al., 2005; Miller, 2005). This relationship between the degree of saturation of lipids in the diet and serum TAGs likely accounts in part for the differences observed between butter, margarine, and the canola oil-based sample treatments. It has also been shown that the physicochemical state of dietary fats influences postprandial lipid metabolism (Mekki et al., 2002; Michalski et al., 2005). For example, the emulsification of dietary fats decreases the size and increases the specific surface area of the fat droplets. Furthermore, the adsorption of lipase is affected by the stability of the fat droplet as well as the presence and type of surfactant (Armand et al., 1996; Michalski et al., 2005). As a result, the kinetics of lipid hydrolysis and absorption are affected by the emulsification of dietary fats. Butter and margarine are water-in-oil emulsions, so their lipid components contain emulsified water droplets, while the lipids of the canola oil and the canola oil organogel are present in a simple bulk state. It is possible that this difference contributed in part to the variation in lipid metabolism that was observed between these two sample groups.

In general, the free fatty-acid response following test-meal and sample ingestion featured an initial decrease in serum FFA concentrations. This decrease corresponds to the uptake of circulating free fatty acids that is triggered by the postprandial spike in glucose and insulin levels resulting from the consumption of the test meal. After one to two hours, serum FFA levels gradually increased for the remaining four hours of the testing period. This steady increase corresponds to the release of the digested dietary lipids from the test meal into the bloodstream.

Figs. 14.4.A, 14.4.B, and 14.4.C clearly demonstrate the differences between the two canola oil-based samples (canola oil and organogel) and butter and margarine. In general, the canola oil and organogel treatments led to lower acute serum levels of free fatty acids than the two commercial spreads. The area under the organogel serum free

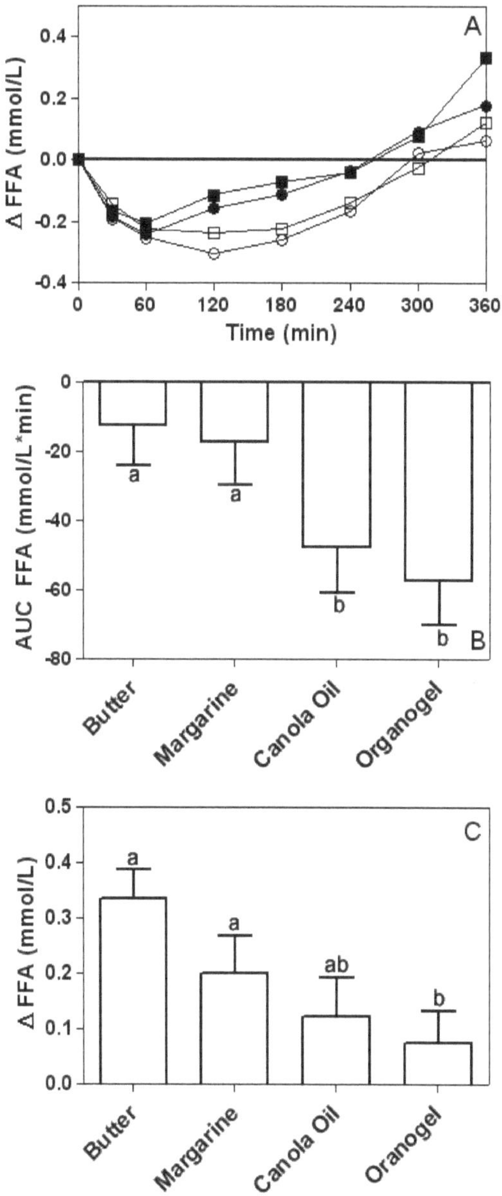

Fig. 14.4. A: Baseline Corrected Postprandial Serum Free Fatty Acid Response Curves for Butter (■), Margarine (●), Canola Oil (□) and Organogel (○); Canola Oil gelled with 2 wt % 12-HSA). Note: values represent the mean of twelve subjects and error bars have been excluded for clarity; B: A Comparison of the Area under the Curve Values; and C: A Comparison of Maximum Serum Free Fatty Acid Concentrations (t = 360 minutes). Values represent the mean and standard error of twelve subjects (Adapted from Hughes et al., 2009 and Hughes, 2009).

fatty acid response curve (–57.02 ± 12.93 mmol/L*min) was statistically significantly lower than those of both butter (–12.35 ± 11.61 mmol/L*min) and margarine (–17.23 ± 12.24 mmol/L*min) but not significantly different from oil (–47.61 ± 13.06 mmol/L*min). Furthermore, the maximum serum concentration of free fatty acids (which occurred at six hours for all treatments) was statistically significantly lower (P value < 0.05) for both the canola oil and the organogel than for butter (See Fig. 14.4.C).

Both acute and chronic elevation of plasma free fatty acid levels have been associated with insulin resistance, which can lead to obesity, type II diabetes, metabolic disorder, and coronary heart disease. Normalizing or decreasing serum free fatty acid levels has been proposed as an approach to the reduction of the risk of these diseases (Boden, 2003). Based on this knowledge, it can be concluded that the consumption of a diet that leads to lower serum free fatty acid levels would be beneficial to public health. The substitution of saturated and *trans* fats, which are solid at room temperature, with highly unsaturated oils, which are liquid at room temperature, lowers the risk for a number of these chronic diseases (Boden, 2003; Woodside & Kromhout, 2005). However, the substitution of conventional spreads like butter or margarine with edible liquid oils is impractical for obvious functional reasons (Toro-Vasquez et al., 2007). By structuring liquid canola oil with the 12-HSA organogel network, the health benefits associated with the consumption of unsaturated oils can be exploited without sacrificing the functionality of a structured, spreadable product. Results from this study demonstrate the exciting potential health benefits associated with the substitution of conventional spreads with vegetable oil organogels (Hughes et al., 2009; Hughes, 2009).

Regardless of the accompanying treatment, the consumption of the test meal led to a sharp spike in serum glucose levels (within the first hour) followed by a gradual decrease for the remainder of the experimental time period. Due to an experimental error in the determination of the fasting values for the margarine treatment, data from this particular treatment was excluded from further analysis.

Based on the observation of Fig. 14.5, it is clear that there is no statistically significant variation between the treatments in terms of glucose response.

The postprandial insulin response to the test meal was similar to the glucose response in that there was a sharp increase within the first hour of testing, followed by a gradual decrease back to baseline values.

Based on the observation of Fig. 14.6, it is clear that there is also no statistically significant variation between the treatments in terms of insulin response.

Despite the close interdependent biochemical relationships between serum triacylglycerols, free fatty acids, glucose and insulin, no significant differences (P value > 0.05) were detected between the four treatments for postprandial glucose or insulin responses. Despite this partial lack of supporting evidence, the statistical significance of the differences in TAG and FFA response levels (P value < 0.05) provides substantial proof of the health benefits that result from the substitution of butter or margarine with an organogel spread that is composed primarily of canola oil.

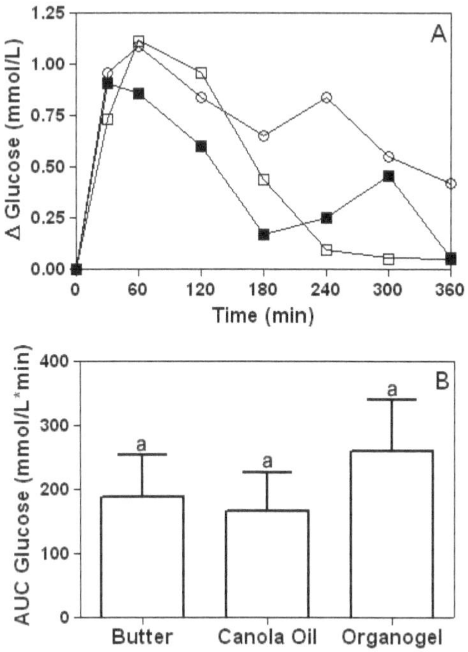

Fig. 14.5. A: Baseline Corrected Postprandial Glucose Response Curves for Butter (■), Canola Oil (□) and Organogel (○); Canola Oil gelled with 2 wt % 12-HSA). Note: values represent the mean of twelve subjects and error bars have been excluded for clarity; B: A Comparison of the Corresponding Area under the Curve Values. Values represent the mean and standard error of twelve subjects. Note: Margarine data has been excluded due to experimental error (Adapted from Hughes et al., 2009 and Hughes, 2009).

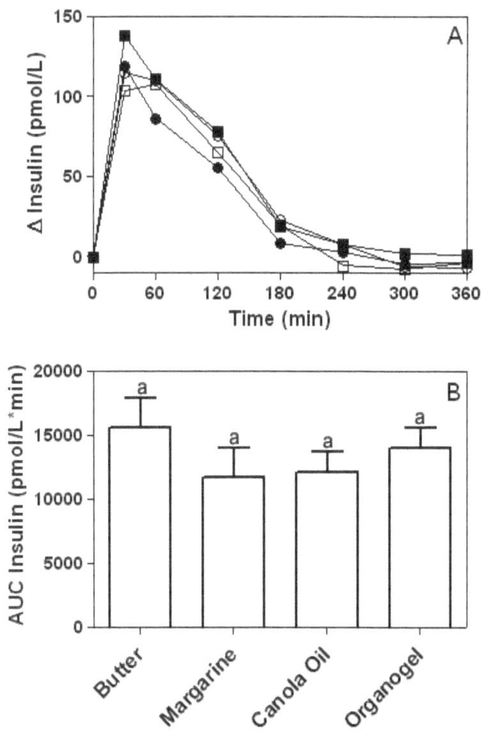

Fig. 14.6. A: Baseline Corrected Postprandial Insulin Response Curves for Butter (■), Margarine (●), Canola Oil (□) and Organogel (○); Canola Oil gelled with 2 wt % 12-HSA). Values represent the mean of twelve subjects. Note: Error bars have been excluded for clarity; B: A Comparison of the Corresponding Area under the Curve Values. Values represent the mean and standard error of twelve subjects (Adapted from Hughes et al., 2009 and Hughes, 2009).

Overview

Based on this review, it is clear that the organogelation of edible oils has great potential to be used within the fields of human health and nutrition. However, more research needs to be done to optimize the functionality of edible organogels in both food and pharmaceutical systems. A more in-depth study of the rate of release of lipophilic compounds from the gel network under the conditions of the human digestive system should be conducted. More frequent sampling throughout digestion will be required to quantify the kinetics of release. Also, to fully characterize the nutritional health benefits associated with the substitution of conventional fat spreads with edible oil organogels, a longer term intervention should be conducted. Furthermore, to reduce variation and increase experimental control, future experiments should include a higher number of participants. A sensory evaluation of the palatability of organogels, which governs the success of new food products, should be conducted to determine the likelihood that this type of product would be acceptable to the consumer. Most importantly, however, the development and optimization of a food grade organogelator system will be required before organogels can be applied in food and/or pharmaceutical products.

References

American Dietetic Association; Position of the American Dietetic Association and Dietitians of Canada: Dietary Fatty Acids. J. Am. Diet. Assoc. 2007, 107, 1599–1611.

Armand, M.; Borel, P.; Pasquier, B.; Dubois, C.; Senft, M.; Andre, M.; Peyrot J.; Salducci J.; Lairon D. Physiochemical characteristics of emulsions during fat digestion in human stomach and duodenum. Am. J. Physiol-Endoc. M. 1996, 271, G172–G183.

Boden, G.; Carnell, L.H. Nutritional effects of fat on carbohydrate metabolism. Best Pract. Red. Cl. En. 2003, 17, 399–410.

Bot A.; Agterof, W.G.M. Structuring of edible oils by mixtures of γ-oryzanol with β-sitosterol or related phytosterols. J. Am. Oil Chem. Soc. 2006, 83, 513–521.

Bot, A.; Pernetti, M.; van Malssen, K.F.; Flöter, E. Structuring of edible oils by alternatives to crystalline fat. Curr. Opin. Colloid Interface Sci. 2007, 12, 221–231.

Cooper, D.A. Carotenoids in Health and Disease: Recent Scientific Evaluations, Research Recommendations and the Consumer. J. Nutr. 2004, 134, 221–224.

Decker, K.J. Health/Nutrition: Fats to Feel Good About. F.P.D. 2006 http://www.foodproductdesign.com/articles/2006/08/fats-to-feel-good-about.aspx# (October, 2008).

Felt, O.; Buri, P.; Gurny, R. Chitosan: A Unique Polysaccharide for Drug Delivery. Drug Dev. Ind. Pharm. 1998, 24. 11, 979–993.

Furr, H.C.; Clark, R.M. Intestinal absorption and tissue distribution of carotenoids. J. Nutr. Biochem. 1997, 8, 364–377.

Garrett, D.A.; Failla, M.L.; Sarama, R.J. Development of an in vitro digestion method to assess carotenoid bioavailability from meals. J. Agric. Food. Chem. 1999, 47, 10, 4301–4309.

Ghosh, S. Recent research and development in synthetic polymer-based drug delivery systems. J. Chem. Res. 2004, 4, 241–246.

Health Canada. Environmental and Workplace Health. http://www.hc-sc.gc.ca/ewh-semt/contaminants/person/impact/list/person-cas-eng.php (September, 2008).

Health Canada. Food and Nutrition. Appendix 4. Current Regulatory Context. http://www.hc-sc.gc.ca/fn-an/nutrition/gras-trans-fats/tf-ge/tf-gt_app4-eng.php. (September, 2008).

Hedren, E.; Diaz, V.; Svanberg, U. Estimation of carotenoids accessibility from carrots determined by an in vitro digestion method. Eur. J. Clin. Nutr. 2002, 56, 5, 425–430.

Hughes, N.E.; Marangoni, A.G.; Wright, A.J.; Rogers, M.A.; Rush, W.E. Potential food applications of edible oil organogels. Trends Food Sci. Technol. 2009, 20, 470–480.

Hughes, N.E. Potential applications of edible oil organogels. M.Sc. Thesis; University of Guelph: Guelph, Ontario, Canada N1G2W1, 2009; 1–241.

Ishi-i, T.; Shinkai, S. Dye-Based Organogels: Stimuli-Responsive Soft Materials Based on One-Dimensional Self-Assembling Aromatic Dyes. Top. Curr. Chem. 2005, 258, 119–160.

Jackson, K.G.; Wolstencroft, A.J.; Bateman, P.A.; Yaqoob, P.; Williams C.M. Greater enrichment of triacylglycerol-rich lipoproteins with apolipoproteins E and C-III after meals rich in saturated fatty acids than after meals rich in unsaturated fatty acids. Am. J. Clin. Nutr. 2005, 81, 1, 25–34.

Lairon, D.; Lopez-Miranda, G.; Williams, C. Methodology for studying postprandial lipid metabolism. Eur. J. Clin. Nutr. 2007, 61, 1145–1161.

Maljaars, J.; Peters, H.P.F.; Masclee, A.M. Aliment. Pharmacol. Ther. 2007, 26, 241–250.

Mekki, N.; Charbonnier, M.; Borel, P.; Leonardi, J.; Juhel, C.; Portugal, H.; Lairon, D. Butter Differs from Olive Oil and Sunflower Oil in Its Effects on Postprandial Lipemia and Triacylglycerol-Rich Lipoproteins after Single Mixed Meals in Healthy Young Men. J. Nutr. 2002, 132, 3642–3649.

Michalski, M.C.; Briard, V.; Desage, M.; Geloen, A. The dispersion state of milk fat influences triglyceride metabolism in the rate. A $^{13}CO_2$ breath test study. Eur. J. Nutr. 2005, 44, 436–444.

Miller, M. Triglyceride Management in Cardiovascular Risk Reduction; Diabetes & Cardiovascular Risk. US Endocrine Review 2005 http://www.touchbriefings.com/pdf/1479/Miller.pdf (accessed December, 2010).

Mozaffarian, D.; Katan, M.B.; Ascherio, A.; Stampfer, M. J.; Willet, W.C. Trans-fatty acids and cardiovascular disease. N. Engl. J. Med. 2006, 354, 1601–1613.

Peppas, N.A. Is there a future in glucose-sensitive, responsive insulin delivery systems? J. Drug Del. Sci. Tech. 2004, 14, 247–256.

Pernetti, M.; van Malssen, K.; Kalnin, D.; Flöter, E. Structuring edible oil with lecithin and sorbitan tri-stearate. Food Hydrocolloids 2007, 21, 855–861.

Rao, A.V.; Rao, L.G. Carotenoids and human health. Pharmacol. Res. 2007, 55, 207–216.

Ritter, H.; Van De, S.R.L.; Muller, V. Liquid fatty component containing composition. Lipton, division of Conopco, Inc. (NJ, USA). US Patent 684650 http://www.freepatentsonline.com/6846507.html"2005 (accessed September, 2008).

Rogers, M.A.; Wright, A.J.; Marangoni, A.G. Engineering the oil binding capacity and crystallinity of self-assembled fibrillar networks of 12-hydroxystearic acid in edible oils. Soft Matter 2008, 4, 1483–1490.

Rogers, M.A.; Wright, A.J.; Marangoni, A.G. Nanostructuring fiber morphology and solvent inclusions in 12-hydroxystearic acid / canola oil organogels. Curr. Opin. Colloid Interface Sci. 2009, 14, 33–42.

Ryan, L.; O'Connell, O.; O'Sullivan, L.; Aherne, S.A.; & O'Brien, N.M. Micellarization of carotenoids from raw and cooked vegetables. Plant Foods Hum. Nutr. 2008, 63, 127–133.

Rich, G.T.; Fillery-Travis, A.; Parker, M.L. Low pH enhances the transfer of carotene from carrot juice to olive oil. Lipids 1998, 33, 985–992.

Sundram, K.; Basiron, Y. The Trans-Fat Dilemma: Health vs. Functionalities. Official Report 4, Malaysian Palm Oil Council. http://www.malaysiapalmoil.org/ (accessed July, 2008).

Schaink, H.M.; van Malssen, K.F.; Morgado Alves, S.; Kalnin, D.; van der Linden, E. Crystal network for edible oil organogels: possibilities and limitations of the fatty acid & fatty alcohol systems. Food Res. Int. 2007, 40, 1185–1193.

Terech, P.; Pasquier, D.; Bordas, V.; & Rossat, C. Rheological properties and structural correlations in molecular organogels. Langmuir 2000, 16, 4485–4494.

Terech, P.; Weiss, R.G. Low molecular mass gelators of organic liquids and the properties of their gels. Chem. Rev. 1997, 97, 3133–3159.

Toro-Vazquez J.F.; Morales-Rueda J.A.; Dibildox-Alvarado E.; Charó-Alonso M.; Alonzo-Macias, M.; González-Chávez, M.M. Thermal and textural properties of organogels developed by candelilla wax in safflower oil. J. Am. Chem. Soc. 2007, 84, 989–1000.

Tsutomu, I.; Shinkai, S. Dye-Based Organogels: Stimuli-Responsive Soft Materials Based on One-Dimensional Self-Assembling Aromatic Dyes Top. Curr. Chem. 2005, 258, 119–160.

Turner, S.; Hite, M.; Federici, C. Controlled Delivery Technologies Applied to the Nutraceutical Industry. PMPS 2002, Autumn, 1–3.

Van Esch, J.H.; Feringa, B.L. New Functional Materials Based on Self-Assembling Organogels: From Serendipity towards Design. Angew. Chem. Int. Ed. 2000, 39, 2263–2266.

Vogelson, C.T. Advances in drug delivery. Mod. Drug Discovery 2001, 4, 49–50, 52.

Woodside, J.V.; Kromhout, D. Fatty acids and CHD. Proc. Nutr. Soc. 2005, 64, 554–564.

Wright, A.J.; Pietrangelo, C.; MacNaughton, A. Influence of simulated upper intestinal parameters on the efficiency of beta carotene micellarization using an in vitro model of digestion. Food Chem. 2008, 107, 1253–1260.

Editors and Contributors

Editors

Alejandro G. Marangoni Alejandro Marangoni is a professor and Canada Research Chair in Food and Soft Materials Science at the University of Guelph. His work concentrates on the physical properties of foods, particularly fat crystallization and structure. He has published over 200 refereed research articles and six books. He is the recipient of many awards including a 1999 Premier's Research Excellence Award, the first Young Scientist Award from the AOCS (2000), a Canada Research Chair (2001, renewed in 2006), two Distinguished Researcher Awards from the Ontario Innovation Trust (2002), a Career Award from the Canadian Foundation for Innovation (2002), an E.W.R. Steacie Memorial Fellowship (2002)—given to the top six Canadian scientists from all disciplines—and the T.L. Mounts Award from AOCS in 2004. Dr. Marangoni is a past chair of the Natural Sciences and Engineering Research Council of Canada's Plant Biology and Food Science Grant Selection Committee, member of NSERC's E.W.R. Steacie Memorial Fellowship selection committee, Editor-in-Chief of Food Research International (Elsevier), an Associate Editor of the JAOCS (Springer), and an editorial board member of Food and Function (RSC), Food Digestion (Springer) and CyTA—The Journal of Food (Taylor and Francis). Dr. Marangoni has co-founded three high-technology companies and is the co-recipient of the 2008 Guelph Partners of Innovation "Innovator of the Year" award for his discovery of a platform technology for the manufacture of structured oil-in-water emulsions. Dr. Marangoni has recently been appointed a Tier 1 Canada Research Chair in Food, Health, and Aging.

Nissim Garti Nissim Garti obtained his B.Sc., M.Sc., and Ph.D. from the Hebrew University of Jerusalem. He has been a full professor since 1990 and holds the Ratner Chair of Chemistry in the Department of Chemistry and Applied Chemistry. He also serves as a Board Member Elect and Director of the Hebrew University Governors Executive Board since January 2011. Nissim is the recipient of numerous prestigious awards including the Rockefeller Award, the Israel President Award for one of the most innovative inventions in 60 years of the existence of the country, Life-Time Achievement Award of the Food Society, the Chang Award of the AOCS, the Corporate Research Achievement Award of the AOCS for 2011, and many others. His achievements include publishing over 380 original (research) refereed papers in peer reviewed journals; writing over 60 review chapters in scientific books; granted over 80 patents; edited 7 books and additional 4 in preparation; invited to over 180 conferences as keynote, session, and invited speaker; and educated and tutored 38 Ph.D. students and 84 M.Sc. students. Nissim is a member of the board of directors of several academic institutions in Israel and consults for several Israeli and global industries. Nissim's expertise, competence, and active research is in colloid chemistry, emulsion technology, dispersed systems, delivery new vehicles, microemulsions and lyotropic liquid crystals, crystallization phenomena, interfacial reactions and reactivity, amphiphilic proteins, hydrocolloids, dendrimers, nutraceuticals, and food science.

Contributors

Abraham Aserin Dr. Aserin is a researcher at the Casali Institute of Applied Chemistry at the Hebrew University of Jerusalem. His research is focused on many practical facets of surfactant chemistry, such as emulsions, microemulsions, lyotropic liquid crystals, dispersions, and improvement of drug bioavailability. He received B.Sc., M.Sc., and Ph.D. from the Hebrew University of Jerusalem, Israel, in 1978, 1981, and 1987, respectively. Dr. Aserin is the author of more than 150 peer reviewed scientific articles, 40 reviews, 2 edited books and holds 12 patents. He has educated and tutored 45 M.Sc. and 15 Ph.D. students. He is the co-founder of NutraLease Ltd.

Arjen Bot Arjen Bot is a physical chemist, who received his Ph.D. at the University of Amsterdam (The Netherlands) for a thesis on light scattering of binary gas mixtures in 1990. Subsequently, he worked for a number of years as a post-doc at the FOM-Institute for Atomic and Molecular Physics in Amsterdam on a subject in surface physics. He joined Unilever R&D Vlaardingen (The Netherlands) in 1993, working in product development of dairy-type products with vegetable fat, mayonnaises, and margarines for the last decade—alternating between group leader and technology leader roles. Some of his work has been published externally in the form of papers or patents.

Eckhard Flöter Eckhard Flöter is a trained Chemical Engineer of the TU Berlin (Germany) and received his Ph.D. degree in Natural Sciences related to the thermodynamics of mineral oils from the TU Delft (The Netherlands). Subsequent to his Ph.D., he changed from mineral oils to edible oils when he joined Unilever R&D in 1996. In the past 14 years, he has held several positions in Unilever. During this time he has been working on numerous projects relating to fat-based food products, mainly spreads. His activities span from science to the application of technology in the factory. Next to being a frequent contributor to scientific conferences, he is a regular member of the scientific committee of the Euro Fed Lipids congresses.

Naomi Hughes Naomi Hughes graduated from Mount Allison University in 2006 with a B.Sc. Honors in Chemistry, and from the University of Guelph in 2008 with a M.Sc. in Food Science. She is currently working as a Food Technologist at CSM Bakery Products in Tucker, Georgia, USA.

Dima Libster Dr. Libster is a leading expert in the field of formation and characterization of new liquid crystalline carriers for delivery and controlled release of peptide and protein drugs. Dr. Libster holds interdisciplinary education both in chemical engineering (from the Technion) and chemistry (Hebrew University). Dr. Libster is the author of more than 40 peer reviewed publications and holds 2 patents.

Raffaele Mezzenga Raffaele Mezzenga received his Master's degree (Summa Cum Laude) from University of Perugia, Italy, in Materials Science and Engineering, while actively working for the European Center for Nuclear Research (CERN), Geneva, and NASA, Houston, in problems related to the interaction between elementary particles and polymer-based structures (NASA Space Shuttle Discovery mission STS91). In

2001 he obtained a Ph.D. in Polymer Physics (with honors), from the Swiss Federal Institute of Technology, Lausanne (EPFL), focusing on the thermodynamics of thermoset-hyperbranched polymer reactive blends. He then spent 2001-2002 as a postdoctoral scientist at University of California, Santa Barbara, working on the self-assembly of polymer colloids for the design of new semiconductive organic materials. In 2003 he joined as a research scientist the Nestlé Research Center (NRC), in Lausanne, working on the self-assembly of surfactants, natural amphiphiles and lyotropic liquid crystals. In 2005 he was hired as Associate Professor in the Physics Department of the University of Fribourg where he has been a board and founding member of the Fribourg Center for Nanomaterials (Frimat). He was appointed Full Professor at ETH Zurich in the Fall 2009 to start a new group on food and soft materials science. Since 2010 he is also an Affiliated Professor of the Materials Department. His research focuses on the fundamental understanding of self-assembly processes in liquid crystalline polymers, supramolecular polymers, lyotropic liquid crystals, biological and food colloidal systems. Prof. Mezzenga has been a Visiting Professor from Helsinki University of Technology and a Nestlé Distinguished Scientist. He is a board member of the Polymer and Colloid Division of the Swiss Chemical Society, winner of the 2004 Swiss National Science Foundation Professorship and recipient of both the 2011 Young Scientist Research Award (AOCS) and the 2011 John H. Dillon Medal (American Physical Society).

Michael A. Rogers Michael A. Rogers graduated from the Ph.D. program in Food Science from the University of Guelph in 2008 after which he took an appointment as an Assistant Professor in the Department of Food and Bioproduct Science at the University of Saskatchewan from 2008-2011. In 2011, Dr. Rogers joined the Food Science Department at Rutgers University where he is currently an Assistant Professor. He has published 25 peer-reviewed publications, 5 textbook chapters and has presented at numerous conferences. In 2008 he was awarded the Top Young Scientist though the International Union of Food Science and Technology (IUoFST).

James W.E. Rush James Rush is Professor and Chair of the Department of Kinesiology at the University of Waterloo. In this role, Dr. Rush leads the department in advancing its mission to discover mechanisms to prevent illness and injury by understanding movement- and lifestyle-related aspects of health.

Dr. Rush holds the Canada Research Chair in Integrative Vascular Biology. His research expertise is in cardiovascular physiology, with a particular focus on identifying mechanisms affecting blood vessel endothelial function in healthy individuals and in cardiovascular disease. His research group focuses on characterizing vascular cell signaling pathways responsible for endothelial control of blood vessel function, and the influences of pharmacological and dietary factors as well as physical activity on vascular function.

Prior to joining the University of Waterloo in 2000, Dr. Rush was a Post-Doctoral Fellow at the Dalton Cardiovascular Research Center at the University of Missouri. He earned his Ph.D. in Physiology from the State University of New York Health Sciences Center at Syracuse.

Eugene Terentjev Eugene Terentjev is a Professor of Polymer Physics at University of Cambridge. He received his Ph.D. in 1985 in Moscow for theoretical research in liquid crystals and did postdoctoral work on modeling liquid crystalline polymers in Case Western Reserve University in Cleveland, Ohio. After moving to Cambridge in 1992, he worked extensively on theory and experiment in liquid crystalline elastomers and is an author of the monograph on that subject. He is interested and involved in a wide range of experimental and theoretical problems in soft-matter and biological physics, researching structure, optical and mechanical properties, and the rheology of complex materials and nanocomposites.

Jorge F. Toro-Vazquez Jorge Fernando Toro-Vazquez, Ph.D. in Food Chemistry from Cornell University, Department of Food Science (1989), is currently working as a Full Professor in the Department of Chemistry, Laboratory of Physical and Chemistical Properties of Food at the Universidad of San Luis Potosi (San Luis Potosi, Mexico). He has published more than 70 peer-reviewed papers and book chapters, mainly in the area of chemical and physical properties/behavior of food using simple and complex model systems. His current research projects are associated with phase transitions and structural organization of lipids (i.e., crystallization, organogelation), and their application in vegetable oil processing and the development of novel lipid-based *trans*-free food products. Dr. Toro-Vazquez is an associate editor of the JAOCS, reviewer of several scientific journals, and a member of the Mexican Academy of Sciences, the AOCS, the Institute of Food Technology. The CONACYT (National Council for Science and Technology of Mexico) has awarded him twice with the National Award in Food Science and Technology (privately funded by an international soft-drink and beverage company) and granted the highest level (SNI-III) as a national researcher by the Mexican Ministry of Education.

Amanda Wright Amanda Wright completed her Ph.D. in Food Chemistry at the University of Guelph, Ontario, Canada in 2001. She then worked as a Post Doctoral Fellow with the Food Engineering group in the Department of Chemical Engineering & Applied Chemistry at the University of Toronto. In 2005, Amanda was appointed to the faculty of the Department of Human Health & Nutritional Sciences at the University of Guelph, where she is presently an Associate Professor. Amanda leads a research group focused on establishing the relationships between food material properties and the biochemical and biophysical aspects of digestion which impact nutritional outcomes. Amanda teaches in the area of functional foods and natural health product human testing and regulations. She also serves as Director of the Human Nutraceutical Research Unit at the University of Guelph. This educational and research unit has a core emphasis on human clinical studies of foods and natural health products pursued on behalf of, or in collaboration with, industrial, government, or other academic partners. Amanda has been an AOCS member since joining the association in 1998, as a student. Since then, she has been active with the Edible Applications Division and the Canadian Section.

Alexander Zetzl Alexander Zetzl is completing his Ph.D. in Food Science at the University of Guelph, only shortly after he received a B.Sc. in Food Science (2009) at the University of Guelph, Ontario, Canada. He is a member of the AOCS and has presented at various international conferences. This young food scientist also has a publication in the prestigious American Chemical Society Journal, Crystal Growth and Design (2009), to his credit.

Index

E

F

G

H

Kolibri 360 GmbH
Ferdinand-Jühlke-Straße 7
99095 Erfurt, Deutschland
produktsicherheit@kolibri360.de